THE
HUMAN
BODY

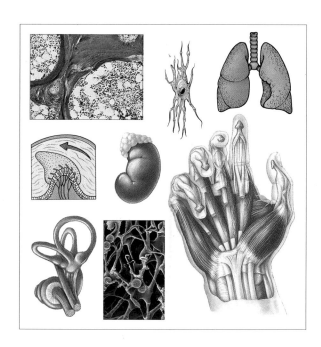

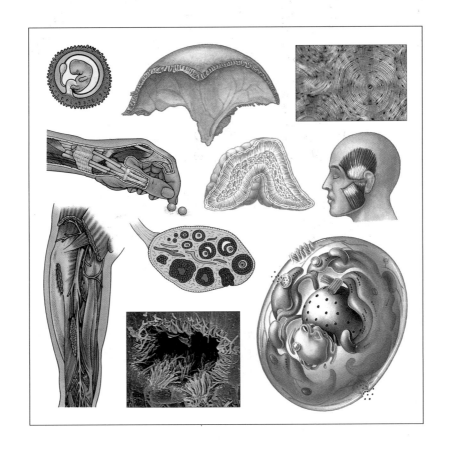

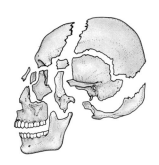
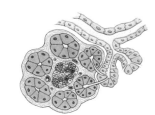
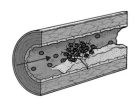
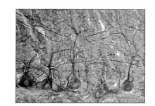
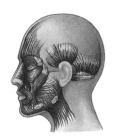
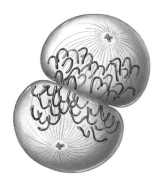
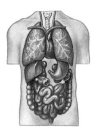
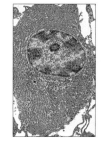

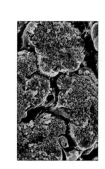
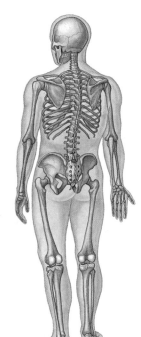
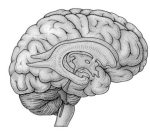

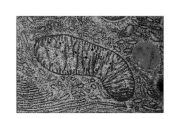
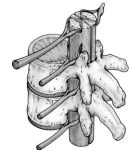
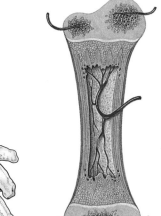

THE
HUMAN
BODY

AN ILLUSTRATED GUIDE TO
ITS STRUCTURE, FUNCTION,
AND DISORDERS

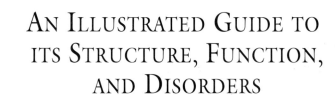

Editor-in-Chief
CHARLES CLAYMAN MD

DK

DORLING KINDERSLEY
LONDON · NEW YORK · STUTTGART · MOSCOW

A DORLING KINDERSLEY BOOK

CONCEIVED, EDITED, AND DESIGNED BY DK DIRECT LIMITED

Editor-in-Chief Charles B. Clayman MD

Senior Editor Andrea Bagg

Art Editor Marianne Markham

Project Editors Christine Murdock, Nance Fyson

US Editor Jill Hamilton

Designers Poppy Jenkins, Donna Askem, Tim Mann, Stephen Cummiskey

Commissioning Designer Jenny Hobson

Visualizers Lydia Umney, Joanna Cameron, Peter Dawson

Illustrators Joanna Cameron, Halli Verrinder, Tony Graham, Philip Wilson,
Sandie Hill, Janos Marffy, Andrew Green, Deborah Maizels

Picture Researchers Sharon Southren, Christina Bankes, Christine Murdock

Text Contributors Dr. Frances Williams, Duncan Brewer, Dr. Tony Smith,
Dr. Robert M. Youngson, Dr. Amanda Jackson,
Dr. Fiona Payne, Dr. Pinkinder Sahota

Indexer Sue Bosanko

Production Manager Ian Paton

Design Director Ed Day

Publisher Jonathan Reed

First American Edition, 1995
2 4 6 8 10 9 7 5 3 1

Published in the United States
by Dorling Kindersley Publishing, Inc.,
95 Madison Avenue
New York, New York 10016

Library of Congress Cataloging-in-Publication Data

The human body / editor-in-chief, Charles B. Clayman.
p.   cm.
ISBN 1-56458-992-7
1. Human physiology--Popular works. 2. Human anatomy--Popular
works. 3. Medicine, Popular. I. Clayman, Charles B.
QP38.H7933   1995   94-37165
612--dc20   CIP

Reproduced by Colourscan, Singapore
Printed and bound in Italy by A. Mondadori Editore, Verona

PREFACE

Throughout history, surgeons and scientists have relied on pictures of the human body to understand its structures and teach their students. Early anatomical investigations by artists such as Leonardo da Vinci are recognized as esthetic masterpieces, but they are also amazingly accurate. Drawing the body from living models and post-mortem specimens has largely been replaced by technological innovation: the inner organs and structures can now be portrayed with astonishing clarity by the computer analysis of X-ray, magnetic resonance, ultrasound, and other data, often producing images of stunning beauty.

The great advances in medical imaging techniques of the past two decades form the background to our unique project: a visual exploration of the human body that takes into full account the extended boundaries of current scientific knowledge. *The Human Body* features beautiful illustrations produced by medical artists who, having benefited from this knowledge, can now render details of human anatomy with greater accuracy than ever before. Complementing the drawings are vivid computer and microscope images created with today's advanced technology.

The Human Body is the most ambitious and detailed presentation of anatomy and function yet offered to the general reader. It contains over 1,000 images, which reveal the body's structures and illustrate the seemingly simple processes we take for granted, such as breathing and heartbeat. In addition to normal anatomy and function, we illustrate and describe the causes, symptoms, diagnosis, and treatment – including many common surgical procedures – of a wide range of diseases and disorders.

As medical science advances, pictures still remain the easiest way to show how the body is constructed, and how it is affected by disease. We believe the illustrations in *The Human Body* will, like the compelling works of earlier artists, stand the test of time.

Charles B. Clayman MD

Charles B. Clayman MD

CONTENTS

ILLUSTRATING *the* HUMAN BODY

TO REVEAL THE INTRICATE STRUCTURES of the human body, this book uses many types of illustration. Some show whole areas, such as the nervous system at right, with its complex branching system of nerve trunks, while others concentrate on smaller parts. Also incorporated are microscope and computer-generated images produced using the latest techniques; most of these are color-enhanced.

THE NERVOUS SYSTEM

BODY PARTS

After an initial overview, each of the body systems is explored in greater depth by showing smaller sections or individual organs. Shown at left is the portal system, the special blood supply to the liver (which processes nutrients and eliminates any toxic substances). Each of the illustrations peels back layers or uses cutaways to expose deeper parts in greater detail.

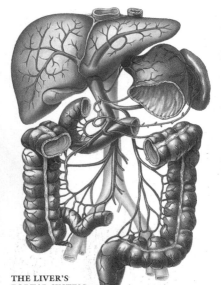

THE LIVER'S
PORTAL SYSTEM

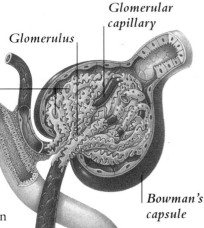

Glomerular capillary

Glomerulus

Podocytes (cells)

Bowman's capsule

FILTERING UNIT
OF KIDNEY

DETAILED STRUCTURE

Some illustrations zoom in on a very tiny part of the body. Seen at right is one of the one million microscopic filtering units found in each kidney; it is so greatly enlarged that even individual cells can be seen. Labels identify the important parts of each illustration.

DYNAMIC PROCESSES

A diagrammatic approach helps clarify the many complex processes that take place inside the body. These may be normal functions, such as digestion, or disease processes, such as how tobacco smoking causes cancer. In some cases, text and illustrations are organized into "steps." Those at the right show how blood clots form in an artery that has been damaged by the accumulation of fatty deposits (atheroma).

1 If the inner wall of an artery is damaged by deposits of atheroma, platelets in the blood may release chemicals.

Fibrin strands

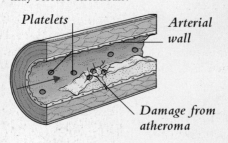

Platelets

Arterial wall

Platelets

Damage from atheroma

2 These chemicals convert a blood protein into fibrin strands that trap blood cells and form a clot, blocking the artery.

The BODY SYSTEMS

THE HUMAN BODY, LIKE THAT OF EVERY OTHER LIVING CREATURE, has one prime biological function – to reproduce itself and ensure the survival of its offspring. However, this is possible only when all the body systems work together efficiently to maintain health. In this book, we describe how every system works as a separate entity, and also how each system is dependent on the others for physical and biochemical support, a true functioning cooperative.

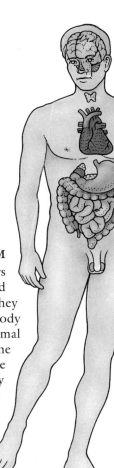

SKELETAL SYSTEM

The skeleton is the framework on which the rest of the body is built. Bones also play a role in the other body systems: red and white blood cells grow and develop in fatty inner tissue known as red marrow. The minerals stored in the bones, especially calcium, are released when the body needs them.

CARDIOVASCULAR SYSTEM

The cardiovascular system's most basic function is to pump blood around the body, and a pause of more than just a few seconds will result in loss of consciousness. All body organs and tissues need a supply of oxygenated blood and the removal of waste products. This delivery system can adapt swiftly to changes in demand.

NERVOUS SYSTEM

The brain is the seat of both consciousness and creativity. Through the spinal cord and nerve branches, the brain also controls all body movement. The nervous system works with endocrine glands to monitor and maintain other systems.

MUSCULAR SYSTEM

Muscles make up about half the body's bulk. Working with the skeleton, they generate the energy to move, make precise and intricate hand movements, lift objects, and even speak. Involuntary muscles, including the specialized cardiac muscle and all smooth muscles, provide the essential power for the respiratory, cardiovascular, and digestive systems. Muscles are dependent on a healthy nerve and blood supply.

ENDOCRINE SYSTEM

Hormones are chemical messengers secreted by the endocrine glands and specialized tissue in other organs. They circulate in the blood and other body fluids, and help maintain an optimal internal environment. The endocrine system initiates the changes that take place at puberty, and governs many of those that are associated with our overall metabolism.

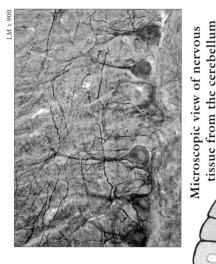

LM x 900

TISSUES: *The* FABRIC *of the* BODY

THE BODY IS MADE UP OF TISSUES – collections of similar cells that are specialized to perform a particular function. Tissues fall into five main categories: muscle; nervous tissue; blood; epithelial tissue, which covers or lines surfaces and forms glands; and connective tissue (including bone), which supports or covers organs. All categories subdivide into a variety of individual types, with their own specialized functions. The microscopic study of tissues is called histology.

Microscopic view of nervous tissue from the cerebellum
In this sample taken from the cerebellum, a number of flask-shaped nerve cells (Purkinje cells) are seen. Thin projecting fibers extend through the cerebellum from the Purkinje cell bodies.

Nervous tissue
Nervous tissue consists of a network made up of two basic types of cell. Neurons transmit and receive electrical impulses via fibers that project from the cell body. Glial cells play a supporting role. The gray matter of the brain and spinal cord consists mainly of neuron bodies; the white matter is mainly nerve fibers.

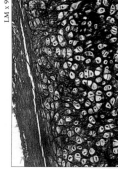

LM x 90

CROSS-SECTION OF THE BRAIN

White matter

Gray matter

Cerebellum

Brain stem

SIDE VIEW OF THE LARYNX

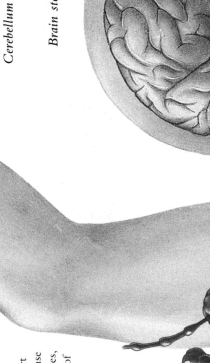

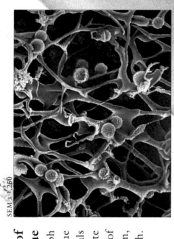

SEM x 1260

Microscopic view of lymphatic tissue
This is an electron micrograph of a piece of lymphatic tissue inside a lymph node. It reveals a fine, fibrous network. White blood cells, a key feature of the body's immune system, are entangled within its mesh.

Lymphatic tissue
This type of tissue has features of both connective tissue and blood. It forms part of the immune system – the body's defense against infection. It occurs in lymph nodes, the spleen, the thymus, and in the walls of the digestive and respiratory tracts.

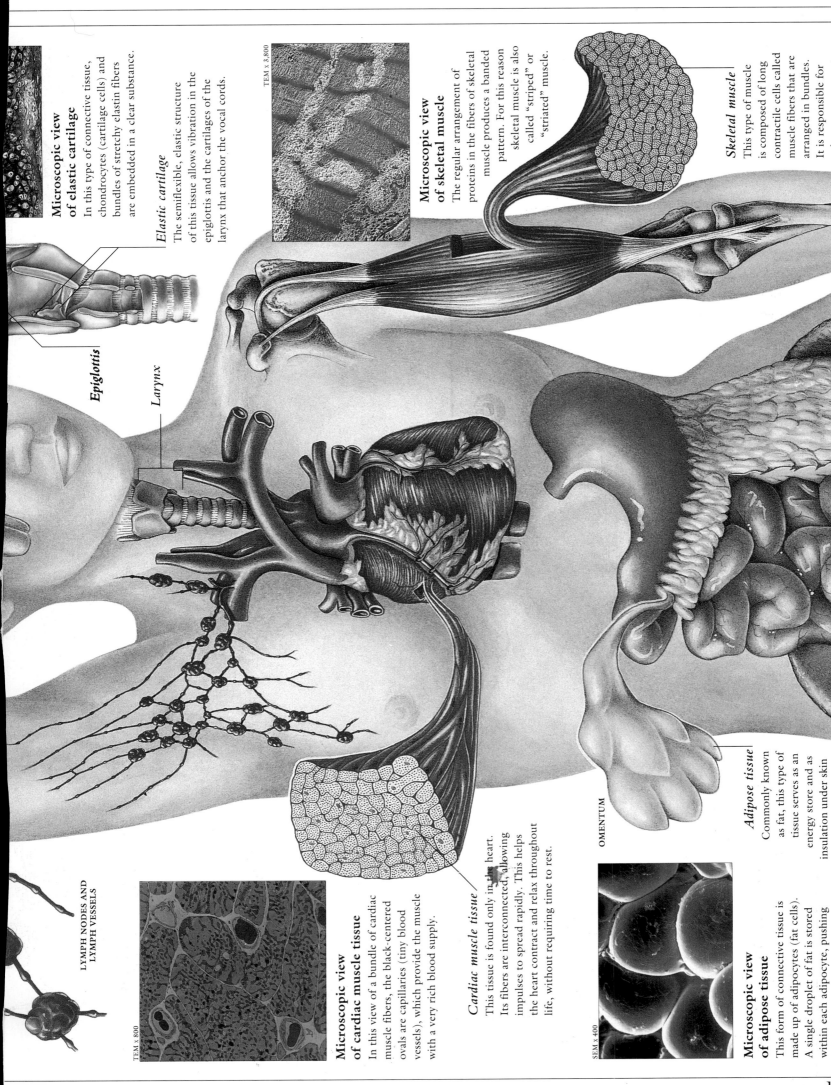

Microscopic view of elastic cartilage
In this type of connective tissue, chondrocytes (cartilage cells) and bundles of stretchy elastin fibers are embedded in a clear substance.

Elastic cartilage
The semiflexible, elastic structure of this tissue allows vibration in the epiglottis and the cartilages of the larynx that anchor the vocal cords.

TEM x 3,800

Microscopic view of skeletal muscle
The regular arrangement of proteins in the fibers of skeletal muscle produces a banded pattern. For this reason skeletal muscle is also called "striped" or "striated" muscle.

Skeletal muscle
This type of muscle is composed of long contractile cells called muscle fibers that are arranged in bundles. It is responsible for voluntary movement.

Epiglottis

Larynx

LYMPH NODES AND
LYMPH VESSELS

Microscopic view of cardiac muscle tissue
In this view of a bundle of cardiac muscle fibers, the black-centered ovals are capillaries (tiny blood vessels), which provide the muscle with a very rich blood supply.

TEM x 800

Cardiac muscle tissue
This tissue is found only in the heart. Its fibers are interconnected, allowing impulses to spread rapidly. This helps the heart contract and relax throughout life, without requiring time to rest.

OMENTUM

Adipose tissue
Commonly known as fat, this type of tissue serves as an energy store and as insulation under skin and around organs.

Microscopic view of adipose tissue
This form of connective tissue is made up of adipocytes (fat cells). A single droplet of fat is stored within each adipocyte, pushing the nucleus and cytoplasm aside.

SEM x 400

MICROSCOPY

The simplest microscopic technique uses focused light rays and magnifying lenses. Specimens viewed through a light microscope – labeled in this book as LM – can be enlarged about 1,500 times. Higher magnifications can be achieved with techniques that use beams of electrons rather than light, either transmission electron microscopy (TEM) or scanning electron microscopy (SEM).

TEM of skin section

For transmission electron microscopy, the specimen is sliced extremely thinly. An electron beam focused by electromagnets is passed through the specimen and onto a photographic plate or fluorescent screen. The images can be magnified up to five million times.

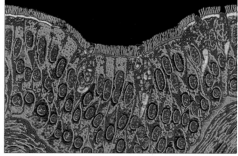

TEM x 700

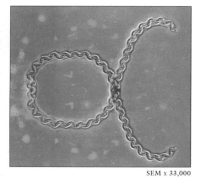

SEM x 33,000

SEM of *Leptospira* bacterium

For scanning electron microscopy, the surface of the specimen is coated with a very thin gold film, over which an electron beam is scanned. Secondary electrons, whose intensity varies with the surface contours, bounce off. The emission patterns create a 3-D image, magnified up to 100,000 times.

PLAIN X-RAYS

X-rays are short electromagnetic waves. When passed through the body to strike photographic film, they create shadow images. Dense structures such as bone absorb more X-rays and show up white, while soft tissues such as muscle appear as shades of gray.

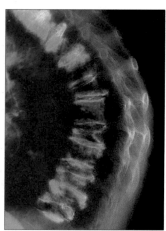

X-ray of deformed spine

This color-enhanced X-ray of the thoracic spine shows severe deformity due to osteoporosis.

X-RAY

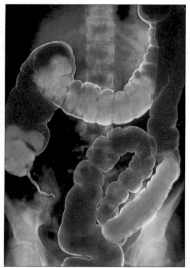

BARIUM X-RAY

CONTRAST X-RAYS

To view hollow or fluid-filled structures clearly, they must first be filled with a contrast medium, a substance that is opaque to X-rays. Angiography involves injecting contrast into blood vessels. For examining the digestive tract, a barium sulfate mixture is swallowed, or is passed into the rectum via a tube (barium enema).

Barium X-ray of colon

The image at left of a healthy colon was taken after a barium enema.

CT SCANNING

Computerized tomography (CT) scanning uses an X-ray scanner that rotates around the patient, and a computer that records the varying proportions of X-rays absorbed by tissues of different densities. This information is used to construct cross-sectional images ("slices") of the body.

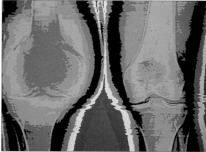

CT scan of sarcoma

CT scans are an especially useful means of investigating bleeding, assessing injuries, or diagnosing the cause of abnormal swellings. The image shows a soft-tissue tumor, called a sarcoma, in the right knee (left of image).

CT SCAN

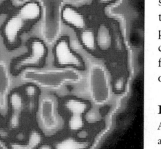

PET SCANNING

The technique of positron emission tomography (PET) scanning relies on chemical tracers that emit radioactive particles called positrons. It can provide data about the function as well as structure of organs such as the brain.

PET scan of the brain

A radioactive tracer reveals areas of brain activity from high (yellow) to low (blue).

PET SCAN

ULTRASOUND SCANNING

Sound waves of an extremely high frequency are emitted by a device called a transducer as it is passed back and forth over the part of the body being examined, such as the uterus. The sound waves echo back to the transducer and are analyzed by a computer; it creates an image that is displayed on a screen.

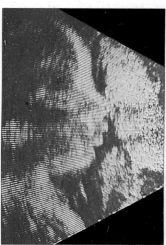

Ultrasound scan of a fetus

Because no radiation is used, this technique is a safe and reliable method for examining the fetus.

ULTRASOUND SCAN

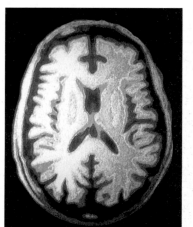

MRI

In magnetic resonance imaging (MRI), the person lies inside a magnetic chamber, which causes the nuclei of hydrogen atoms in the body to line up. A pulse of radio waves is released, throwing the atoms out of alignment. As they realign, they produce radio signals that are analyzed by a computer to create an image.

MRI scan of the brain

MRI scans are particularly useful for studying the brain and spinal cord.

MRI SCAN

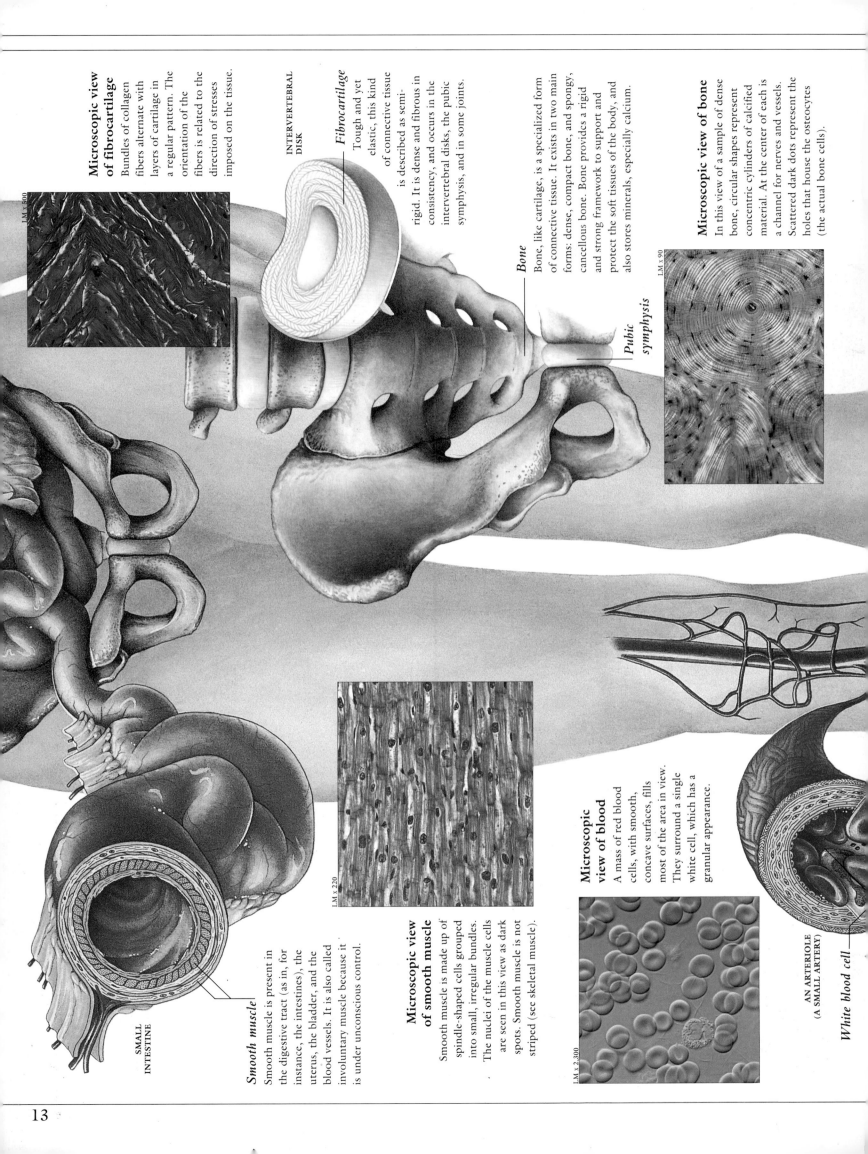

Microscopic view of fibrocartilage

Bundles of collagen fibers alternate with layers of cartilage in a regular pattern. The orientation of the fibers is related to the direction of stresses imposed on the tissue.

LM x 500

INTERVERTEBRAL DISK

Fibrocartilage

Tough and yet elastic, this kind of connective tissue is described as semi-rigid. It is dense and fibrous in consistency, and occurs in the intervertebral disks, the pubic symphysis, and in some joints.

Bone

Bone, like cartilage, is a specialized form of connective tissue. It exists in two main forms: dense, compact bone, and spongy, cancellous bone. Bone provides a rigid and strong framework to support and protect the soft tissues of the body, and also stores minerals, especially calcium.

Microscopic view of bone

In this view of a sample of dense bone, circular shapes represent concentric cylinders of calcified material. At the center of each is a channel for nerves and vessels. Scattered dark dots represent the holes that house the osteocytes (the actual bone cells).

LM x 90

Pubic symphysis

Smooth muscle

Smooth muscle is present in the digestive tract (as in, for instance, the intestines), the uterus, the bladder, and the blood vessels. It is also called involuntary muscle because it is under unconscious control.

SMALL INTESTINE

Microscopic view of smooth muscle

Smooth muscle is made up of spindle-shaped cells grouped into small, irregular bundles. The nuclei of the muscle cells are seen in this view as dark spots. Smooth muscle is not striped (see skeletal muscle).

LM x 220

Microscopic view of blood

A mass of red blood cells, with smooth, concave surfaces, fills most of the area in view. They surround a single white cell, which has a granular appearance.

LM x 2,300

AN ARTERIOLE (A SMALL ARTERY)

White blood cell

CHAPTER 1

CELLS, SKIN, *and* EPITHELIUM

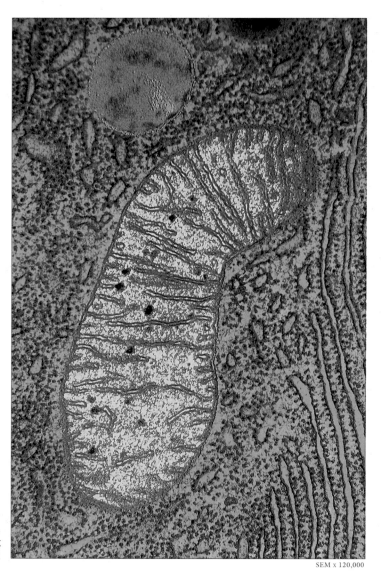

A mitochondrion, an energy-producing unit in the cell

SEM x 120,000

IMMUNE SYSTEM

The immune system's defenses help provide vital protection against infectious disease and malfunctions of the internal systems of the body. In a healthy person, the intricate interrelationship of physical, cellular, and chemical defenses can help protect against many threats, but poor general health lowers the body's resistance.

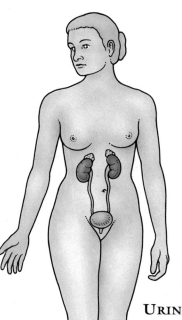

RESPIRATORY SYSTEM

The respiratory tract, working together with breathing muscles, carries air into and out of the lungs, where gases are exchanged. The cardiovascular system transports these gases to and from all body tissues, supplying vital oxygen and removing waste carbon dioxide. A variety of viruses, bacteria, and chemicals contaminate most of the air we breathe; overcoming these threats to our health is a vital role played by the immune system.

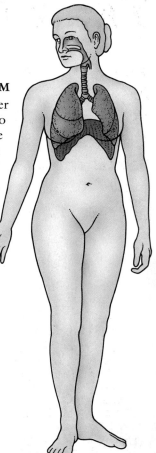

DIGESTIVE SYSTEM

The 30ft (9m) of tubing between the mouth and anus has a complex range of functions. It is needed to store food, digest it, eliminate the wastes, and make optimal use of nutrients. A healthy digestion depends on the proper functioning of the immune and nervous systems. Psychological health is an aid to efficient digestion.

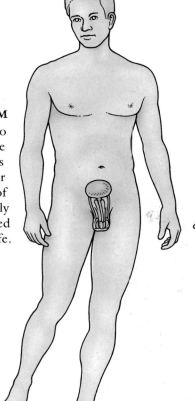

REPRODUCTIVE SYSTEM

Although small when compared to other body systems, the reproductive system is without doubt the body's biological centerpiece. Unlike other systems, it functions for only part of the human lifespan. It is also the only system that can be surgically removed without threatening a person's life.

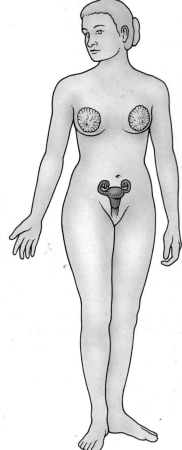

URINARY SYSTEM

The formation of urine by the kidneys eliminates wastes and helps maintain the body's water and chemical balance. Production of urine is influenced by blood flow, blood pressure, hormones, and various general rhythms and cycles of the body, such as sleeping and waking.

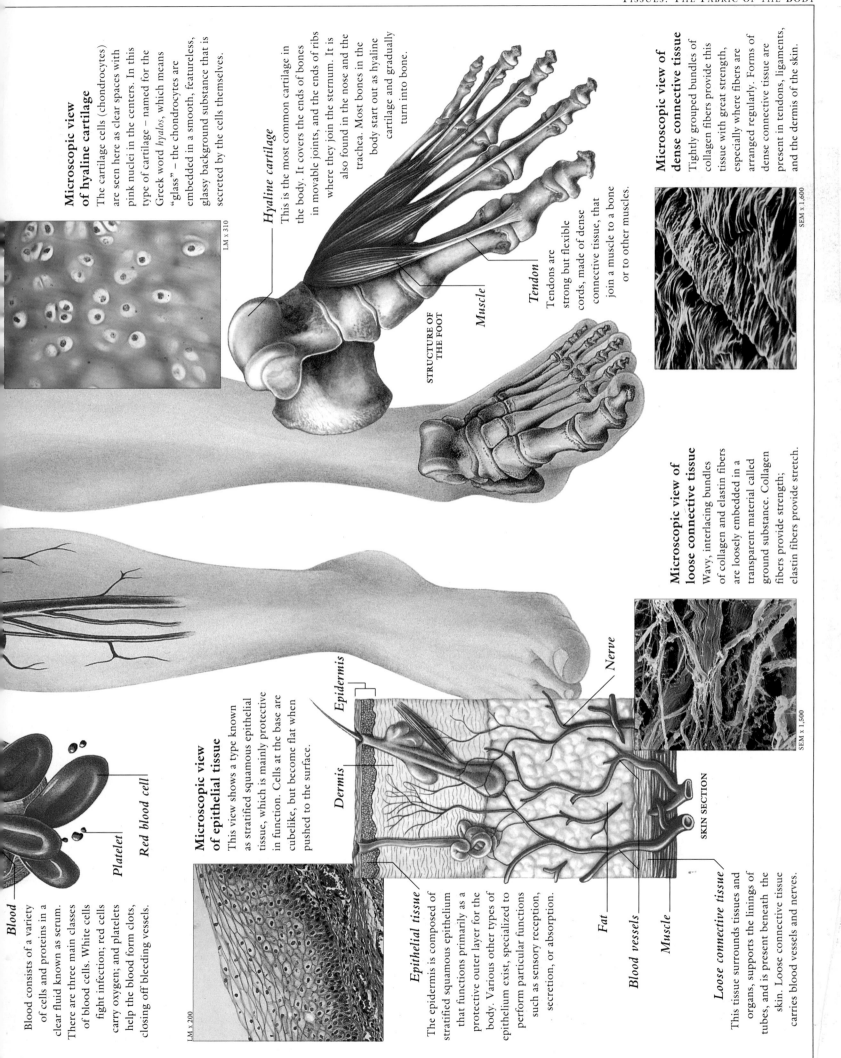

Microscopic view of hyaline cartilage

The cartilage cells (chondrocytes) are seen here as clear spaces with pink nuclei in the centers. In this type of cartilage – named for the Greek word *hyalos*, which means "glass" – the chondrocytes are embedded in a smooth, featureless, glassy background substance that is secreted by the cells themselves.

LM x 310

Hyaline cartilage
This is the most common cartilage in the body. It covers the ends of bones in movable joints, and the ends of ribs where they join the sternum. It is also found in the nose and the trachea. Most bones in the body start out as hyaline cartilage and gradually turn into bone.

Microscopic view of dense connective tissue

Tightly grouped bundles of collagen fibers provide this tissue with great strength, especially where fibers are arranged regularly. Forms of dense connective tissue are present in tendons, ligaments, and the dermis of the skin.

SEM x 1,600

Muscle

Tendon
Tendons are strong but flexible cords, made of dense connective tissue, that join a muscle to a bone or to other muscles.

STRUCTURE OF THE FOOT

Microscopic view of loose connective tissue

Wavy, interlacing bundles of collagen and elastin fibers are loosely embedded in a transparent material called ground substance. Collagen fibers provide strength; elastin fibers provide stretch.

SEM x 1,500

Nerve

Epidermis

Dermis

Microscopic view of epithelial tissue

This view shows a type known as stratified squamous epithelial tissue, which is mainly protective in function. Cells at the base are cubelike, but become flat when pushed to the surface.

LM x 200

Blood
Blood consists of a variety of cells and proteins in a clear fluid known as serum. There are three main classes of blood cells. White cells fight infection; red cells carry oxygen; and platelets help the blood form clots, closing off bleeding vessels.

Platelet

Red blood cell

Epithelial tissue
The epidermis is composed of stratified squamous epithelium that functions primarily as a protective outer layer for the body. Various other types of epithelium exist, specialized to perform particular functions such as sensory reception, secretion, or absorption.

Fat

Blood vessels

Muscle

Loose connective tissue
This tissue surrounds tissues and organs, supports the linings of tubes, and is present beneath the skin. Loose connective tissue carries blood vessels and nerves.

SKIN SECTION

INTRODUCTION

Human beings begin life as a single, newly fertilized cell. Like every cell that contains a nucleus, the fertilized cell holds all the instructions for its future growth and development. The term "cell" was first applied by Robert Hooke, a seventeenth-century English scientist, who compared the internal structure of a piece of cork to the "cells" inhabited by monks in a monastery. The characteristics common to all living cells include the ability to reproduce, exchange gases,

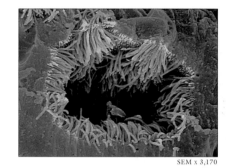

Epithelial tissue lining the trachea

SEM x 3,170

move, react to external stimuli, and create or utilize energy to perform their tasks. During the course of evolution, the cells of the body have become ever more specialized. One example is in the retina of the eye, where there are two kinds of cone cell, some reacting to red light and some to blue or green. When similar cells are grouped together, they form tissues, such as the epithelial cells that make up the protective coverings of body surfaces and the linings of the lungs and the intestines. The outer layer of the skin (the epidermis) is another type of epithelial tissue. This is subject to continual wear and tear: although it is capable of a fair amount of self-repair, it is also vulnerable to a wide range of disorders, from minor rashes to cancer.

A boil – a common disorder of the skin

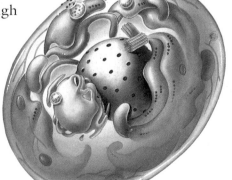

Features of a human cell

CELL STRUCTURE

THE CELL IS THE FUNDAMENTAL UNIT OF LIFE. It is the smallest structure of the body capable of performing all the processes that define life, including respiration, movement, digestion, and reproduction – although not every cell can perform all these functions. Most cells are invisible to the naked eye. Even the female germ cell, the largest in the body, is only as large as the period at the end of this sentence. Size and shape vary with cell functions.

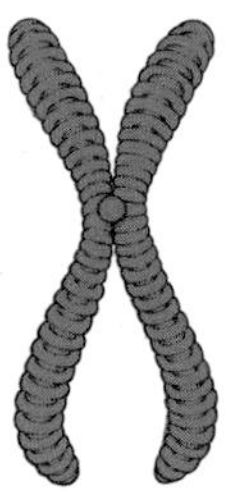

Chromosome
Just before dividing, a cell's genetic material, chromatin, duplicates and coils into rodlike bodies; these link to form X-shaped chromosomes.

FEATURES OF CELLS

Most human cells contain smaller substructures known as organelles ("little organs"); each performs a specialized task, and most are surrounded by a membrane. Organelles float in cytoplasm, a jellylike substance of which 90 percent is water. The cell also contains enzymes, amino acids, and other molecules needed for cell functions.

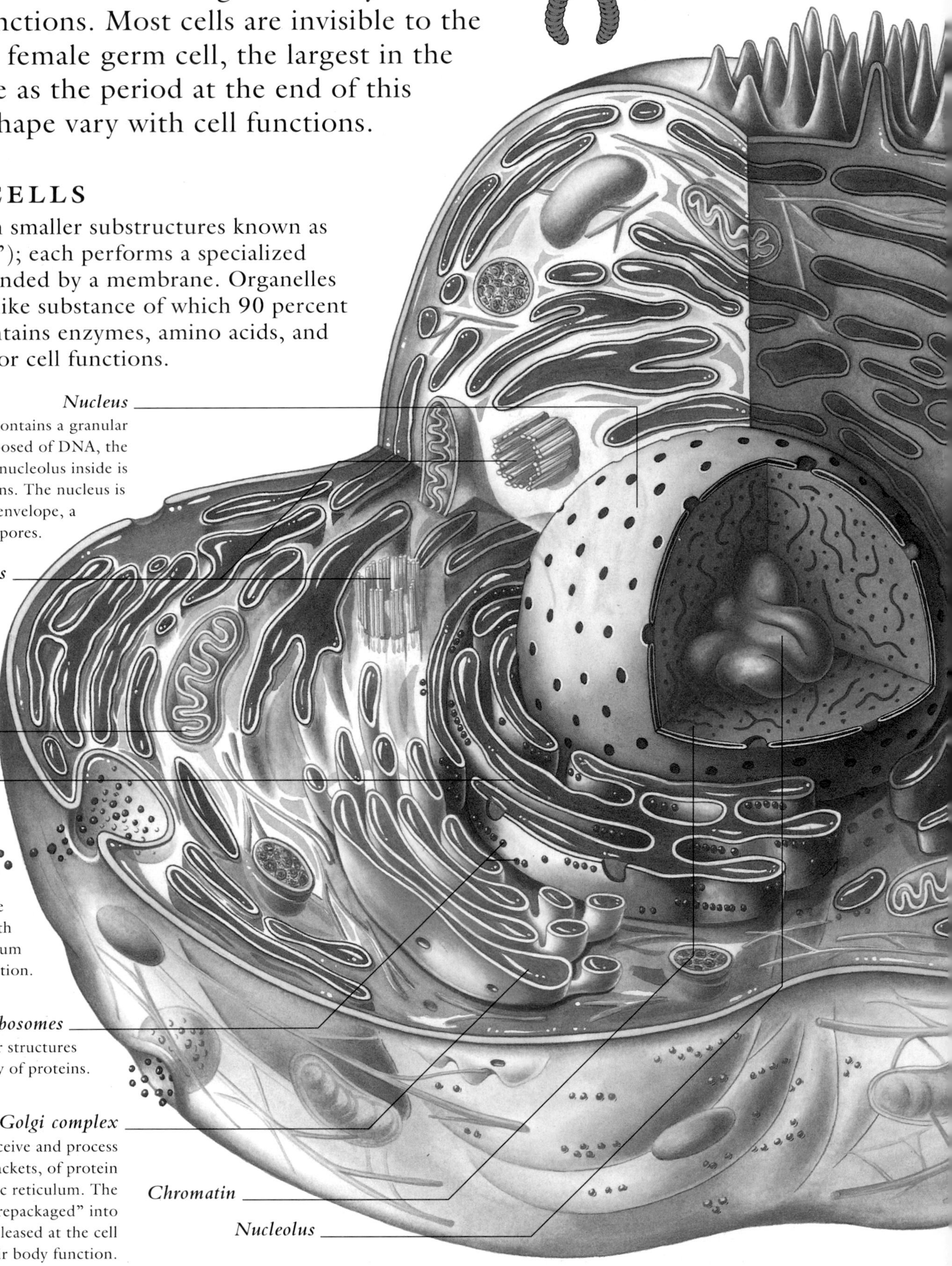

Nucleus
The cell's control center contains a granular material, chromatin, composed of DNA, the cell's genetic material. The nucleolus inside is made up of RNA and proteins. The nucleus is surrounded by the nuclear envelope, a two-layered membrane with pores.

Centrioles
Located near the center of the cell are these two cylinders; each is made up of nine pairs of hollow tubules. The centrioles play an important role in cell division.

Mitochondrion

Endoplasmic reticulum
This network of tubules and thin, curved sacs helps transport materials through the cell. Rough reticulum is the site of attachment for ribosomes, which play a role in protein formation; smooth reticulum is the site of calcium storage and fat production.

Ribosomes
These small, granular structures function in the assembly of proteins.

Golgi complex
Stacks of flattened sacs receive and process small vesicles, or sealed packets, of protein from the rough endoplasmic reticulum. The proteins are modified and "repackaged" into larger vesicles and then released at the cell membrane to perform their body function.

Chromatin

Nucleolus

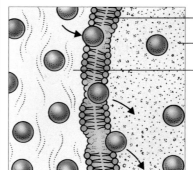

Mitochondrion
This organelle, the cell's powerhouse, is the site for both respiration and the breakdown of fats and sugars in order to produce energy. The inner folds contain enzymes that produce the energizing chemical called adenosine triphosphate (ATP), which provides energy needed for many cell functions.

TEM x 12,000

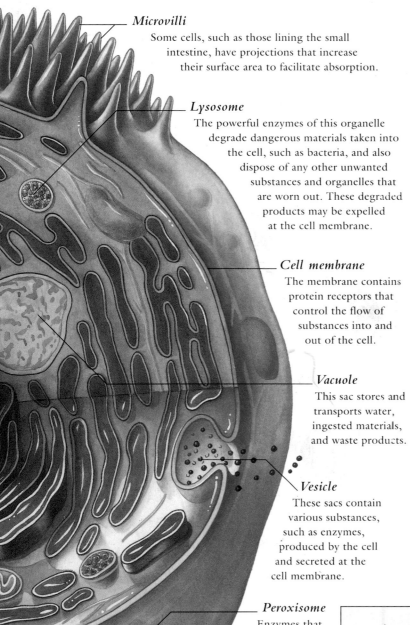

Microvilli
Some cells, such as those lining the small intestine, have projections that increase their surface area to facilitate absorption.

Lysosome
The powerful enzymes of this organelle degrade dangerous materials taken into the cell, such as bacteria, and also dispose of any other unwanted substances and organelles that are worn out. These degraded products may be expelled at the cell membrane.

Cell membrane
The membrane contains protein receptors that control the flow of substances into and out of the cell.

Vacuole
This sac stores and transports water, ingested materials, and waste products.

Vesicle
These sacs contain various substances, such as enzymes, produced by the cell and secreted at the cell membrane.

Peroxisome
Enzymes that are made here oxidize some cell substances.

Cytoskeleton
The internal framework of the cell is made up of two main types of structure. Prominent in all cells are filaments, which are thought to provide support for the cell and are sometimes associated with the plasma membrane. Hollow microtubules are thought to aid movement of substances through the cell's watery cytoplasm.

TRANSPORT MECHANISMS

The cell membrane regulates the substances that flow in and out of the cell. Because cell membranes allow only certain substances to pass through, determined in part by the cell's role in the body, they are called selectively permeable. Cell membranes may contain several types of receptor protein, each responding to a specific molecule. Some membrane proteins bind to each other, forming connections between cells.

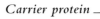

Cell membrane

Cell interior

Fluid outside cell

Diffusion
Diffusion describes the random movement of molecules from areas of high concentration to areas of low concentration. Movement of liquids or gases generally occurs by diffusion.

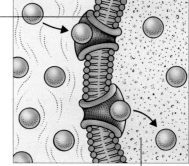

Carrier protein

Facilitated diffusion
A carrier protein temporarily binds with large molecules, such as glucose, outside of the cell membrane. It then changes its shape, opening up to the interior of the cell. Every substance is carried by a specific protein.

Cell interior

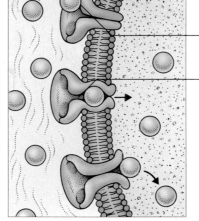

Molecule at receptor site

Protein forms channel

Active transport
If substances must be moved from areas of low to high concentration, energy is supplied by ATP. Molecules bind to a receptor site on the cell membrane, triggering a protein to change into a channel through which molecules are squeezed in and out.

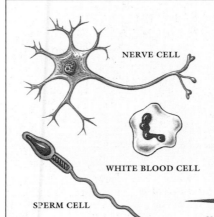

NERVE CELL

WHITE BLOOD CELL

SPERM CELL

SMOOTH (INVOLUNTARY) MUSCLE CELL

TYPES OF CELL
Each cell has a shape, size, and life span adapted for its function. Nerve signals are dispatched along the axons of nerve cells. The flexible membrane of the white blood cells allows them to squeeze through the tiny spaces between capillaries. Sperm have whiplike tails to propel them up into the genital tract. Muscle cells can change their length in order to vary contractile force.

DNA: CONTROLLER *of* CELL ACTIVITY

KEY TO BASES
Adenine
Guanine
Thymine
Cytosine

DNA (DEOXYRIBONUCLEIC ACID), the material from which the chromosomes of a cell's nucleus are formed, governs cell growth and inheritance. Shown on these pages is the process by which DNA instructs chemical compounds to form, or synthesize, the proteins that control specific cell functions. Protein synthesis starts when DNA temporarily unwinds at specific points.

Cell nucleus

Cell membrane

Cytoplasm

Mitochondrion

Endoplasmic reticulum studded with ribosomes

Nucleotide bases
The large DNA molecule is classed as a polymer because it comprises several smaller molecules. These subunits, known as nucleotide bases, always pair in specific ways: adenine with thymine and cytosine with guanine.

Paired nucleotide bases

Nucleosome
DNA is wrapped around a core of binding proteins in beadlike bodies seen when the chromosomes unwind.

1 The nucleus of every human cell contains 46 chromosomes; each is a long, coiled molecule of DNA, and together they contain about 100,000 genes. Every gene is a tiny segment of DNA that controls a specific cell function by governing the synthesis, or manufacture, of a specific protein.

2 When the threadlike chromosome is unraveled, DNA structure is seen to be two intertwined strands: a double helix. Each strand is composed of four types of subunits called nucleotide bases projecting from a backbone of sugar and phosphate.

DNA double helix unwinds

Base triplet

Sugar-phosphate backbone

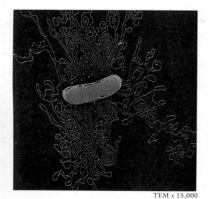

TEM x 15,000

The DNA molecule
Seen at left is *Escherichia coli*, a normally harmless bacterium found in the intestinal tract. It is surrounded by its DNA, whose total length is 1,000 times that of the bacterium. In humans, a DNA molecule is longer than the body itself.

3 A unit of three successive pairs of nucleotide bases is called a triplet. Each triplet carries the code for one of the 20 amino acids, the building blocks that form proteins. The sequence of the pairs in each segment of DNA – or gene – determines which protein is synthesized under the control of that particular gene.

THE MANY ROLES OF PROTEINS

Proteins are needed for development and growth as well as for carrying out vital chemical functions in the body. Some proteins form structures such as hair and muscle. Others serve as antibodies, hormones, or enzymes, or, like oxygen-carrying hemoglobin, transport substances in the body.

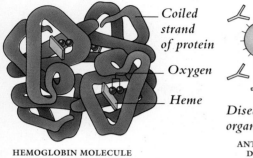

Coiled strand of protein

Oxygen

Heme

HEMOGLOBIN MOLECULE

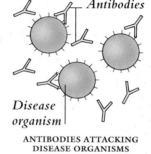

Antibodies

Disease organism

ANTIBODIES ATTACKING DISEASE ORGANISMS

MITOSIS

Mitosis is a simple copying process that organizes and redistributes DNA during cell division in all normal growth. It occurs continuously during growth and also as the body replaces old, worn-out cells. During this copying process, one cell produces two daughter cells that are identical to each other and to the parent cell. Cells of the growing embryo multiply by mitosis, as do adult tissues such as skin and intestinal lining.

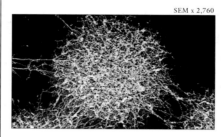

SEM x 2,760

Chromatin

The stage that precedes actual cell division is called interphase. During this time, molecules of DNA are loosely organized into a network of extended filaments, which is known as chromatin.

4 For protein synthesis, the DNA strands temporarily separate along the length of the gene that governs the production of that protein. Only one strand carries the genetic code and acts as a template for the formation of messenger ribonucleic acid (mRNA). The process of creating a molecule of mRNA from DNA is known as transcription.

Molecule of messenger RNA (shown here as a 6-based strand)

Pore in nuclear membrane

5 Once the mRNA has formed, the strands of DNA reunite and mRNA leaves the cell nucleus and enters the cytoplasm. Here it attaches to structures known as ribosomes; using raw materials of the cell, ribosomes produce the protein by following the sequence of nucleotide bases in the mRNA.

Molecule of messenger RNA

Ribosome

Amino acids

Amino acid chain growing into a protein

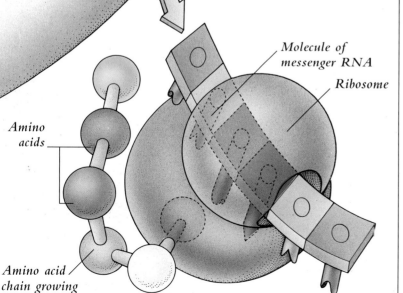

Prophase

DNA strands replicate and coil up, forming spiral filaments called chromatids that join at the centromere. These filaments then condense and form 46 X-shaped pairs of chromosomes.

Cytoplasm

Nucleus

Chromatin

Paired chromosomes

Centromere

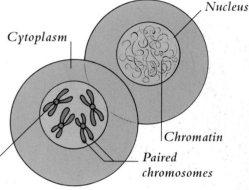

Threads of the spindle

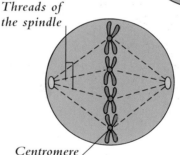

Centromere

Metaphase

Chromosome pairs line up in the center of the cell. Threadlike fibers create the spindle that connects the centromere of each chromosome pair to opposite poles of the dividing cell.

Anaphase

Each centromere splits, dividing the paired chromosomes so that there are 92 single chromosomes. One-half, or 46, of these daughter chromosomes move toward each side of the cell.

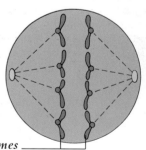

Daughter chromosomes

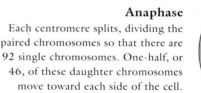

Telophase

Spindle fibers disappear, and a nuclear membrane forms around each group of 46 daughter chromosomes. The cell becomes pinched in the middle, and the chromosomes start to uncoil.

Late telophase

The cytoplasm begins to divide, a cell plate forms between the two groups of chromosomes, and the cell splits into two. Each cell has a set of 46 chromosomes; these revert to chromatin filaments.

Chromatin

Nucleus

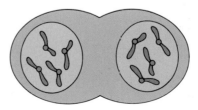

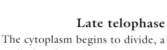

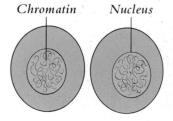

SKIN STRUCTURE *and* EPITHELIAL TISSUES

SKIN IS THE BODY'S PROTECTIVE OUTER BOUNDARY. It is one of the largest organs of the body, and is self-repairing. Skin also plays an essential role in regulating body temperature. A variety of sensory receptors make many parts of this organ sensitive to the lightest touch. Because its appearance can alter with both emotional states and general health, skin reveals signs of a wide range of disorders.

SEM x 100

Hair on the scalp
Hair grows from follicles located in the dermis. The scaly cells covering each hair are lubricated by oil glands.

THE STRUCTURE OF SKIN

Skin is composed of two main layers. The outer layer is the stratified squamous epithelial tissue, and consists of sheets of cells that are flatter and more scaly near the surface. The inner dermis consists of fibrous and elastic tissue pierced by blood vessels, nerves, hair follicles, and sweat glands; its deepest layer anchors skin to the underlying tissues.

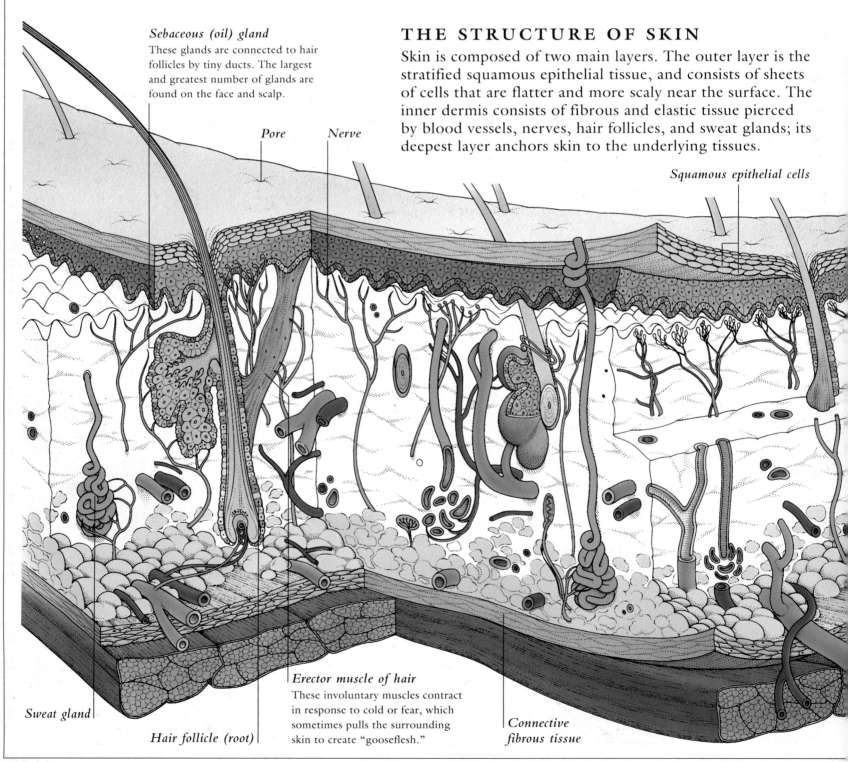

Sebaceous (oil) gland
These glands are connected to hair follicles by tiny ducts. The largest and greatest number of glands are found on the face and scalp.

Pore

Nerve

Squamous epithelial cells

Sweat gland

Hair follicle (root)

Erector muscle of hair
These involuntary muscles contract in response to cold or fear, which sometimes pulls the surrounding skin to create "gooseflesh."

Connective fibrous tissue

NAIL STRUCTURE

Nails are made of keratin, a hard, fibrous protein, which is also the main constituent of hair. The nails rest on a bed served by blood vessels, which creates the pink color; they grow from a matrix of active cells under skin folds at their base and sides.

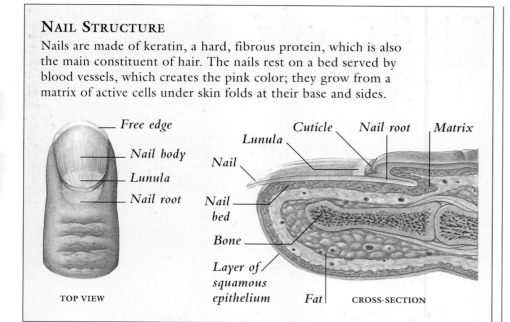

TOP VIEW
- Free edge
- Nail body
- Lunula
- Nail root

CROSS-SECTION
- Lunula
- Nail
- Cuticle
- Nail root
- Matrix
- Nail bed
- Bone
- Layer of squamous epithelium
- Fat

PSEUDOSTRATIFIED EPITHELIUM

This type of columnar epithelial tissue appears to be stratified, but actually consists of a single layer of cells of different heights. Sometimes the taller cells are specialized; they are either goblet cells, which secrete mucus, or ciliated cells, which have tiny surface hairs to trap or move foreign particles.

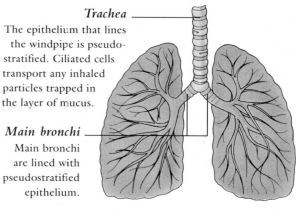

Trachea
The epithelium that lines the windpipe is pseudo-stratified. Ciliated cells transport any inhaled particles trapped in the layer of mucus.

Main bronchi
Main bronchi are lined with pseudostratified epithelium.

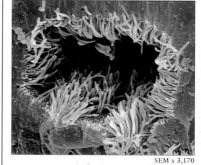

SEM x 3,170

Tracheal epithelium
Cilia that project from tracheal epithelial cells are seen here in green. Mucus-secreting goblet cells between the cilia possess tiny microvilli (seen here in yellow).

Basal cell layer
In this layer, cells constantly divide. The new cells push toward the upper surface, replacing dead or worn-out cells of the outermost layer.

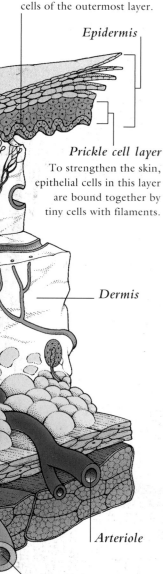

- Epidermis
- Prickle cell layer
  To strengthen the skin, epithelial cells in this layer are bound together by tiny cells with filaments.
- Dermis
- Arteriole
- Venule

EPITHELIAL TISSUES

Epithelial tissue, also called epithelium, is an important structural element that acts as a lining or covering for other body tissues. Epithelium can be classified according to cell type and arrangement of cells into one or more layers. These tissues are specialized for protection, absorption, or secretion.

SIMPLE AND STRATIFIED EPITHELIUM

Simple epithelium consists of a single layer of cells, which are called squamous if platelike or flattened, cuboidal if cube-shaped, or columnar if tall and thin. Simple epithelium is usually found in areas where substances need to pass through easily. Stratified epithelium has more than one layer, and is thus well adapted for protection.

The eye
There are two types of epithelium in the eye. Simple cuboidal epithelium occurs in the pigmented layer of the retina, whereas stratified squamous epithelium is found in the outer cornea.

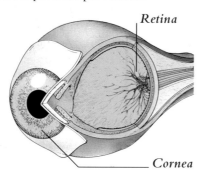

- Retina
- Cornea

The cornea
The cornea is covered by about five layers of stratified squamous epithelial cells, which form a transparent coating that permits light rays to enter the eye. Tiny ridges, called microplicae (seen on the surface of the cell at lower center), hold fluid, which helps refract the incoming rays of light.

SEM x 1,170

TRANSITIONAL EPITHELIUM

This kind of epithelial tissue is similar to stratified squamous epithelium, but has the ability to stretch without tearing. It is particularly well suited to the urinary system. Covering the columnar cells at the base are progressively rounder surface cells that flatten, or become more squamous, as they stretch.

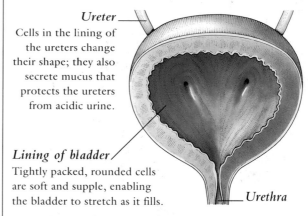

Ureter
Cells in the lining of the ureters change their shape; they also secrete mucus that protects the ureters from acidic urine.

Lining of bladder
Tightly packed, rounded cells are soft and supple, enabling the bladder to stretch as it fills.

- Urethra

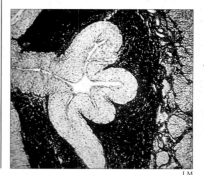

LM

Inside a ureter
When the muscles of the ureters are relaxed, epithelial cells in the lining appear folded; here they are seen to curve around the star-shaped white interior of the relaxed ureter.

SKIN DISORDERS

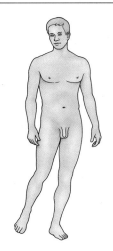

SKIN MAY BE AFFECTED BY A WIDE RANGE OF DISORDERS – from physical injury to chemical damage. It is susceptible to infection by viruses, bacteria, fungi, and protozoa, and to infestation by mites and other parasites. Skin may also develop tumors and may be affected by an inadequate blood supply. Its own glands, blood vessels, and nerves may become disordered, and skin may show various allergic reactions. Rashes and skin discoloration may reflect general body disorders.

COMMON SWELLINGS

Skin swellings may be small, inflamed, pus-filled spots known as pustules, or larger ones called boils. A cluster of boils may link and enlarge, forming a carbuncle. Other swellings may be caused by a local increase in the number of cells, as in warts, moles, or tumors (benign or malignant). Swellings may also be due to acne, cysts, and allergies.

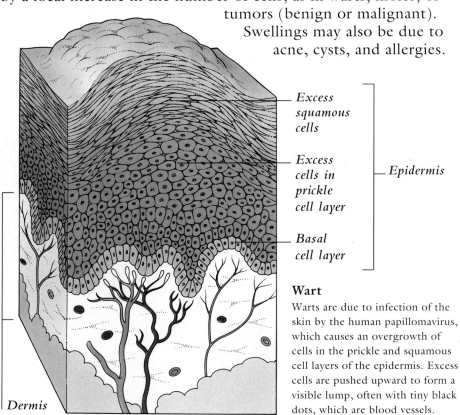

Excess squamous cells

Excess cells in prickle cell layer

Basal cell layer

Epidermis

Dermis

Mole

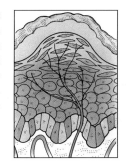

A mole, also called a nevus, is an aggregate of pigment cells (melanocytes) or of tiny blood vessels. Moles rarely become malignant, but any change in size, shape, or color should be discussed with a doctor.

Boil

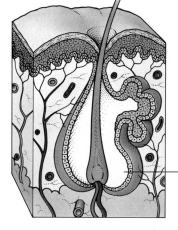

A boil is a collection of pus inside a hair follicle or a sebaceous gland. It usually results when staphylococcal bacteria infect the area, causing acute inflammation.

Pus-filled follicle

Wart

Warts are due to infection of the skin by the human papillomavirus, which causes an overgrowth of cells in the prickle and squamous cell layers of the epidermis. Excess cells are pushed upward to form a visible lump, often with tiny black dots, which are blood vessels.

Cyst

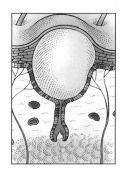

Cysts are saclike structures containing fluid or semisolid material. Most have a strong capsule. The most common type of cyst is a "wen"; it is full of sebaceous secretions.

ACNE VULGARIS

In acne vulgaris, the sebaceous (oil-secreting) glands of the skin produce excessive amounts of their secretion, known as sebum. The sebum oxidizes and forms a blackened plug in the skin pore. Trapped sebum, dead cells, and infection by bacteria inflame the area and can cause a pustule to form. Acne can be treated with skin medications, for example tretinoin, or drugs that can be taken orally, such as tetracycline.

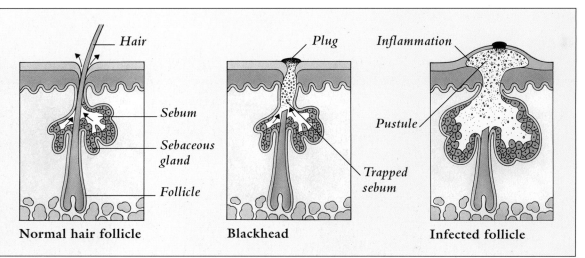

Hair

Plug

Inflammation

Sebum

Sebaceous gland

Follicle

Pustule

Trapped sebum

Normal hair follicle　　　**Blackhead**　　　**Infected follicle**

WOUNDS

Damage to or removal of part of the skin surface to variable depths is called a wound, and may occur by accident or as a result of an operation. How well a wound heals depends upon the alignment of the edges and the prevention of infection. Well-closed, clean wounds usually heal in a few weeks, while open wounds heal more slowly and usually result in a puckered scar.

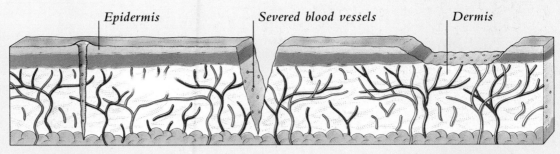

Epidermis   *Severed blood vessels*   *Dermis*

Puncture wounds
Punctures usually heal quickly but infection is a risk. Because tetanus sometimes follows puncture wounds, immunization may be necessary.

Cuts
Deep cuts often need stitching to avoid scarring. Clean cuts will heal with only minimal scarring if they are closed well.

Abrasions
Abrasions involving only the epidermis usually heal without a scar. Any deeper abrasion is likely to leave a scar.

SKIN RASHES

Rashes or groups of spots are areas of inflamed skin. They may occur in small patches, or cover a large part of the body. Among the main causes of rashes are skin conditions such as eczema and psoriasis, infectious diseases, and allergic reactions. Some types of skin rashes are accompanied by fever or itching.

Eczema

The term eczema refers to various skin inflammations with common features; these include itching, red patches, and small blisters that burst, making the skin become moist and crusty. The most common type, atopic eczema, is an allergic reaction that often initially appears in the first year of life.

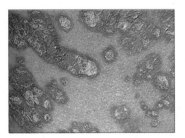

Psoriasis

This common, noninfectious skin disease of unknown cause features sharply outlined, bright red or pink, dry, nonitchy plaques with silvery, scaly surfaces. These occur mainly on the skin surface over the elbows, knees, shins, scalp, and lower back.

INFECTIOUS RASHES

Common infectious diseases such as measles, rubella (German measles), and chickenpox – as well as other less common diseases, including typhoid and scarlet fever – have a toxic effect on the skin that produces a characteristic, temporary rash. The rashes are due to organisms or their toxins circulating in the skin.

Chickenpox rash

The photograph seen at right is a close-up view of a typical chickenpox rash. The spots are itchy, fluid-filled blisters that dry out and form scabs after a few days. The rash, which is fullest on the trunk, may be accompanied by a low fever.

SKIN CANCER

Skin cancers are related to excessive exposure to sunlight. Ultraviolet light can damage DNA, and change the cell's genetic material. The type that occurs most frequently is basal cell carcinoma, which spreads only locally and not to other parts of the body. Squamous cell carcinoma and the rarer malignant melanoma are more dangerous. Skin cancers may be treated by excising (cutting away), radiation, or cryocautery (freezing).

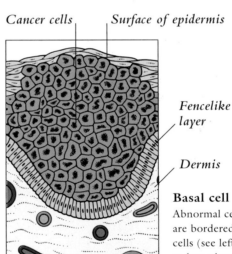

Cancer cells   *Surface of epidermis*

Fencelike layer

Dermis

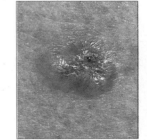

TYPICAL APPEARANCE

Basal cell carcinoma

Abnormal cells grow in the epidermis and are bordered by a fencelike basal layer of cells (see left). The tumor, also called a rodent ulcer, is firm and pearly looking, and occurs most frequently on the face.

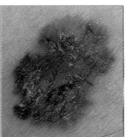

TYPICAL APPEARANCE

Cancer cells in epidermis

Cancer cells spreading into dermis

Arteriole

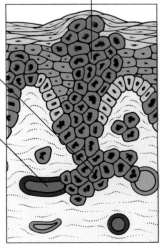

Malignant melanoma

Malignant melanoma occurs in pigment cells, called melanocytes. Cancerous cells spread through the skin layers (see right). This type of tumor is often very dark and asymmetrical, with an indistinct border. It can grow rapidly and spread to distant sites.

CHAPTER TWO

C H A P T E R   2

The SKELETAL SYSTEM

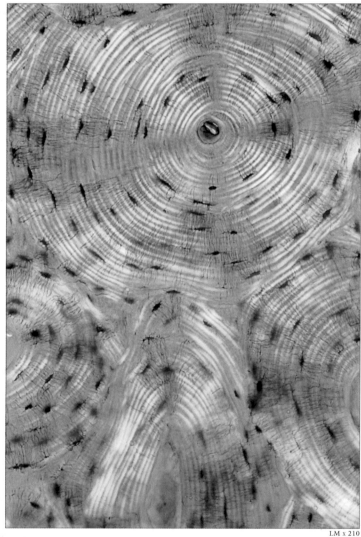

The structure
of hard bone

LM x 210

INTRODUCTION

There is more life in a bone than most people think. The living skeleton, although tough, is a flexible structure with blood flowing through every part, and it is in a state of constant growth and remodeling. Even bones discovered by archaeologists tell a story: they often reveal a great deal about the dead person's age, sex, height and weight, activities, and whether he or she was a meat-eater or a vegetarian. Fossil bones also provide clear evidence that bone disorders such as rickets and arthritis have a long history. Diseases of the bones and joints have always been among the most common causes of poor health and disability, especially in older people. However, many are preventable: the bulk and strength of bones in later life depends critically on a person's health in earlier adult life. Being overweight, for example, increases the risk of developing osteoarthritis, while a calcium-rich diet combined with moderate, regular exercise reduces the risk of a variety of bone diseases. An important recent advance in treating arthritis and some other bone disorders has been the development of artificial replacements for many joints in the body. Scientists are also exploring the complex interactions between human genes and environmental factors to find out why some bone disorders affect only a small fraction of the population.

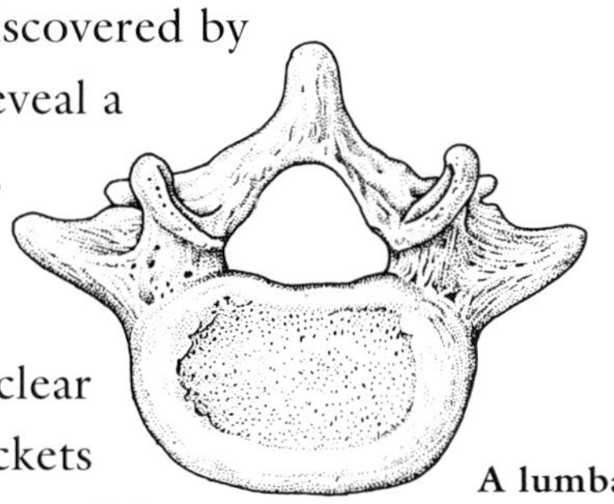

A lumbar vertebra

X-RAY

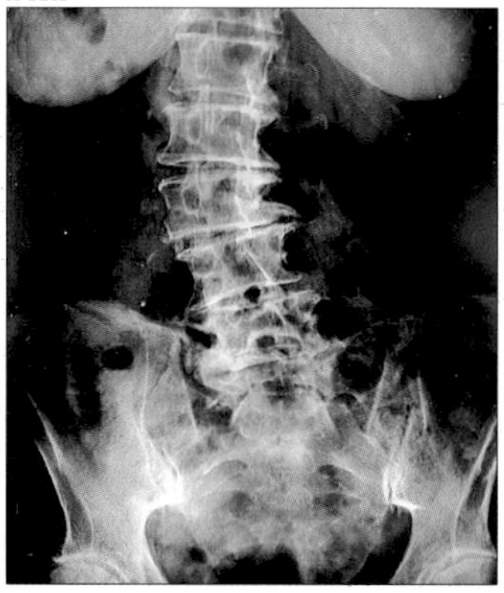

Osteoarthritis of the lumbar spine

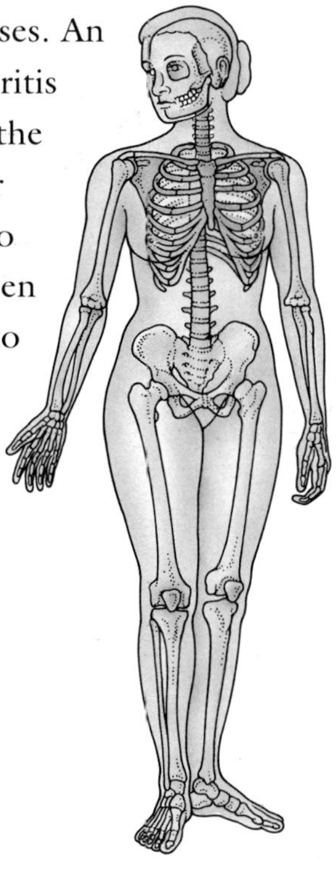

THE SKELETAL SYSTEM

BONES *of the* BODY I

BOTH STRONG ENOUGH to support weight and light enough to enable movement, bones also offer protection to internal organs and store most of the calcium, phosphorus, and other minerals, such as magnesium salts, needed by the body. While the bones seen in a museum appear dry and rigid, living bones are actually a moist hive of activity. The marrow in bones is the site where red blood cells and some types of white blood cells are produced.

THE SKELETON

The precise number of bones in the adult human skeleton varies from one person to another, but on average there are 206 bones of varying shapes and sizes. The skeleton is divided into two main parts. The central bones of the skull, ribs, vertebral column, and sternum form the axial skeleton. The bones of the arms and legs, along with the scapula, clavicle, and pelvis, make up the appendicular skeleton.

SPINAL LINK

The cylindrical linked vertebrae of the spinal column offer strong, bony protection for the spinal cord. Aided by muscles and ligaments, they support the skull, hold the body upright, and permit twisting and bending of the trunk.

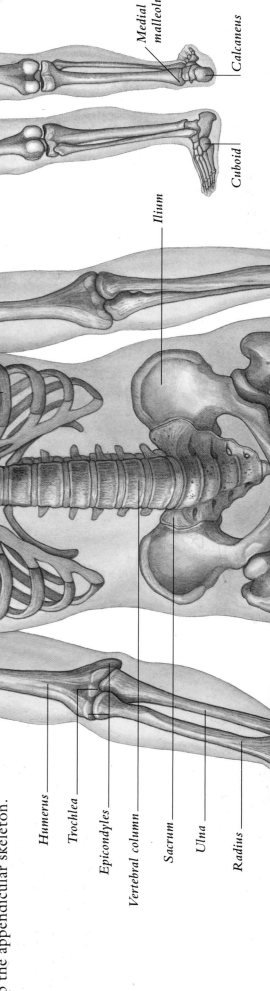

Parietal bones

Occipital bone

Coccyx

Medial malleolus

Calcaneus

Cuboid

Acromion

Scapula

Coracoid process

Ribs

Clavicle

Ilium

Sternum

Humerus

Trochlea

Epicondyles

Vertebral column

Sacrum

Ulna

Radius

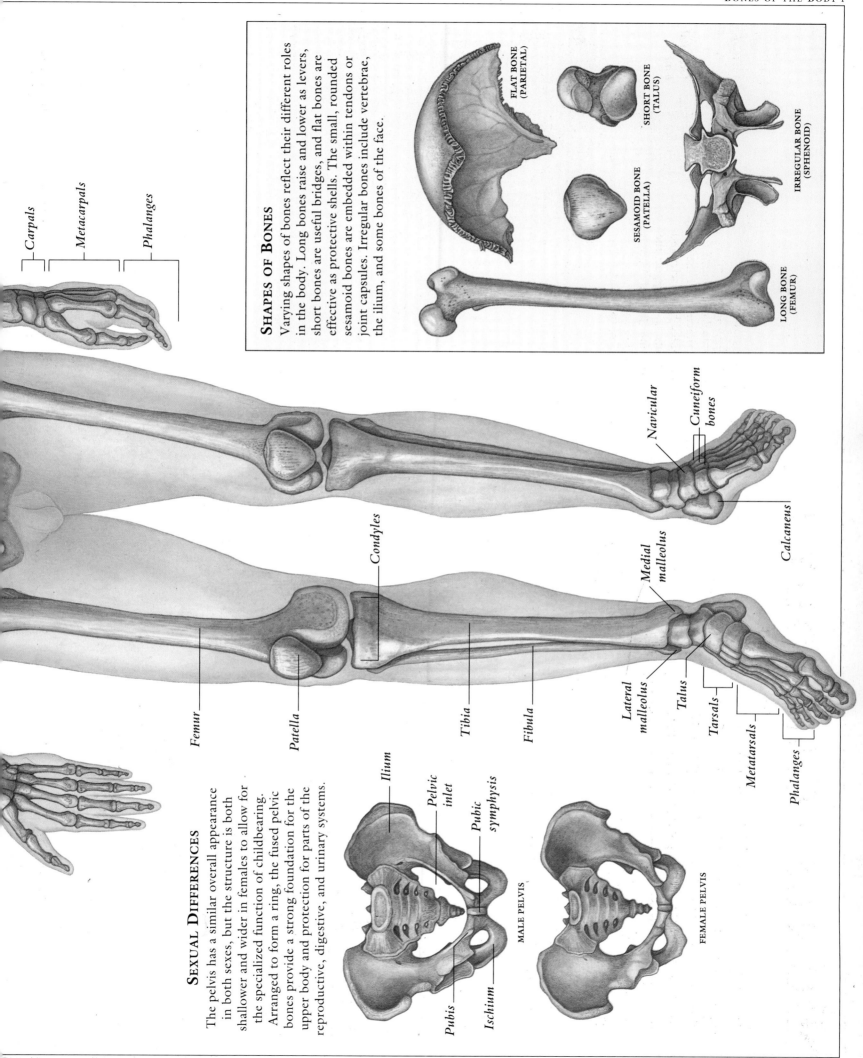

Carpals

Metacarpals

Phalanges

SHAPES OF BONES

Varying shapes of bones reflect their different roles in the body. Long bones raise and lower as levers, short bones are useful bridges, and flat bones are effective as protective shells. The small, rounded sesamoid bones are embedded within tendons or joint capsules. Irregular bones include vertebrae, the ilium, and some bones of the face.

FLAT BONE (PARIETAL)

SHORT BONE (TALUS)

IRREGULAR BONE (SPHENOID)

SESAMOID BONE (PATELLA)

LONG BONE (FEMUR)

Navicular

Cuneiform bones

Calcaneus

Condyles

Medial malleolus

Femur

Patella

Tibia

Fibula

Lateral malleolus

Talus

Tarsals

Metatarsals

Phalanges

SEXUAL DIFFERENCES

The pelvis has a similar overall appearance in both sexes, but the structure is both shallower and wider in females to allow for the specialized function of childbearing. Arranged to form a ring, the fused pelvic bones provide a strong foundation for the upper body and protection for parts of the reproductive, digestive, and urinary systems.

Ilium

Pelvic inlet

Pubic symphysis

MALE PELVIS

FEMALE PELVIS

Pubis

Ischium

BONES *of the* BODY II

A CAGE OF BONES within the thorax, or chest, demonstrates both the supportive and protective functions of the skeleton. The ribcage and its muscles form the chest wall while shielding vital internal organs such as the heart, lungs, and liver. Most people have 12 pairs of ribs, but about 5 percent are born with one or more extra ribs. Some people, such as Down's syndrome sufferers, have one pair of ribs less than is usual.

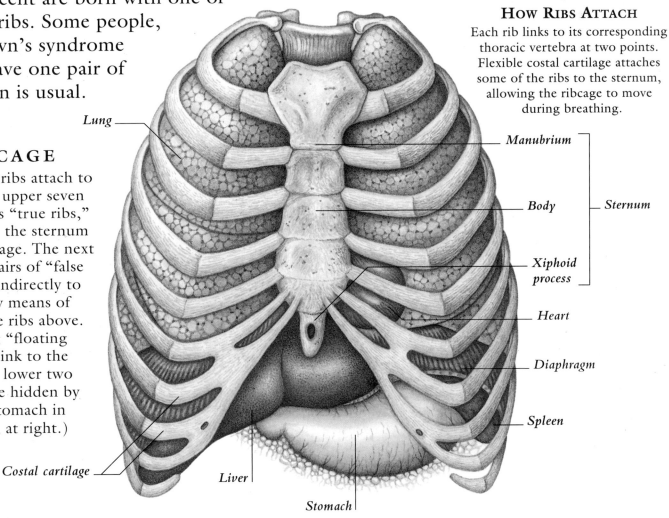

Thoracic vertebra

Rib

Costal cartilage

Points of attachment

Sternum

HOW RIBS ATTACH
Each rib links to its corresponding thoracic vertebra at two points. Flexible costal cartilage attaches some of the ribs to the sternum, allowing the ribcage to move during breathing.

THE RIBCAGE

All 12 pairs of ribs attach to the spine. The upper seven pairs, known as "true ribs," link directly to the sternum by costal cartilage. The next two or three pairs of "false ribs" connect indirectly to the sternum by means of cartilage to the ribs above. The remaining "floating ribs" have no link to the sternum. (The lower two pairs of ribs are hidden by the liver and stomach in the illustration at right.)

Lung

Manubrium

Body

Sternum

Xiphoid process

Heart

Diaphragm

Spleen

Costal cartilage

Liver

Stomach

BONES OF THE HAND AND FOOT

A similar arrangement of bones is repeated in the hand and foot, but with some key differences. For example, the phalanges in toes are generally shorter than those in fingers.

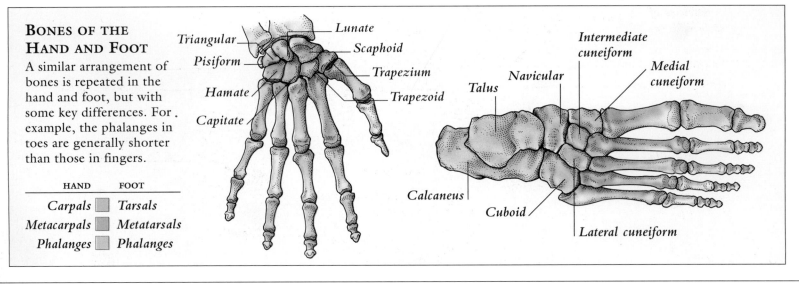

Triangular

Pisiform

Hamate

Capitate

Lunate

Scaphoid

Trapezium

Trapezoid

Talus

Navicular

Intermediate cuneiform

Medial cuneiform

Calcaneus

Cuboid

Lateral cuneiform

HAND		FOOT
Carpals		Tarsals
Metacarpals		Metatarsals
Phalanges		Phalanges

THE BONES OF THE SKULL

Two separate sets of bones form the intricate structure of the skull. The eight bones that enclose and protect the brain are called the cranial vault, while the other 14 bones compose the skeleton of the face. The wavy lines on the skull surface are actually a type of joint known as a suture. While sutures are flexible in very young children to allow for growth, with age they ossify and become virtually fixed. The mandible bone of the lower jaw is the only skull bone secured by a truly movable joint. The ossicles of the middle ear, which conduct sound waves from the eardrum to the inner ear, are not technically part of the skull.

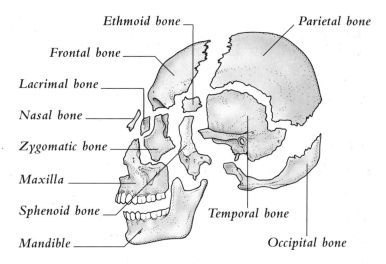

Ethmoid bone — Frontal bone — Lacrimal bone — Nasal bone — Zygomatic bone — Maxilla — Sphenoid bone — Mandible — Parietal bone — Temporal bone — Occipital bone

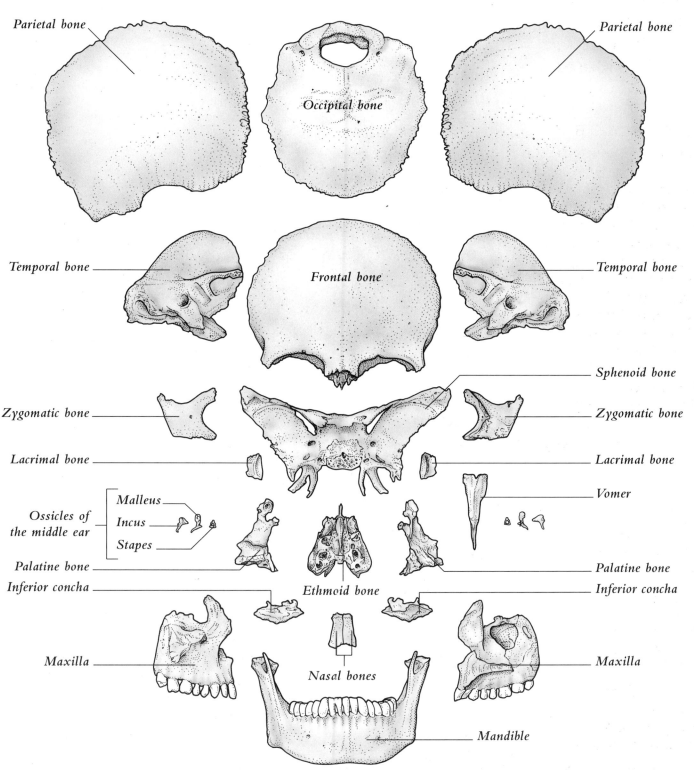

Parietal bone — Occipital bone — Parietal bone

Temporal bone — Frontal bone — Temporal bone

Sphenoid bone

Zygomatic bone — Zygomatic bone

Lacrimal bone — Lacrimal bone

Vomer

Ossicles of the middle ear — Malleus — Incus — Stapes

Palatine bone — Palatine bone

Inferior concha — Inferior concha

Maxilla — Ethmoid bone — Maxilla

Nasal bones

Mandible

BONE STRUCTURE *and* GROWTH

BONE IS A TYPE OF CONNECTIVE TISSUE that is as strong as steel but as light as aluminum. It is made up of specialized cells and protein fibers interwoven in a gel-like matrix composed of water, mineral salts, and carbohydrates. Bone tissue is not completely rigid. It continually breaks down and rebuilds, renewing its shape and proportion during the growing process and after an injury.

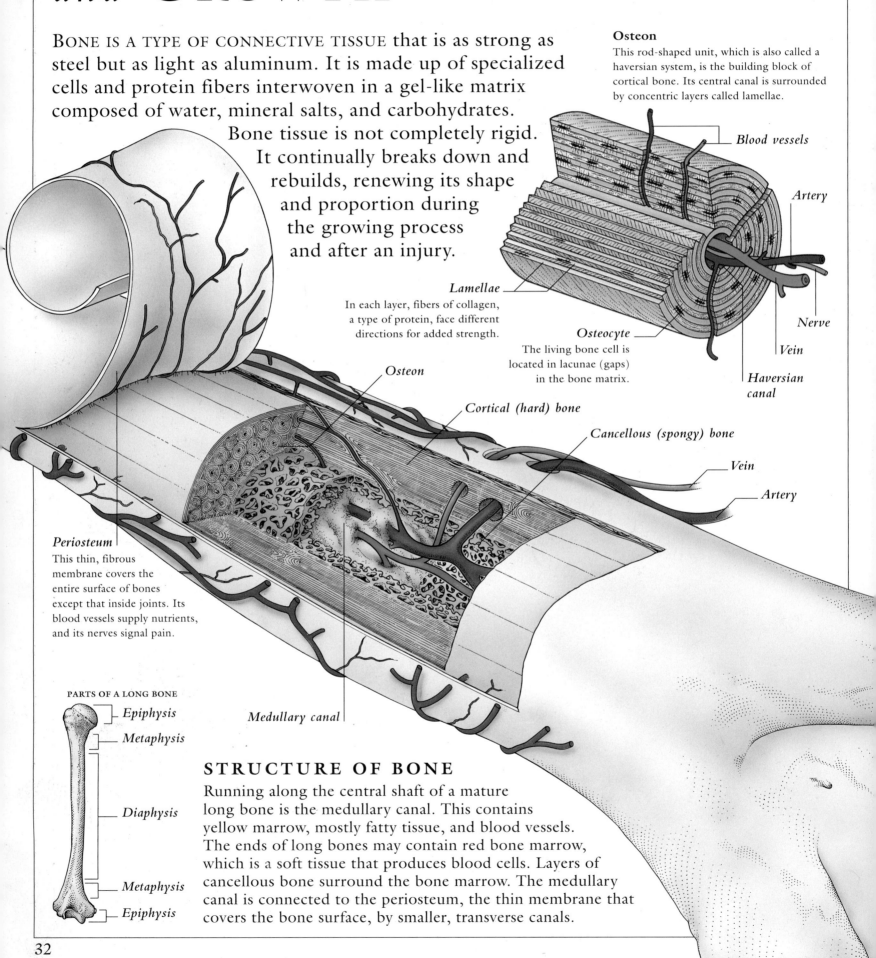

Osteon
This rod-shaped unit, which is also called a haversian system, is the building block of cortical bone. Its central canal is surrounded by concentric layers called lamellae.

Blood vessels

Artery

Nerve

Vein

Haversian canal

Lamellae
In each layer, fibers of collagen, a type of protein, face different directions for added strength.

Osteocyte
The living bone cell is located in lacunae (gaps) in the bone matrix.

Osteon

Cortical (hard) bone

Cancellous (spongy) bone

Vein

Artery

Periosteum
This thin, fibrous membrane covers the entire surface of bones except that inside joints. Its blood vessels supply nutrients, and its nerves signal pain.

PARTS OF A LONG BONE

├ *Epiphysis*

├ *Metaphysis*

├ *Diaphysis*

├ *Metaphysis*

├ *Epiphysis*

Medullary canal

STRUCTURE OF BONE

Running along the central shaft of a mature long bone is the medullary canal. This contains yellow marrow, mostly fatty tissue, and blood vessels. The ends of long bones may contain red bone marrow, which is a soft tissue that produces blood cells. Layers of cancellous bone surround the bone marrow. The medullary canal is connected to the periosteum, the thin membrane that covers the bone surface, by smaller, transverse canals.

Bone marrow

The microscopic image at right shows red bone marrow dotted with red blood cells, which are (with white blood cells) produced here. At birth, red marrow is present in all bones, but in the long bones it gradually becomes yellow marrow and loses its capacity to produce blood cells.

SEM x 340

Cortical bone

Cortical bone consists of many closely packed osteons. The central canals (black) in each osteon contain blood vessels and nerves. The tiny black dots located between the concentric lamellae are lacunae (gaps), which contain osteocytes (bone cells).

LM x 100

Cancellous bone

Shown at right is the latticework structure of cancellous bone, which consists of bony spikes called trabeculae ("little beams"). The trabeculae are arranged along the lines of greatest pressure or stress, making bones both strong and light.

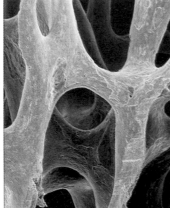

SEM x 40

BONE LENGTHENING

Near the ends of the long bones, which are covered with cartilage instead of periosteum, is an area known as the epiphyseal plate. Cartilage cells (chondrocytes) proliferate here and form columns that push older cells toward the middle of the bone shaft. As the chondrocytes enlarge and die, the space they occupied is filled by new bone cells.

The process of bone growth
Bone growth continues until the age of 17 to 21. If the epiphyseal plate is injured during adolescence, a shortened bone and impaired movement may result.

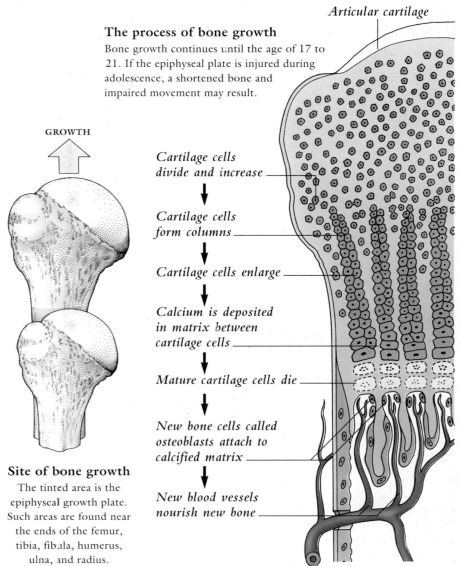

Articular cartilage

GROWTH

Cartilage cells divide and increase

Cartilage cells form columns

Cartilage cells enlarge

Calcium is deposited in matrix between cartilage cells

Mature cartilage cells die

New bone cells called osteoblasts attach to calcified matrix

New blood vessels nourish new bone

Site of bone growth
The tinted area is the epiphyseal growth plate. Such areas are found near the ends of the femur, tibia, fibula, humerus, ulna, and radius.

Epiphysis

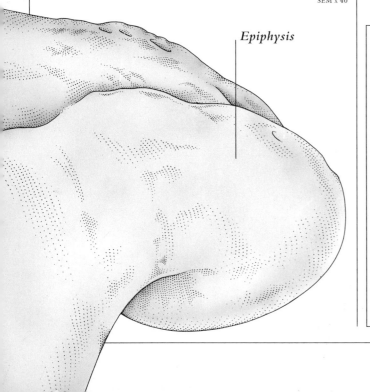

THE BIRTH OF BONES

Most bones are descended from an ancestor cartilage. Ossification is the process by which the cartilage slowly hardens into bone as a result of the deposition of mineral salts, primarily calcium. Bones are not completely ossified until early in adult life, when bone growth has ended.

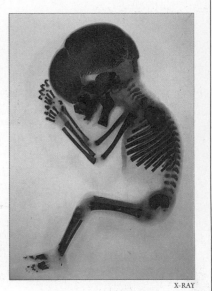

Fetal skeleton

The lighter areas in the hands, feet, and knees of the 12-week-old fetus shown at right are pieces of bone-shaped cartilage that will not harden until after the baby is born. The suture joints in the skull will also not harden until much later.

X-RAY

FRACTURES

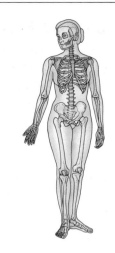

BROKEN BONES – FRACTURES – ARE A COMMON INJURY that can occur at any age. Injuries may range from minor cracks in the surface, known as fissures, to complete breaks through the entire bone. Fractures may be caused by a sudden, unusual impact, or by compression. So-called "stress" fractures are caused by prolonged or repeated force straining the bone and are known to occur during long-distance walks. Nutritional deficiencies or certain chronic diseases, which can weaken bone, may increase the likelihood of fractures.

HOW BONES BREAK

If a broken bone remains beneath the skin, the fracture is described as closed (simple); if the ends of the fractured bones project outside the skin, the injury is described as open (also called compound). A displaced fracture occurs when the bones are forced from their normally aligned position.

TYPES OF FRACTURE

Bones break in different ways, depending on the angle and degree of force to which they are subjected, and on the part affected. From these patterns, a surgeon can usually gauge the probable stability of the bone fragments, as well as the easiest way in which to reposition the injury.

Transverse fracture
A powerful direct or angular force may cause a break straight across the width of a bone. Such fractures are usually stable.

Spiral fracture
A sharp sudden twist may break a bone diagonally across the shaft, sometimes leaving jagged ends.

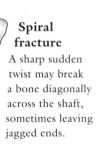

Comminuted fracture
A powerful direct impact can shatter a bone into several fragments. This type of fracture can easily occur as the result of a car accident.

Greenstick fracture
Strong force may cause long bones to bend and crack obliquely on only one side. This fracture is most common in young children and heals well.

BONES COMMONLY INJURED

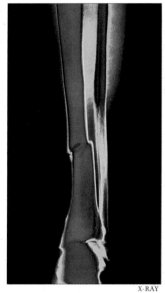

X-RAY

Fractured tibia
A younger person is especially likely to injure a lower leg bone during active movement, such as playing sports. The image above shows a displaced tibial fracture. This type of injury is frequently accompanied by a fracture of the fibula.

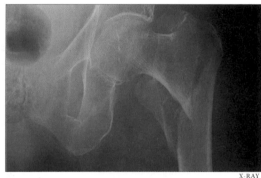

X-RAY

Fractured neck of femur
Bones naturally become thinner and more brittle with age, and are more likely to fracture from only minimal force. The hip joint is especially vulnerable.

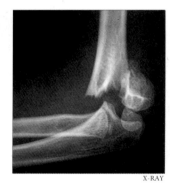

X-RAY

Elbow fracture
An injury that often occurs in childhood, a supracondylar fracture of the humerus (upper arm bone) just above the elbow may damage the brachial artery and affect circulation of the forearm and hand.

Colles' fracture
Flexing a hand to cushion a fall sometimes breaks the end of the radius and occasionally the tip of the ulna. This injury can occur at any age, but it is especially likely to occur in older people with thin bones, who fall due to unsteady balance.

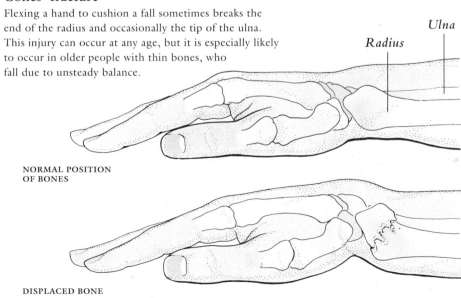

Radius   *Ulna*

NORMAL POSITION OF BONES

DISPLACED BONE

REPAIRING FRACTURES

How fractures are treated depends on the nature of the injury, the bone(s) affected, and the extent of adjacent tissue damage. If the bones are displaced, manipulation to restore their normal position may be performed under general anesthesia. Physical therapy often encourages healing and prevents deformity.

METHODS OF IMMOBILIZATION

Some fractures need to be immobilized in order to ensure that bones unite soundly in a good position. In most cases this is achieved with a splint or a cast made from plaster of paris or plastic. Certain fractures, particularly those with many bone fragments, require an operation to reposition and fix bones securely, so that surrounding tissues heal.

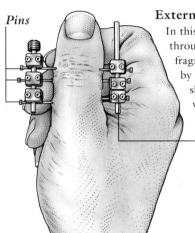

Pins

External fixation

In this form of fixation, pins are inserted through the skin into repositioned bone fragments. These are held firmly in place by a metal frame positioned outside the skin. The pins and frame are removed when the bone has healed.

External metal frame

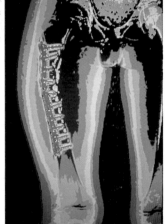

Internal fixation

This type of fixation may be used if a bone is severely fractured in several places. An incision is made in the skin to expose the injured bone, and some combination of plates, wires, screws, rods, and nails is inserted. The image at right shows a fractured femur held in place by a metal plate and screws.

CT SCAN

TRACTION

Pulling on the ends of a broken bone in order to realign them is like straightening a string of beads by tugging at each end. Traction is the force generated by a mechanical system of weights and pulleys to perform this necessary realignment. If rapid repositioning of a fracture is needed, traction may be applied while a patient is anesthetized.

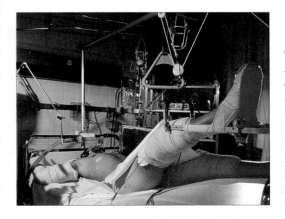

Gradual traction

A fractured femur is often treated by the application of gradual traction, as shown at left. This prevents the large muscles from going into spasm and pulling the correctly repositioned bones out of alignment.

INTERNAL FIXATION: FRACTURED SHAFT OF FEMUR

The strong, weight-bearing femur (thighbone) is the largest bone in the body. If the femoral shaft is severely fractured, it may be necessary to insert a rod into the central medullary canal to hold the repositioned bone in place while it heals. The rod may be either left in place permanently or removed at a later date, and full use of the leg usually follows within 3 months.

Traction pin

1 When the patient is anesthetized, a temporary traction pin is attached to the lower end of the fractured femur just above the knee. The ends of the fractured bone are then repositioned gradually to restore normal alignment.

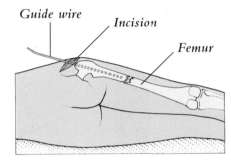

Guide wire

Incision

Femur

2 After an incision is made, a hole is drilled in the top of the exposed femur. A wire is guided down into the medullary canal at the center of the bone, passing the fracture and stopping just above the knee.

3 Using an instrument known as a reamer, the surgeon enlarges the medullary canal and then passes a hollow metal rod of the correct length down over the guide wire. Once this rod is in place, the guide wire is removed.

Screws

Rod

4 The bones are held in position by screws that pass through each end of the femur and the intramedullary rod.

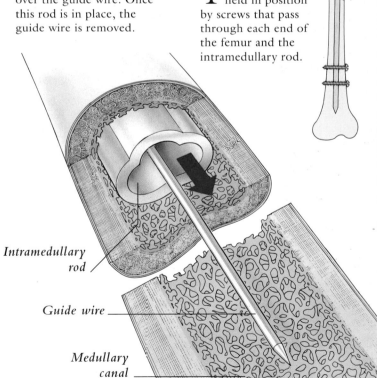

Intramedullary rod

Guide wire

Medullary canal

BONE DISORDERS

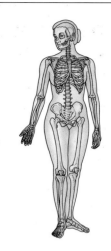

BONE STRENGTH AND STRUCTURE may be affected by nutritional, hormonal, and other disorders throughout life. These include Cushing's syndrome, hyperparathyroidism, rickets, and cancer. Some people are born with a bone deformity such as a shortened or partially absent limb. Although such birth defects affect only a small minority of people, the bone-weakening disorder called osteoporosis occurs gradually with advancing age in virtually everyone. Exercise and calcium supplements may slow the effects of this condition.

OSTEOPOROSIS

After middle age, bones become distinctly thinner and more porous – with bone loss increasing in both sexes. However, because the hormone estrogen helps maintain bone mass, loss of this hormone in females at the menopause leads to much more severe osteoporosis in older women than in men.

EFFECTS OF OSTEOPOROSIS

Because of their decreased density, bones affected by osteoporosis are much more prone to crush fractures; this may lead to spinal curvature. Fractures of the hip or wrist may result from falls.

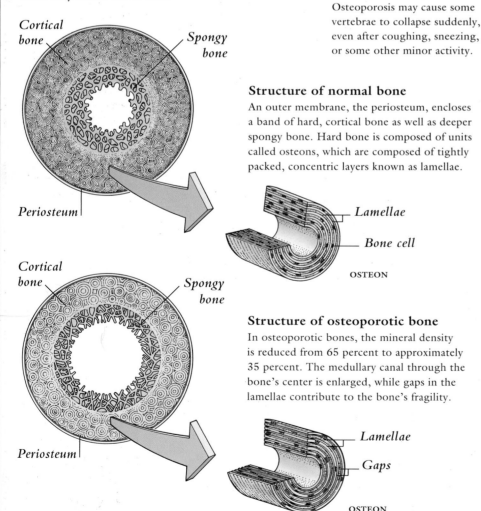

Structure of normal bone
An outer membrane, the periosteum, encloses a band of hard, cortical bone as well as deeper spongy bone. Hard bone is composed of units called osteons, which are composed of tightly packed, concentric layers known as lamellae.

Structure of osteoporotic bone
In osteoporotic bones, the mineral density is reduced from 65 percent to approximately 35 percent. The medullary canal through the bone's center is enlarged, while gaps in the lamellae contribute to the bone's fragility.

X-RAY

Osteoporotic spine
Osteoporosis may cause some vertebrae to collapse suddenly, even after coughing, sneezing, or some other minor activity.

WHY OSTEOPOROSIS OCCURS

Bones are continually being broken down and then rebuilt in order to facilitate growth and repair. In young people, the rate of bone formation exceeds the rate at which cells are reabsorbed. This process begins to change in early adulthood, with the rate of reabsorption becoming greater than formation. Bones gradually become weaker and lighter.

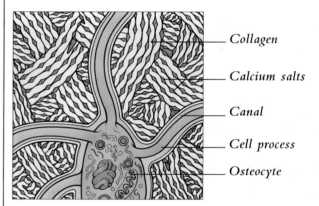

Bone formation
Bone is built up by the deposition of minerals (mainly calcium salts) on an organic matrix of collagen fibers. Osteocytes, or bone cells, form the collagen and also aid in the laying down of calcium. Canals in the bone permit calcium to move in or out of the blood in response to hormones that regulate the body's needs.

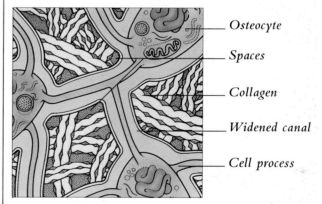

Bone reabsorption
As osteoporosis develops with age, both the collagen framework and deposited minerals are broken down much faster than they are formed. The canals that connect the osteocytes become wider and new spaces appear in the collagen matrix. These changes weaken the bone.

OSTEOMALACIA

In osteomalacia, bones are weakened by the loss of calcium and phosphorus. The condition differs from osteoporosis in that there is not any loss of the bone's protein matrix. In children, this is called rickets. A primary cause is a shortage of vitamin D, essential to enable the body to deal with calcium and phosphorus. Inadequate sunlight may be a factor.

Pelvic deformity
When the pelvic bones are softened as a result of osteomalacia, they may become weakened and severely deformed. Someone with such an abnormal pelvis would probably find walking difficult and painful.

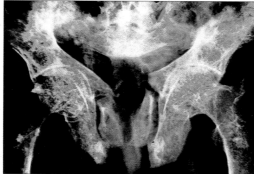

X-RAY

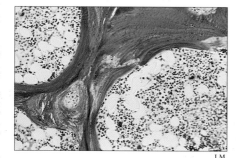

LM

Bone biopsy
This stained sample of bone tissue reveals distinctive signs of osteomalacia. The areas that have reduced deposits of calcium are shown as brown, while normal areas are green. This irregular structure may result in tiny fractures on the surface of affected bones.

PAGET'S DISEASE

The normal formation of bone is disturbed in Paget's disease, also called osteitis deformans. Bone is broken down at an increased rate and is replaced rapidly by abnormal bone. This condition occurs less frequently in younger people, but it affects up to 3 percent of those over the age of 40. Paget's disease occurs most commonly in the skull, spine, pelvis, and leg bones.

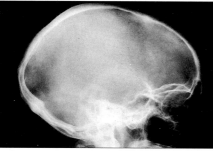

X-RAY

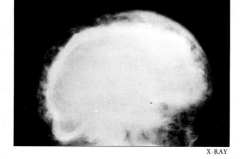

X-RAY

Bone enlargement and thickening
One effect of Paget's disease may be the enlargement and thickening of certain bones. Contrast the normal skull (top) with an affected skull (below), in which areas of increased bone density are seen as white patches. Bone enlargement has a variety of effects, such as hearing loss as the cranial nerves become compressed or greater heat loss from the head due to the increased blood supply.

BONE CANCER

Cancer that originates in a bone is known as primary. More frequently, a malignant tumor in a bone is the result of cancer cells that have spread from a primary tumor elsewhere in the body; this type of bone cancer is referred to as secondary or metastatic.

PRIMARY BONE CANCER

Cancers that start in bone are most likely to occur in younger people. Osteosarcoma is the most common type, and affects long bones such as the femur. Another primary bone cancer, chondrosarcoma, occurs mainly in the pelvis, ribs, and breastbone. Treatment for primary bone cancer may include replacement by a bone graft.

Osteosarcoma
The most common site of an osteosarcoma is just above the knee at the lower end of the femur. The first sign of the tumor is usually a painful, visible swelling (shown as the blue area at left center).

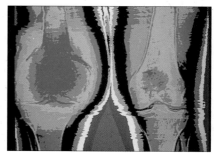

CT SCAN

SECONDARY BONE CANCER

Secondary bone cancer is more likely to occur in older people. Commonly affected areas are the skull, sternum, pelvis, vertebrae, ribs, and, less often, the upper ends of the femur and humerus. Treatment for secondary bone cancer may include anticancer drugs and radiotherapy in order to reduce tumor size, and painkillers.

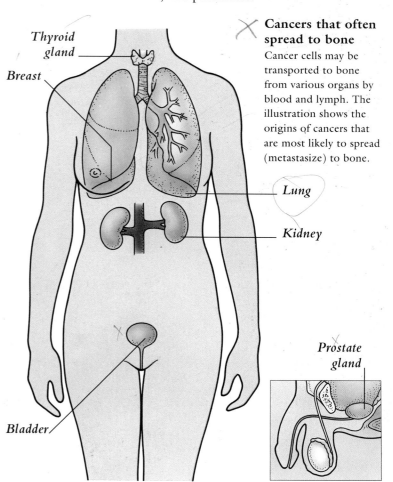

Thyroid gland

Breast

Cancers that often spread to bone
Cancer cells may be transported to bone from various organs by blood and lymph. The illustration shows the origins of cancers that are most likely to spread (metastasize) to bone.

Lung

Kidney

Bladder

Prostate gland

The SPINE

THE SPINE IS A STRONG but flexible support that holds the head and body upright, and allows the upper body to bend and twist. Thirty-three ringlike bones called vertebrae are linked by a series of mobile joints. Sandwiched between the vertebrae are springy disks of tough cartilage that compress when under pressure to absorb shocks. These disks are subjected to enormous forces, sometimes as great as several hundred pounds per square inch during strenuous movements. Strong ligaments and muscles that surround the spine stabilize the vertebrae and help control movement.

STRUCTURE OF THE SPINE

Three main types of individual vertebrae, each a different shape, make up the spine: cervical in the neck, thoracic in the upper back, and lumbar in the lower back. Both the wedge-shaped sacrum and the tail-like coccyx at the base of the spine consist of several fused vertebrae. Cartilage disks separate these bony components of the spine and cushion them from pressure during twisting, weight-bearing, or jumping.

MOVEMENT OF SPINAL JOINTS
Individual spinal joints do not have a wide range of movement, but working together they give the spine great flexibility, letting it arch backward, twist around, or curve forward for toe-touching.

Facet joint
Helps determine the degree of movement between vertebrae.

Vertebral body

Ligament
Stabilizes vertebrae and holds them in alignment during movement.

Intervertebral disk
Absorbs forces directed through its axis and acts like a ball-bearing during bending or twisting.

Flexibility
The body is capable of bending farther forward than backward, due to the shape of the vertebrae. The top seven vertebrae (cervical spine) are the most flexible.

Spinal cord
This vital cable of nerve tissue, which relays messages between the brain and different parts of the body, is protected by the 33 vertebrae of the spinal column.

Spinal nerve
Connected to the spinal cord are 31 pairs of nerves that emerge through gaps between the vertebrae and travel out to body tissues and organs.

Sacrum

Coccyx

Vertebral body
Vertebral bodies become progressively larger toward the base of the spine to support increasing weight.

Vertebral processes
These bony knobs extend from the back of each vertebra. Three processes serve as anchor points for muscles; the other four form the linking facet joints between adjacent vertebrae.

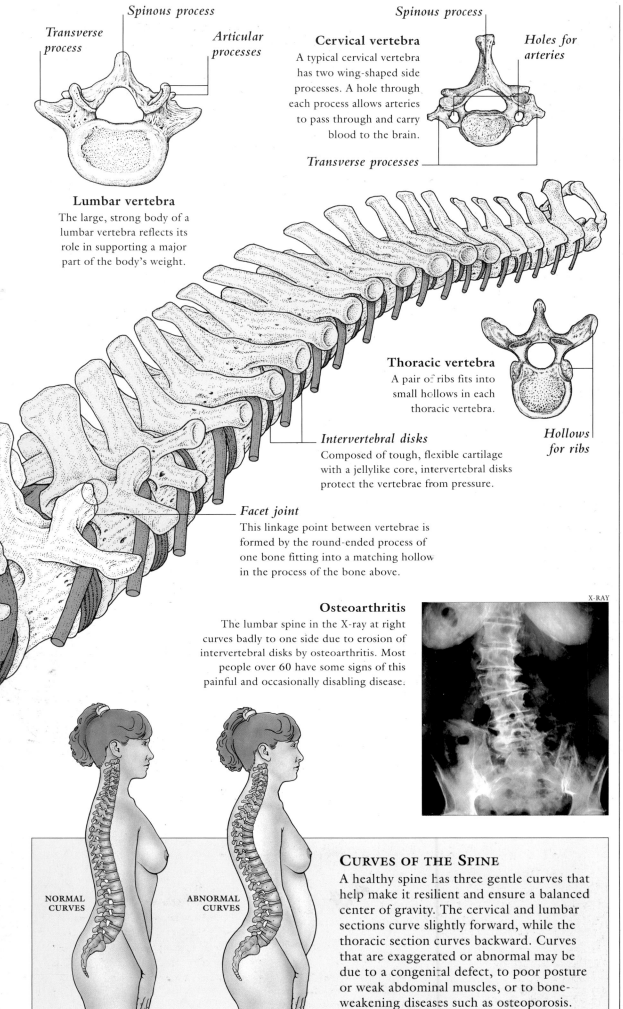

Transverse process

Spinous process

Articular processes

Lumbar vertebra
The large, strong body of a lumbar vertebra reflects its role in supporting a major part of the body's weight.

Spinous process

Cervical vertebra
A typical cervical vertebra has two wing-shaped side processes. A hole through each process allows arteries to pass through and carry blood to the brain.

Holes for arteries

Transverse processes

Thoracic vertebra
A pair of ribs fits into small hollows in each thoracic vertebra.

Hollows for ribs

Intervertebral disks
Composed of tough, flexible cartilage with a jellylike core, intervertebral disks protect the vertebrae from pressure.

Facet joint
This linkage point between vertebrae is formed by the round-ended process of one bone fitting into a matching hollow in the process of the bone above.

Osteoarthritis
The lumbar spine in the X-ray at right curves badly to one side due to erosion of intervertebral disks by osteoarthritis. Most people over 60 have some signs of this painful and occasionally disabling disease.

X-RAY

NORMAL CURVES

ABNORMAL CURVES

CURVES OF THE SPINE
A healthy spine has three gentle curves that help make it resilient and ensure a balanced center of gravity. The cervical and lumbar sections curve slightly forward, while the thoracic section curves backward. Curves that are exaggerated or abnormal may be due to a congenital defect, to poor posture or weak abdominal muscles, or to bone-weakening diseases such as osteoporosis.

REGIONS OF THE SPINE
Each section of the spine is adapted to its function. The cervical vertebrae support the head and neck, the thoracic vertebrae anchor the ribs, and the strong, weight-bearing regions of the lower spine give a stable center of gravity during any movement.

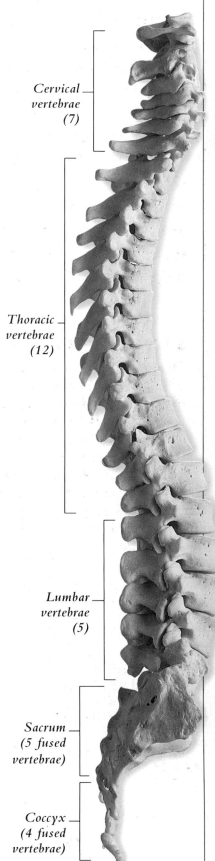

Cervical vertebrae (7)

Thoracic vertebrae (12)

Lumbar vertebrae (5)

Sacrum (5 fused vertebrae)

Coccyx (4 fused vertebrae)

SPINAL INJURIES *and* DISORDERS

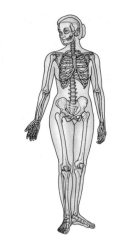

MANY INJURIES TO THE SPINE ARE MINOR and cause only slight bruising, but a severe fall or an accident may dislocate or fracture vertebrae. If the spinal cord or specific nerves are damaged, a loss of bodily sensation or function, or even paralysis, may result. Bone diseases, for example osteoporosis, and deformities can affect the spine and may increase the likelihood of fractures.

SPINAL FRACTURES

Most major spinal injuries occur as a result of severe forces of compression or of rotation or bending beyond the spine's normal range of movement. The most important consideration in assessing a spinal injury is whether a fracture is stable (unlikely to shift) or unstable, in which case damage to the spinal cord or nerves is more likely.

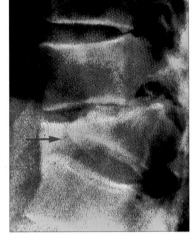

X-RAY

Compression fracture
A compression force through the longitudinal axis of the spine may cause the front part of a vertebral body to collapse (shown by arrow).

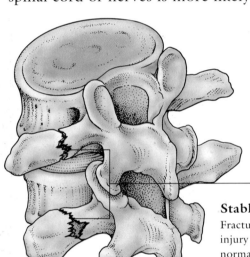

Fractures of transverse processes

Stable fracture
Fracture of a transverse process is usually a minor injury because the vertebra does not shift from its normal position. Most commonly affecting lumbar vertebrae, these injuries are often the result of a direct blow. They rarely cause nerve damage.

SPINAL FUSION

An operation to fuse vertebrae may be needed to treat unstable fractures or to correct spinal distortions or deformity. Bone taken from the back of the pelvis is placed on either side of the spine so that it bridges the two vertebrae to be fused. The bone is held in its new position by muscles in the back, and a metal plate and wire may also be used if greater stability is needed.

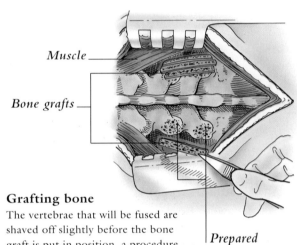

Muscle

Bone grafts

Prepared vertebrae

Grafting bone
The vertebrae that will be fused are shaved off slightly before the bone graft is put in position, a procedure that stimulates bone growth so that the graft is incorporated more easily into the vertebrae. Bone from the person's own body is most readily compatible with the new site.

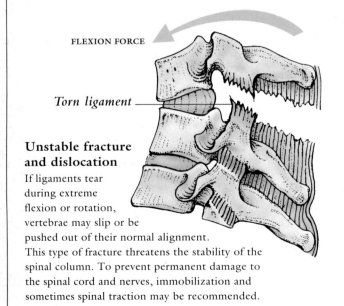

FLEXION FORCE

Torn ligament

Unstable fracture and dislocation
If ligaments tear during extreme flexion or rotation, vertebrae may slip or be pushed out of their normal alignment. This type of fracture threatens the stability of the spinal column. To prevent permanent damage to the spinal cord and nerves, immobilization and sometimes spinal traction may be recommended.

WHIPLASH INJURY

Suddenly forcing the neck forward and then backward often sprains the ligaments and/or partially dislocates a cervical joint. Following a whiplash injury, often the result of a car accident, an orthopedic neck collar may be worn for several weeks until the neck moves freely and without pain. Physical therapy, analgesics, and muscle-relaxants are often prescribed. The recovery is usually complete.

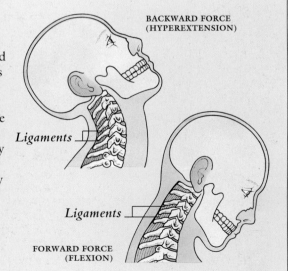

BACKWARD FORCE (HYPEREXTENSION)

Ligaments

Ligaments

FORWARD FORCE (FLEXION)

DISK PROLAPSE

The cartilage disks that separate adjacent vertebrae have a hard outer covering and a jellylike center. Wear and tear or pressure may rupture the outer layer, forcing the center through and causing pressure on a spinal nerve root. If the symptoms are not relieved by rest, the protruding disk can be removed by opening the vertebral canal or by a less invasive procedure (see right).

Protruding part of disk

Vertebra

Disk

Nerve root

Pressure on a nerve root

A so-called "slipped disk" may cause severe back pain if it presses on a nerve root. The pain may be aggravated by even simple activities such as coughing, bending, or sitting for a long time.

SCIATICA

Pain that affects the buttock and back of the thigh can be caused by pressure on the spinal roots of the sciatic nerve. The source of this pressure is usually a prolapsed intervertebral disk, but it may be due to a tumor, a blood clot, a muscle spasm, or just sitting in an awkward position.

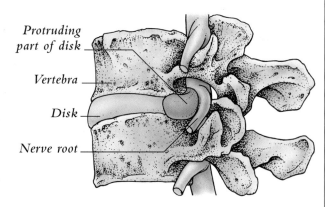

Gluteus maximus muscle

Sciatic nerve

Sciatic nerve

Hamstring muscles

Tibial nerve

Peroneal nerve

Pain from sciatica

Pressure on the sciatic nerve, the largest nerve in the body, may cause pain to radiate down the entire leg. In some severe cases, numbness and weakness of the muscles also occur.

OPERATION

MICRODISKECTOMY

In this operation for a disk prolapse, the surgeon makes only a tiny incision in the patient's back. An operating microscope and special instruments are used to remove the protruding intervertebral disk tissue. Other disk operations are also effective.

1 The anesthetized patient is placed face down. A needle is inserted into the spine at the prolapse level, and an X-ray is taken to confirm the level. The confirmed disk level and site of incision are marked on the patient's skin.

Site of incision

Level of prolapse | *Needle*

2 After making an incision, the surgeon locates the ruptured disk using an operating microscope. A guide instrument is then inserted into the intervertebral space, right next to the prolapsed disk.

Guide

Vertebra

Prolapsed disk

Nerve root

3 The compressed nerve root is pulled aside using a retractor to reveal the underlying prolapsed intervertebral disk. A tiny, crosslike incision is made in its fibrous outer covering, the dura mater.

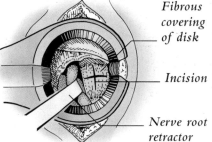

Fibrous covering of disk

Incision

Nerve root retractor

4 A rongeur, a scissorlike instrument, is inserted into the guide instrument, and then pushed through the incision to the protruding disk tissue that will be removed.

Rongeur

5 The disk tissue is trimmed away using the rongeur and is removed through the guide. The instruments are withdrawn, and the incision is closed. The patient is usually able to get out of bed the next day.

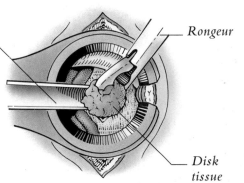

Instrument to protect nerve root

Rongeur

Disk tissue

JOINTS *of the* BODY

THE PLACE WHERE TWO BONES MEET is known as a joint, or articulation. Joints are classified either by their structure or by the way they move. In the freely movable synovial joints, the surfaces in contact slide over each other with almost no friction. Less mobile joints, such as those in the skull, are more firmly linked by fibrous tissue or cartilage to permit growth or provide stability.

OTHER JOINTS

Not all joints have a wide range of movement. Some types allow for growth or for limited flexibility where greater stability is needed.

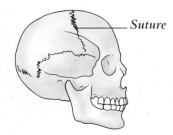

Suture

Fixed joints

After growth is complete, all the separate bone plates of the skull are securely connected by interlocking fibrous tissue, forming so-called suture joints.

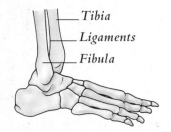

Tibia
Ligaments
Fibula

Slightly movable joints

Bones with limited movement may be stabilized by pads of cartilage, as between spinal vertebrae, or by slightly flexible ligaments, as in the lower leg.

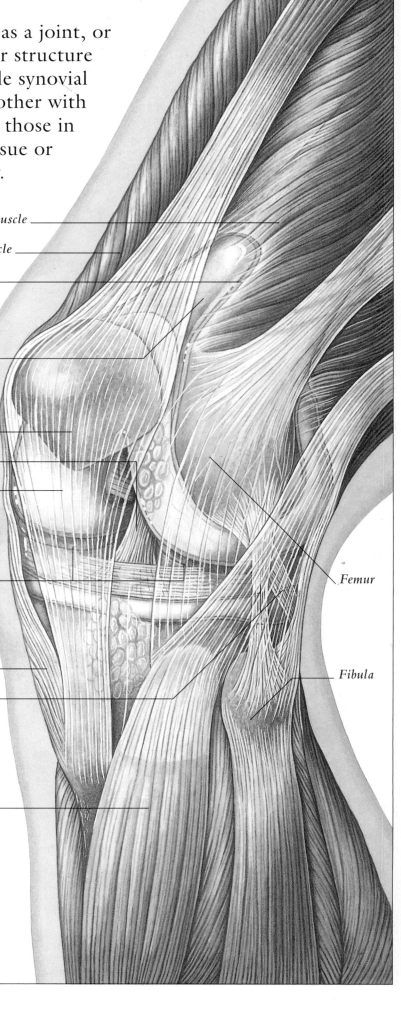

Vastus lateralis muscle

Vastus medialis muscle

Synovial membrane
Tissue that lines the noncontact surfaces within the joint capsule secretes lubricating synovial fluid.

Synovial fluid
This clear fluid lubricates and nourishes all the tissues inside the joint capsule.

Patella

Internal ligaments

Articular cartilage
Where the bone ends are in contact, connective tissue provides a smooth, protective surface for ease of movement.

Menisci (articular disks)
Unique to the knee and wrist are pads of fibrous cartilage. Called menisci, they help the weight-bearing bones absorb shock.

Joint capsule (cut away)

External ligaments
Thickenings of the capsule form these fibrous cords. They stabilize the joint, especially during movement. Some joints, such as the knee, have internal ligaments for added stability.

Tibialis anterior muscle

Femur

Fibula

SYNOVIAL JOINT STRUCTURE

Most of the joints in the body are synovial joints. These are movable, highly versatile, lubricated joints. Articular cartilage covers and protects the bone ends, ligaments help provide stability, and a fibrous capsule encloses the structure. Muscles around the joint contract to produce movement. The knee, shown here, is the largest joint.

TYPES OF SYNOVIAL JOINT

The shape of articular cartilage surfaces in a synovial joint and the way they fit together determine the range and the direction of the joint's movement. Hinge and pivot joints move in only one plane (from side to side, for example, or up and down), while ellipsoidal joints are able to move in two planes at right angles to each other. Most joints in the body can move in more than two planes, which allows for a wide range of motion.

RANGE OF MOVEMENT

The shoulder is one of the most mobile and most complex joints of the body: it moves up and down, forward and backward, and can rotate in a complete circle at the side of the body. Joints like this that move in more than two planes are called multiaxial.

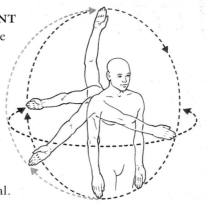

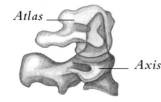

Atlas

Axis

Pivot joint

A projection from one bone turns within a ring-shaped socket of another bone, or the ring turns around the bony projection. A pivot joint formed by the top two cervical vertebrae allows the head to turn from side to side, as when shaking the head "no."

Humerus

Ulna

Radius

Hinge joint

In this simplest of joints, the convex surface of one bone fits into the concave surface of another bone. This allows for movement like a hinged door in only one plane. Both the elbow and the knee are modified hinge joints: they bend up and down in one plane quite easily, but are also capable of very limited rotation.

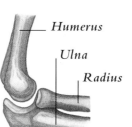

Radius

Scaphoid

Ellipsoidal joint

An ovoid, or egg-shaped, bone end is held within an elliptical cavity. The radius bone of the forearm and the scaphoid bone of the hand meet in an ellipsoidal joint. This type of joint can be flexed or extended and moved from side to side, but rotation is limited.

Scapula

Humerus

Ball-and-socket joint

The rounded head of one bone fits into the cuplike cavity of another bone. Of all joint structures, a ball-and-socket type allows for the greatest range of movement. The shoulder and the hip are both ball-and-socket joints.

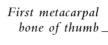

Trapezium of wrist

First metacarpal bone of thumb

Saddle joint

The joint surface of each bone has both concave and convex areas so that the bones can rock back and forth and from side to side, but they have limited rotation. The only saddle joints in the body are at the base of the thumbs.

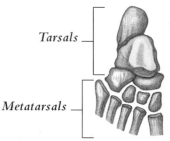

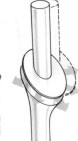

Tarsals

Metatarsals

Gliding joint

The two surfaces of bones that meet in a gliding joint are almost flat, and slide over one another. Movement is limited, however, by strong encasing ligaments. Some joints in the foot and wrist move in this way.

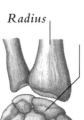

JOINT INJURIES
and DISORDERS

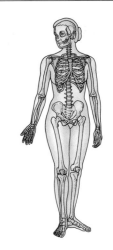

JOINTS ARE MADE TO WORK IN SPECIFIC WAYS, and any movement beyond the normal range, or in an unnatural direction, often causes an injury. Certain injuries may result from a direct blow or fall, or, occasionally, from overuse. Arthritis is a general term used to describe several different disorders that cause painful, swollen joints. The most common disorder is osteoarthritis.

LIGAMENT INJURIES

Ligaments, which are strong bands of fibrous tissue, link bone ends together. If the bones within a joint are pulled too far apart, often as a result of a sudden or unexpected movement, or one that is too forceful, fibers may be overstretched or torn. This commonly results in swelling, pain, or muscle spasm, and, if the injury is severe, joint instability or even dislocation.

Ankle sprain

A sprain is a partial tearing of a ligament. The ankle may be sprained as a result of a fall or stumble that forces the full weight of the body onto the outer edge of the foot. Rest, ice, compression, and elevation are the steps used to treat sprains.

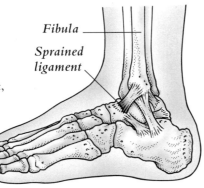

Fibula

Sprained ligament

TREATING PAIN AND INFLAMMATION

Inflamed tissue triggers the release of prostaglandins, which stimulate nerve endings, cause blood vessels to dilate, and attract white blood cells to the area. The resulting pain can be treated by nonsteroidal anti-inflammatory drugs, which block further synthesis of prostaglandins. Corticosteroid drugs can reduce inflammation by inhibiting the proteins (cytokines) that are secreted by white blood cells.

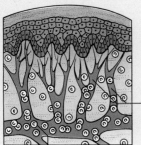

Before medication

Abundant white blood cells flow through dilated blood vessels to the injured area, causing heat, redness, pain, and swelling.

Dilated blood vessels

Reduced swelling

Increased white blood cells

After medication

Drugs inhibit the production of white blood cells, and thus help reduce swelling, by blocking the synthesis of prostaglandins.

TORN CARTILAGE

One type of cartilage found in the body consists of firm, flexible, slightly elastic connective tissue. In the knee, disks called menisci made of this fibrocartilage cushion the bones from excessive force. If a meniscus is torn, which may occur by twisting the knee while playing sports, a meniscectomy to remove part or all of the damaged cartilage may be performed.

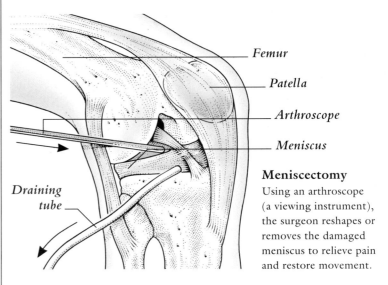

Femur

Patella

Arthroscope

Meniscus

Draining tube

Meniscectomy

Using an arthroscope (a viewing instrument), the surgeon reshapes or removes the damaged meniscus to relieve pain and restore movement.

DISLOCATED JOINTS

A drastic shift of two bone ends out of their normal end-to-end position, particularly of a bone from its complementary socket, is called a dislocation. Often painful, a dislocation frequently can tear ligaments in the joint and fracture one or both bones. This can damage nerves or adjacent blood vessels. The area around the joint may also appear deformed.

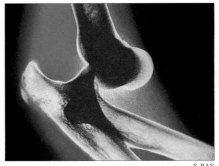

X-RAY

Dislocated elbow

Usually caused by a fall onto an outstretched hand, a dislocated elbow is a common injury that affects both children and adults. It is treated by repositioning the bones and then immobilizing the joint for about 3 to 6 weeks. Exercises are recommended to regain movement and strength.

RHEUMATOID ARTHRITIS

This autoimmune form of arthritis develops when the immune system, usually triggered by an antigen in a genetically predisposed person, begins to attack body tissues. The joints become inflamed, swollen, stiff, and deformed. Early symptoms include fever, pallor, and weakness. If the disease is chronic, tissues of the eyes, skin, heart, nerves, and lungs may be affected.

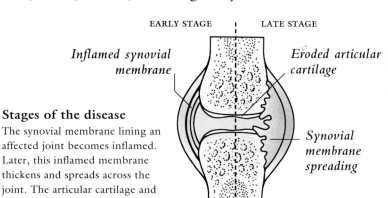

EARLY STAGE | LATE STAGE

Inflamed synovial membrane

Eroded articular cartilage

Synovial membrane spreading

Stages of the disease
The synovial membrane lining an affected joint becomes inflamed. Later, this inflamed membrane thickens and spreads across the joint. The articular cartilage and the bone ends are eroded.

EFFECTS OF RHEUMATOID ARTHRITIS
Characteristically, many of the small joints are affected in a symmetrical pattern; for example, the hands and feet may be inflamed to the same degree on both sides. Stiffness is often worse in the mornings, but eases during the day. A doctor will suspect the disease from these symptoms and signs. The diagnosis is confirmed if a blood test detects an antibody associated with rheumatoid arthritis.

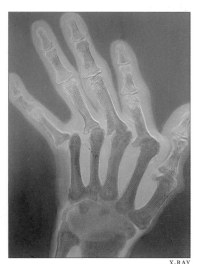

X-RAY

Painful deformity
In severe rheumatoid arthritis, joint spaces disappear and the angle at which bone ends meet changes as a result of ligament laxity. Bone ends are roughened and eroded, and nodules, which are aggregates of inflamed tissue cells, collect around the bone ends. Above the nodules, the skin is thin and fragile. These features all restrict movement.

GOUT
Gout, also called crystal-induced arthritis, may cause sudden and severe pain, swelling, and redness in one joint, usually at the base of the big toe. More common in men than in women, it is caused by excess uric acid in the body. Normally, uric acid is dissolved and excreted in urine. In gout, however, this substance collects in the synovial fluid of a joint, forming needlelike crystals.

LM

Crystals of uric acid

TREATMENT
Although there is no cure yet, it is currently believed that if powerful immunosuppressant drugs are given at an early stage, the disease may go into prolonged remission. Rest and drugs to relieve inflammation and pain are often prescribed during an acute attack, while gentle exercise during remission helps to keep affected joints mobile.

Types of drugs
Certain drugs aim to reduce the swelling that is caused by inflammation of the synovial membrane and the production of excess synovial fluid. Other drugs are used to slow down cartilage and bone destruction.

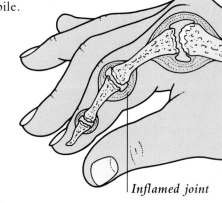

Inflamed joint

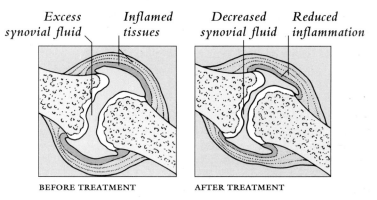

Excess synovial fluid | *Inflamed tissues* | *Decreased synovial fluid* | *Reduced inflammation*

BEFORE TREATMENT | AFTER TREATMENT

SURGERY
When inflammation of the synovial membrane cannot be controlled by drugs, surgery may be carried out to alleviate symptoms and to slow down further joint deterioration. Contracted tendons can be released to permit greater movement, or the inflamed synovial membrane may be removed in a procedure called a synovectomy. Joint replacement surgery is very occasionally carried out for severely deformed, immobile, and painful joints.

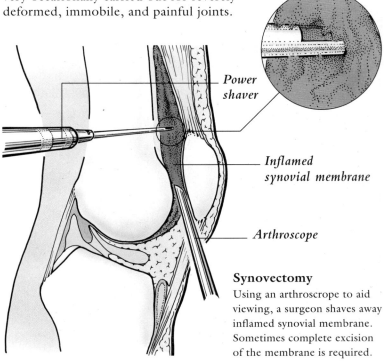

Power shaver

Inflamed synovial membrane

Arthroscope

Synovectomy
Using an arthroscrope to aid viewing, a surgeon shaves away inflamed synovial membrane. Sometimes complete excision of the membrane is required.

OSTEOARTHRITIS

Unlike rheumatoid arthritis, a disease that affects a few body systems at once, osteoarthritis may affect a single joint. Degeneration of the joint may be hastened by a variety of factors, including a congenital defect, injury or infection, obesity, or a familial tendency. Cartilage normally wears away as the body ages, so a mild form of osteoarthritis affects most people by about age 60.

STAGES IN OSTEOARTHRITIS

The process by which articular cartilage begins to break down may be genetically determined. Gradually this cartilage becomes thinner and roughened. When the bone underneath the cartilage eventually erodes, bone surfaces rub directly against one another, causing severe discomfort. However, joints may be painfully inflamed only occasionally; minor degrees are quite common, and be treated. Only a few people have progressive joint damage.

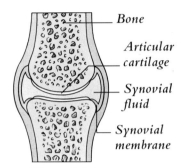

Normal joint structure
Healthy articular cartilage lubricated by synovial fluid allows ease of movement.

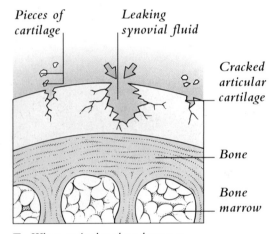

1 When articular chondrocytes (cartilage cells) die, surface cracks appear; this allows synovial fluid to leak in, causing greater cartilage degeneration. Pieces of this weakened cartilage break off, inflaming the synovial membrane.

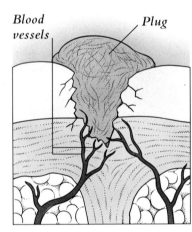

2 Eventually, a gap developing in the cartilage reaches the underlying bone. Blood vessels begin to grow, and a plug made of fibrocartilage fills the gap.

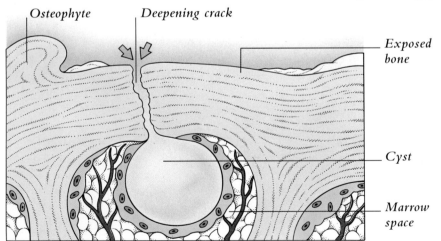

3 The fibrocartilage plug wears away, thus exposing the bone surface. If surface cracks deepen, synovial fluid can leak into the marrow space, and may form a cyst surrounded by weakened bone. Small outgrowths called osteophytes may further deform the bone surface.

OSTEOARTHRITIS OF THE HIP

In weight-bearing or frequently used joints, the articular cartilage is more likely to erode in time. The hip is particularly vulnerable to osteoarthritic changes, which may severely inhibit movement; obesity may also accelerate this process in the hip.

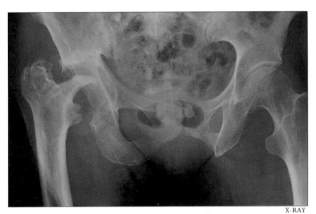

X-RAY

Appearance of an osteoarthritic hip
Here osteoarthritis has eroded the hip socket and almost completely flattened the normally convex curve of the femoral head.

DRUGS FOR OSTEOARTHRITIS

Although drugs cannot repair damaged bone and cartilage, they reduce any inflammation and control pain to help maintain joint mobility and minimize deformity. Nonsteroidal anti-inflammatory drugs (NSAIDs), for example ibuprofen, are prescribed first; these may then be followed by injections of corticosteroid drugs into severely inflamed joints.

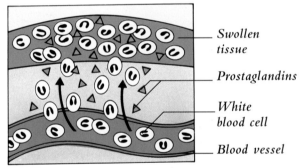

Before corticosteroids
Damaged tissues release prostaglandins to attract numerous white blood cells that repair the injury. The continued secretion of prostaglandins causes swelling and pain.

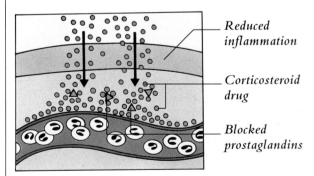

After corticosteroids
Corticosteroids inhibit the production of prostaglandins; this reduces the number and activity of the white blood cells. Inflammation and pain are relieved, and mobility increases.

OPERATION

HIP JOINT REPLACEMENT

In older people, a severe fracture or a stiff, painful hip due to osteoarthritis that cannot be treated successfully with drugs may be replaced by an artificial joint, or prosthesis. The prosthesis at right, composed of a metallic femoral shaft and cuplike pelvic socket, is often cemented in place; newer versions can be fitted without cement because they stimulate bone growth.

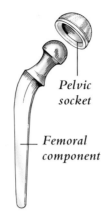

Pelvic socket

Femoral component

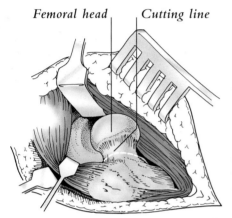

Femoral head   *Cutting line*

1 The surgeon makes an incision over the affected hip. Ligaments and muscles are pushed aside or cut through so that the entire hip joint may be seen. The head of the femur, which is often eroded or broken, is subsequently removed.

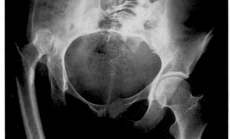

Pelvic bone   *Reamer*

2 The surgeon employs an instrument called a reamer to shave and shape the cavity in the pelvic bone that normally holds the femoral head in place. The plastic pelvic socket is fitted and may be cemented in place.

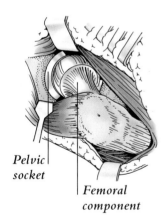

3 After cleaning out the central medullary canal in the shaft of the femur, the surgeon implants the long femoral component and then fits the rounded head of the component into the plastic pelvic socket. Both of these components must be fitted securely to prevent loosening in the future.

Pelvic socket

Femoral component

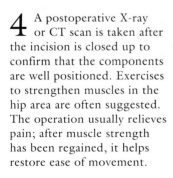

CT SCAN

4 A postoperative X-ray or CT scan is taken after the incision is closed up to confirm that the components are well positioned. Exercises to strengthen muscles in the hip area are often suggested. The operation usually relieves pain; after muscle strength has been regained, it helps restore ease of movement.

HIP PROBLEMS IN CHILDREN

Although most of the bone and joint abnormalities occurring in children are caused by injuries, a painful or deformed hip may also be due to some congenital defect, a bone infection, or an acquired disorder such as Perthes' disease, a disabling childhood condition. In a child who walks, the most obvious sign of a hip malformation is a limp, which may be due either to pain or to shortening of the affected leg.

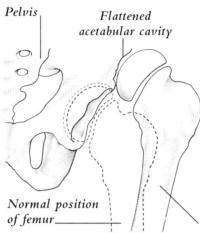

Pelvis   *Flattened acetabular cavity*

Normal position of femur

Displaced femur

CONGENITAL HIP DISLOCATION

This condition is the result of a flattened or misplaced acetabular cavity that fails to enclose the head of the femur. Although normally detected and treated with splints soon after birth, hip dislocation is occasionally missed; it is then detected only when the child begins to walk with a limp.

Appearance of a dislocated hip
A flattened acetabular cavity and an upwardly displaced hip (left side of image) were the result of an untreated congenital hip dislocation. If severe, a misalignment of the hips causes obvious limping.

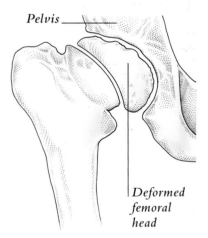

X-RAY

PERTHES' DISEASE

This disorder, which is more common in boys, is thought to be due to abnormal blood circulation in the epiphyseal growth plate. The head of the femur softens and becomes deformed, causing pain in the thigh and groin and a limp. The disease often affects only one hip, and must be treated as soon as possible by rest, splinting, and possibly surgery to prevent osteoarthritis later.

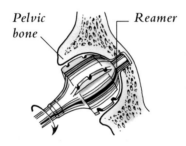

Pelvis

Deformed femoral head

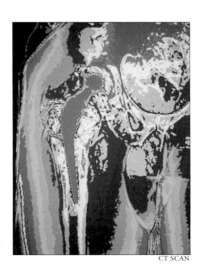

NORMAL POSITION OF EPIPHYSIS   SLIPPED EPIPHYSIS

SLIPPED EPIPHYSIS

The upper growing end of the femur, the epiphysis, may slip out of position as a result of a fall or injury. More often, the displacement occurs gradually, possibly due to tissue softening caused by the growth hormones produced during adolescence. Operative repair repositions the displaced bone, which is then secured with metal pins.

CHAPTER THREE

C H A P T E R   3

The MUSCULAR SYSTEM

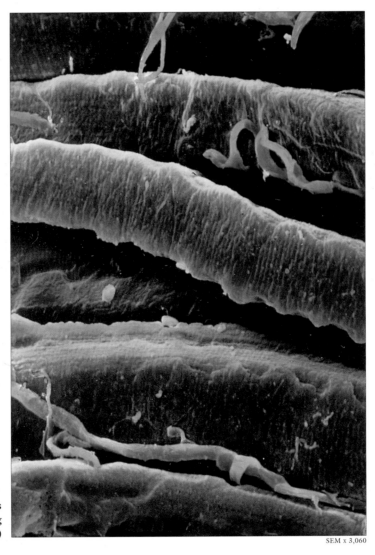

Skeletal muscle fibers
(with accompanying
capillaries seen in blue)

SEM x 3,060

INTRODUCTION

Muscles make up the bulk of the body and account for about half its weight. They are divided into three distinct types: striated skeletal muscle, involuntary (or smooth) muscle, and cardiac muscle. All muscles have in common the ability to contract, to relax, to be excited by a stimulus, and to return to their original size and shape. Involuntary muscles are those that handle the unconscious routine of the body: they are constantly at work performing tasks such as propelling food throughout the alimentary canal, keeping the eyes in focus, and controlling the caliber of the arteries. Cardiac muscle is found only in the heart and is unique for its branched interconnections. Skeletal muscle, the subject of this section of the book, is also called voluntary muscle because we can usually choose to contract or relax it. All muscles in the trunk and limbs are kept in a partly contracted state, known as muscle tone, by a steady flow of nerve impulses from the spinal cord. If a muscle loses its nerve supply, it will shrink to about two-thirds of its bulk within a few months. Many of the diseases affecting muscles, such as poliomyelitis and myasthenia gravis, are really diseases of the nervous system rather than the muscles. Muscles are more often injured than diseased, and are capable of self-repair: if one is partly destroyed, the remaining part will grow larger and stronger in an effort to compensate.

TEM x 4,020

Striated appearance of skeletal muscle

Muscles of the hand and lower arm

THE MUSCULAR SYSTEM

MUSCLES of the BODY I

OVER 600 SKELETAL MUSCLES make up nearly half the weight of the human body. By working with bones, these muscles provide the vital power that enables the body to move under conscious and voluntary control. A skeletal muscle usually attaches to one end of a bone, stretches across a joint, and tapers to attach to another bone. As a muscle contracts, it moves one bone while the other bone remains fairly stable. A muscle's point of attachment on the more stable bone is its origin; where it attaches to the more movable bone is its insertion. Many muscles have more than one point of origin and insertion.

SUPERFICIAL AND DEEP

Layers of skeletal muscles overlap each other in intricate patterns. Those just below the skin and its underlying fat are described as superficial (on right side of illustration); beneath these are the deep muscles (on left side of illustration). The muscles of the abdominal wall form three layers; the fibers of each run in a different direction. These provide a strong barrier that is flexible enough to contract or relax, accommodating changes of volume in the abdominal cavity.

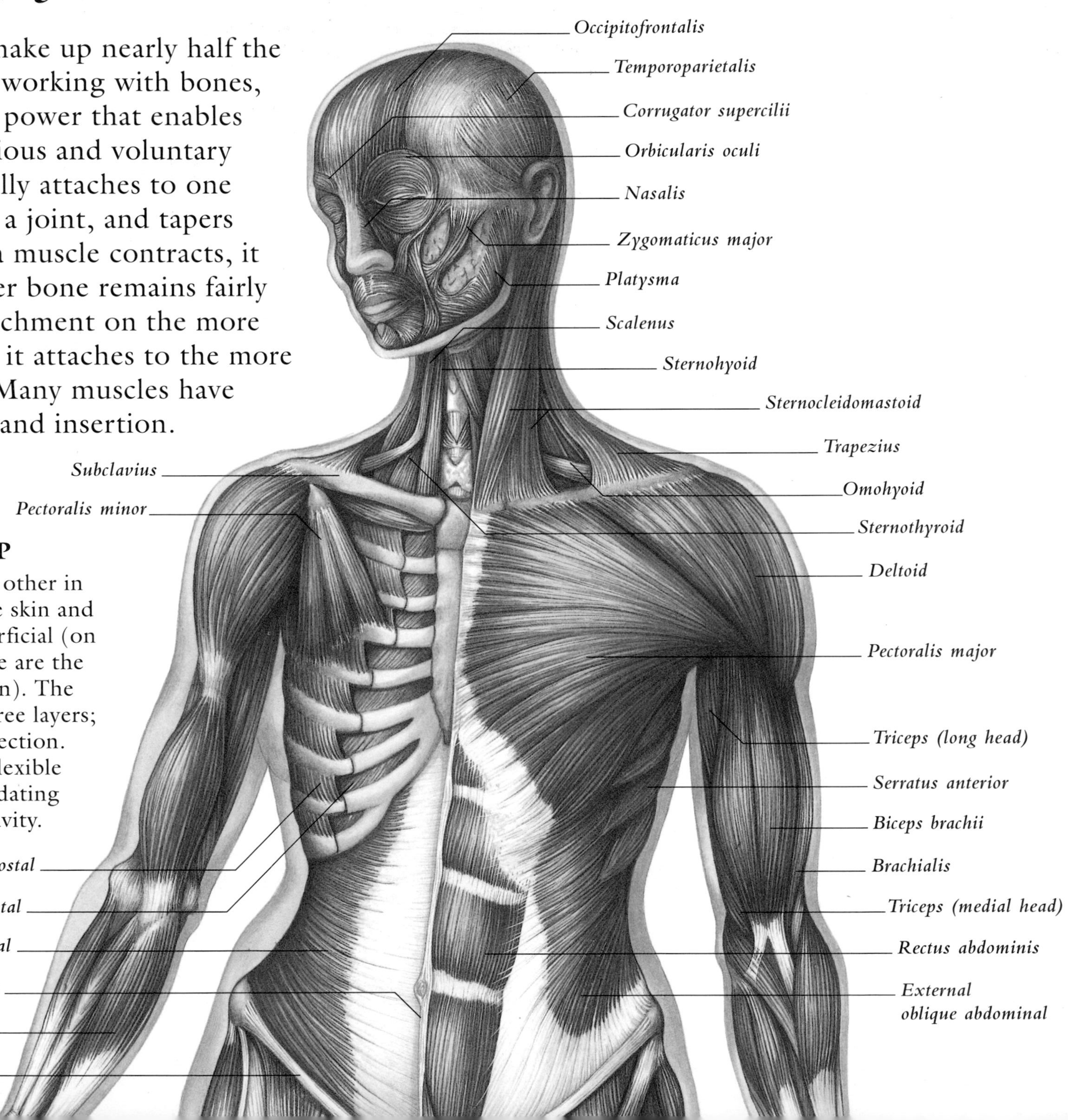

Occipitofrontalis

Temporoparietalis

Corrugator supercilii

Orbicularis oculi

Nasalis

Zygomaticus major

Platysma

Scalenus

Sternohyoid

Sternocleidomastoid

Trapezius

Omohyoid

Sternothyroid

Deltoid

Pectoralis major

Triceps (long head)

Serratus anterior

Biceps brachii

Brachialis

Triceps (medial head)

Rectus abdominis

External oblique abdominal

Subclavius

Pectoralis minor

External intercostal

Internal intercostal

Internal oblique abdominal

Linea alba

Flexor digitorum profundus

Inguinal ligament

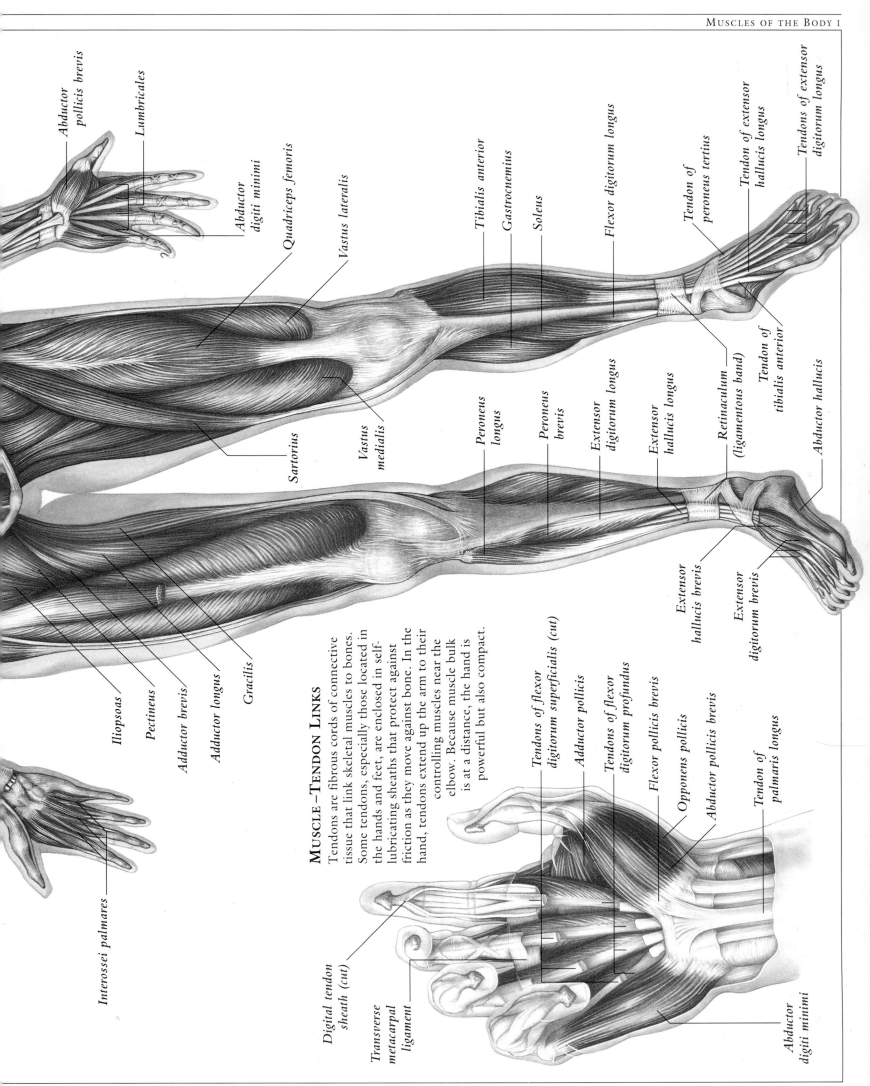

Abductor pollicis brevis

Lumbricales

Abductor digiti minimi

Quadriceps femoris

Vastus lateralis

Tibialis anterior

Gastrocnemius

Soleus

Flexor digitorum longus

Tendon of peroneus tertius

Tendon of extensor hallucis longus

Tendons of extensor digitorum longus

Sartorius

Vastus medialis

Peroneus longus

Peroneus brevis

Extensor digitorum longus

Extensor hallucis longus

Retinaculum (ligamentous band)

Tendon of tibialis anterior

Abductor hallucis

Iliopsoas

Pectineus

Adductor brevis

Adductor longus

Gracilis

Extensor hallucis brevis

Extensor digitorum brevis

Interossei palmares

Muscle–Tendon Links

Tendons are fibrous cords of connective tissue that link skeletal muscles to bones. Some tendons, especially those located in the hands and feet, are enclosed in self-lubricating sheaths that protect against friction as they move against bone. In the hand, tendons extend up the arm to their controlling muscles near the elbow. Because muscle bulk is at a distance, the hand is powerful but also compact.

Digital tendon sheath (cut)

Transverse metacarpal ligament

Tendons of flexor digitorum superficialis (cut)

Adductor pollicis

Tendons of flexor digitorum profundus

Flexor pollicis brevis

Opponens pollicis

Abductor pollicis brevis

Tendon of palmaris longus

Abductor digiti minimi

51

MUSCLES of the BODY II

MUSCLE APPEARANCE VARIES GREATLY, from the massive triangles of the upper back to the slender cables of the small, dexterous hand. The shape of a muscle determines the strength with which it contracts, and thus influences its specific function. The most powerful muscles are those that run along the spine; they maintain posture and provide the strength for lifting and pushing. The smallest is the stapedius inside the ear.

STRONG, STABILIZING MUSCLES

Muscles in the neck and upper back provide strength and permit complex movement. Those in the neck support the head and keep it upright. The upper-back muscles that attach to the winglike scapula stabilize the shoulder, the body's most mobile joint.

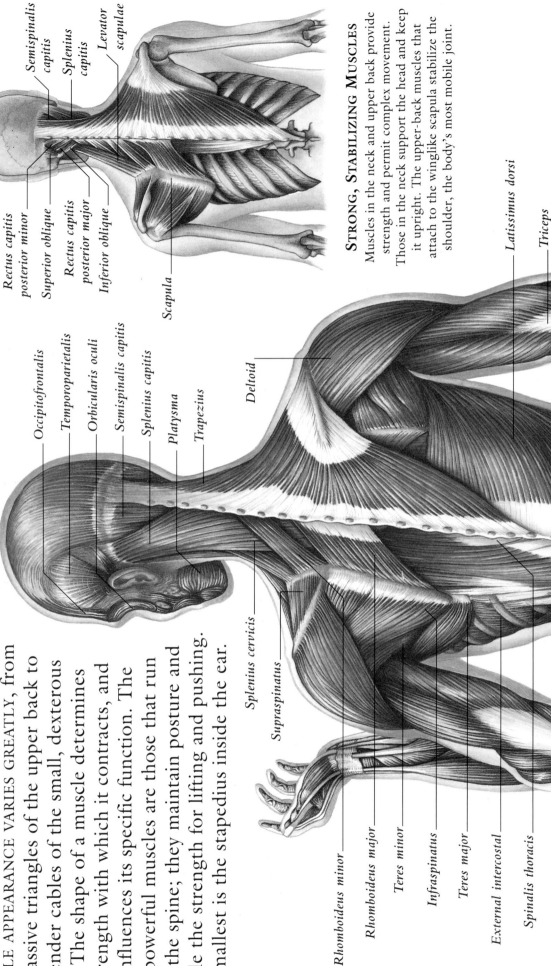

Rectus capitis
posterior minor
Superior oblique
Rectus capitis
posterior major
Inferior oblique

Semispinalis
capitis
Splenius
capitis
Levator
scapulae

Scapula

Occipitofrontalis
Temporoparietalis
Orbicularis oculi
Semispinalis capitis
Splenius capitis
Platysma
Trapezius

Deltoid

Latissimus dorsi
Triceps
External oblique abdominal
Anconeus
Extensor digitorum
Flexor carpi ulnaris
Extensor carpi ulnaris

Splenius cervicis
Supraspinatus

Rhomboideus minor
Rhomboideus major
Teres minor
Infraspinatus
Teres major
External intercostal
Spinalis thoracis
Longissimus thoracis
Erector spinae
Internal oblique abdominal
Gluteus minimus

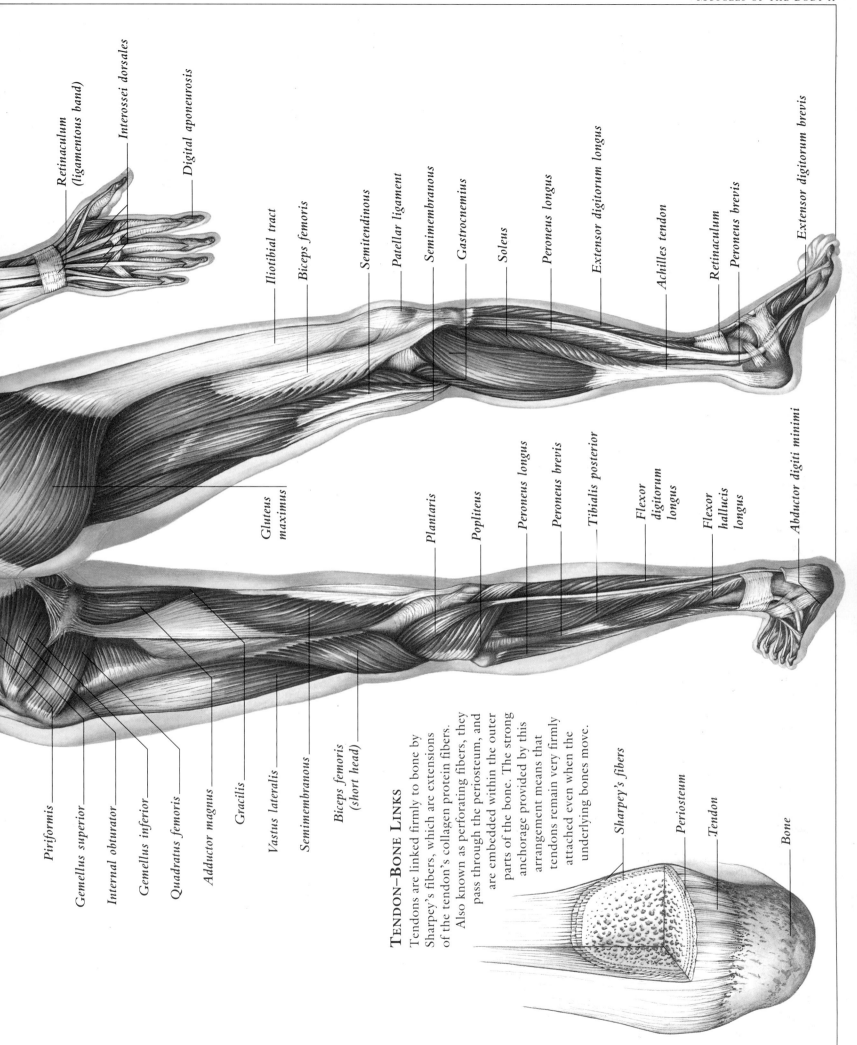

Retinaculum (ligamentous band)

Interossei dorsales

Digital aponeurosis

Iliotibial tract

Biceps femoris

Semitendinous

Patellar ligament

Semimembranous

Gastrocnemius

Soleus

Peroneus longus

Extensor digitorum longus

Achilles tendon

Retinaculum

Peroneus brevis

Extensor digitorum brevis

Gluteus maximus

Plantaris

Popliteus

Peroneus longus

Peroneus brevis

Tibialis posterior

Flexor digitorum longus

Flexor hallucis longus

Abductor digiti minimi

Piriformis

Gemellus superior

Internal obturator

Gemellus inferior

Quadratus femoris

Adductor magnus

Gracilis

Vastus lateralis

Semimembranous

Biceps femoris (short head)

Tendon–Bone Links

Tendons are linked firmly to bone by Sharpey's fibers, which are extensions of the tendon's collagen protein fibers. Also known as perforating fibers, they pass through the periosteum, and are embedded within the outer parts of the bone. The strong anchorage provided by this arrangement means that tendons remain very firmly attached even when the underlying bones move.

Sharpey's fibers

Periosteum

Tendon

Bone

53

MUSCLE STRUCTURE *and* CONTRACTION

THE STRUCTURE OF SKELETAL MUSCLES enables them to contract when stimulated by nerve impulses, pulling some part of the skeleton in the same direction as the contraction. Because muscles can only pull, not push, they are arranged in opposition to each other. This means that the movement produced by a group of muscles can always be reversed by its opposing group.

MUSCLE STRUCTURE

Skeletal muscles consist of densely packed groups of elongated cells known as muscle fibers held together by fibrous connective tissue. Numerous capillaries penetrate the connective tissue to keep muscles supplied with the abundant quantities of oxygen and glucose needed to fuel muscle contraction.

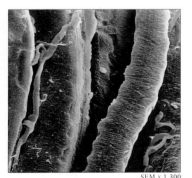

Striated muscle
The alternation of thick and thin myofilaments gives skeletal muscle fibers their striated (striped) appearance, as shown at left. The thin blue tubular structures are capillaries.

SEM x 1,300

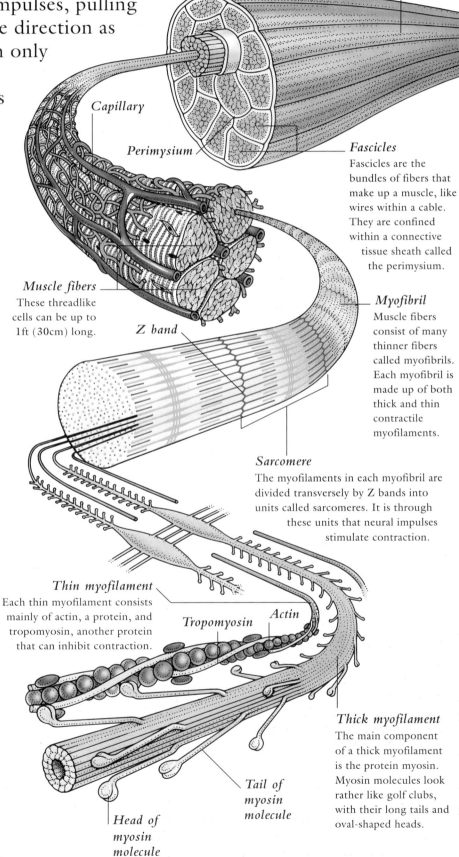

Muscle

Capillary

Perimysium

Fascicles
Fascicles are the bundles of fibers that make up a muscle, like wires within a cable. They are confined within a connective tissue sheath called the perimysium.

Muscle fibers
These threadlike cells can be up to 1ft (30cm) long.

Z band

Myofibril
Muscle fibers consist of many thinner fibers called myofibrils. Each myofibril is made up of both thick and thin contractile myofilaments.

Sarcomere
The myofilaments in each myofibril are divided transversely by Z bands into units called sarcomeres. It is through these units that neural impulses stimulate contraction.

Thin myofilament
Each thin myofilament consists mainly of actin, a protein, and tropomyosin, another protein that can inhibit contraction.

Tropomyosin   *Actin*

Thick myofilament
The main component of a thick myofilament is the protein myosin. Myosin molecules look rather like golf clubs, with their long tails and oval-shaped heads.

Tail of myosin molecule

Head of myosin molecule

HOW MUSCLES CONTRACT

In a relaxed muscle, the thick and thin myofilaments overlap a little. When a muscle contracts, the thick filaments slide farther in between the thin filaments, rather like interlacing fingers, and draw closer to the Z bands. This action shortens both the myofibrils and the entire muscle fiber. The more shortened muscle fibers there are, the greater the contraction in the muscle as a whole.

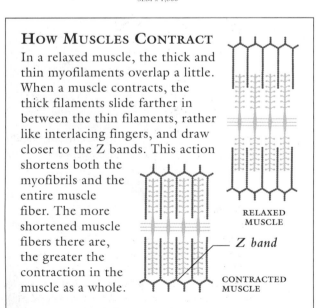

RELAXED MUSCLE

Z band

CONTRACTED MUSCLE

LEVER SYSTEMS

Most bodily movements employ the mechanical principles by which a force applied to one part of a rigid lever arm is transferred via a pivot point, or fulcrum, to a weight elsewhere on the lever. In the body, muscles apply force, bones serve as levers, and joints function as fulcrums in order to move a body part. Lever systems in the body sacrifice mechanical advantage for range of movement.

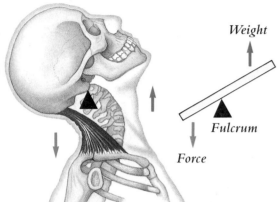

First-class lever

A first-class lever works like a see-saw, with the fulcrum lying between the force and the weight. One rare example in the body is the action of the posterior neck muscles tilting back the head. The lever at the base of the skull pivots on the fulcrum of the atlanto-occipital joint.

Second-class lever

In a second-class lever, the weight lies between the force and the fulcrum. The action of raising the heel from the ground is an example of this type of system in the body. The calf muscles are the force to lift the body weight, the heel and most of the foot form the lever, and the metatarsal-phalangeal joints provide the fulcrum.

Third-class lever

In a third-class lever, which is the most common type in the body, the force is applied to the lever between the weight and the fulcrum. A typical example is flexing the elbow joint (the fulcrum) by contracting the biceps brachii muscle in order to lift the forearm and hand.

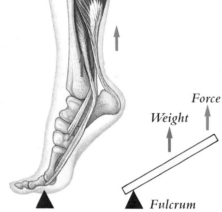

MUSCLES WORK TOGETHER

In order to lift the upper arm away from the trunk, the anterior and posterior sectors of the deltoid muscle balance each other, while the middle sector carries out the work. When a muscle contracts to produce movement it is called the agonist, and its opposite, relaxing muscle is known as the antagonist. Sometimes stabilizing muscles also play an important role in creating this coordinated muscle action.

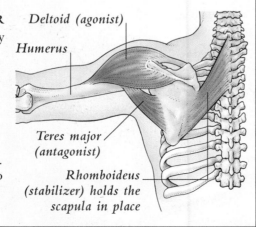

Deltoid (agonist)

Humerus

Teres major (antagonist)

Rhomboideus (stabilizer) holds the scapula in place

FACIAL EXPRESSIONS

For humans and other primates, varying facial expressions are a significant means of communication. The musculature involved is highly complex, allowing for many subtle nuances of expression. Facial muscles have their insertions (attachments to moving parts) within the skin, which means that even a slight degree of muscle contraction can produce movement of facial skin.

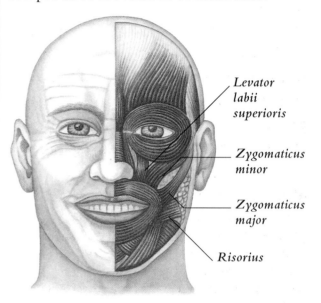

Levator labii superioris

Zygomaticus minor

Zygomaticus major

Risorius

Smiling

The smile is an ambiguous expression, which can convey a number of different emotions apart from simply pleasure. The levator labii superioris lifts the upper lip, while the zygomaticus major, the zygomaticus minor, and the risorius muscles pull the angle of the mouth and the corners of the lips upward as well as sideways.

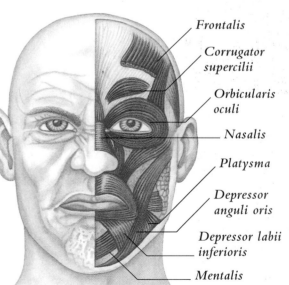

Frontalis

Corrugator supercilii

Orbicularis oculi

Nasalis

Platysma

Depressor anguli oris

Depressor labii inferioris

Mentalis

Frowning

A frown can express various feelings, including disapproval and confusion. The frontalis and corrugator supercilii furrow the brow, the nasalis widens the nostrils, while the orbicularis oculi narrows the eyes. The platysma and depressors pull the mouth and corners of the lips downward and sideways, and the mentalis puckers the chin.

MUSCLE INJURIES *and* DISORDERS

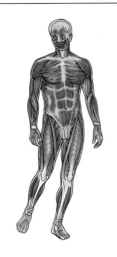

INJURIES TO MUSCLES and their tendon attachments are usually the result of overexertion during daily activities or of sudden pulling or twisting movements, such as those occurring during sports. Various work actions can also damage muscles and tendons. A number of rare muscle disorders may be responsible for muscle weakness and progressive degeneration.

MUSCLE STRAINS AND TEARS

Muscle strain is the term for a moderate amount of damage to muscle fibers. Limited bleeding inside the muscle causes tenderness and swelling, which may be accompanied by painful spasms. Visible bruising may follow. More severe damage that involves a larger number of torn fibers is called a muscle tear.

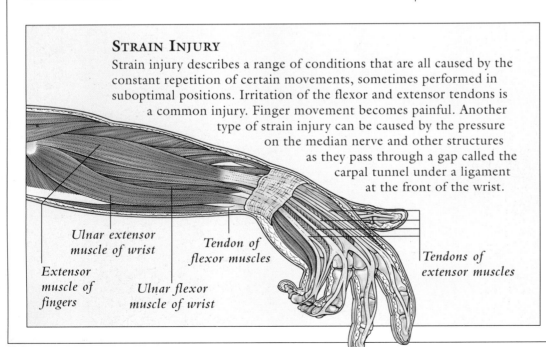

Muscle tear
A torn muscle causes severe pain and swelling. Extensive bleeding may result in the formation of a blood clot, which a doctor may need to remove by needle aspiration or surgical drainage. Vigorous shoulder movements may tear the deltoid or pectoral muscle where it attaches to the humerus.

Deltoid muscle

Tear

Pectoral muscle

Humerus

TENDON INFLAMMATION

Inflammation involving tendons may affect the tendon itself (tendinitis) or the inner lining of fibrous sheaths that enclose some tendons (tenosynovitis). Tendinitis may occur when strong or repeated movement creates excessive friction between the tendon's outer surface and an adjacent bone. Tenosynovitis may be caused by overstretching or repeated movements.

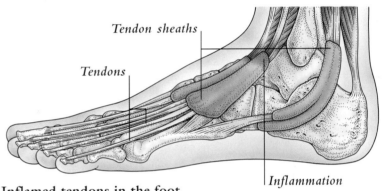

Tendon sheaths

Tendons

Inflammation

Inflamed tendons in the foot
The complexity of the foot makes it susceptible to tendon damage. Activities that involve running, kicking, or any awkward or complex movements (such as dancing) may cause tendon inflammation, as may friction from ill-fitting shoes. Symptoms include pain, swelling, and restricted movement.

STRAIN INJURY
Strain injury describes a range of conditions that are all caused by the constant repetition of certain movements, sometimes performed in suboptimal positions. Irritation of the flexor and extensor tendons is a common injury. Finger movement becomes painful. Another type of strain injury can be caused by the pressure on the median nerve and other structures as they pass through a gap called the carpal tunnel under a ligament at the front of the wrist.

Ulnar extensor muscle of wrist

Extensor muscle of fingers

Ulnar flexor muscle of wrist

Tendon of flexor muscles

Tendons of extensor muscles

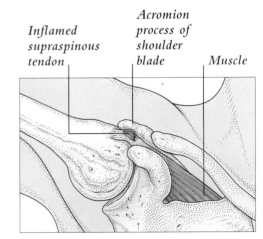

Inflamed supraspinous tendon

Acromion process of shoulder blade

Muscle

Inflamed supraspinous tendon
Anyone who is devoted to racquet sports risks tendinitis of the shoulder. Repeated arm lifting causes the supraspinous tendon in the shoulder to rub against the shoulder blade's acromion process, creating friction.

TENDON TEARS

A sudden, powerful muscle contraction can severely damage a tendon, and can even tear it away from the bone. For example, the strain of lifting a heavy weight, such as a barbell, may tear either the tendons that are attached to the biceps muscle or the main tendon at the front of the thigh (the quadriceps tendon).

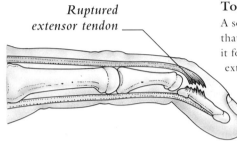

Ruptured extensor tendon

Torn finger tendon
A solid object, such as a hard ball, that strikes a fingertip may bend it forward with the result that the extensor tendon is torn from its attachment. The finger may need to be immobilized for several months, or else be corrected surgically.

OPERATION

TORN ACHILLES TENDON

A minor tear of the Achilles tendon may heal with rest and physical therapy, but a serious injury often requires surgery and months of convalescence. A torn Achilles tendon is a familiar injury to tennis players when they rise abruptly onto their toes to serve. It is also a risk for sprinters, who frequently subject their calf muscles to explosive bursts of muscle contraction.

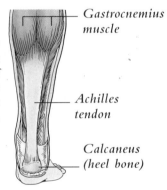

Gastrocnemius muscle

Location of tendon
The Achilles tendon runs from the base of the gastrocnemius muscle of the calf down to the calcaneal bone of the heel. If the tendon is torn, it becomes impossible to raise the heel.

Achilles tendon

Calcaneus (heel bone)

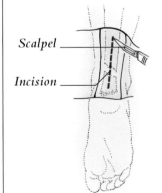

Scalpel

Incision

1 To reattach the torn ends of the tendon, the surgeon first applies a tourniquet around the thigh, to keep the injured area free of blood. An incision over the area of the tear exposes the separated tendon ends.

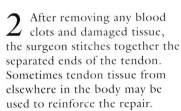

Stitches

2 After removing any blood clots and damaged tissue, the surgeon stitches together the separated ends of the tendon. Sometimes tendon tissue from elsewhere in the body may be used to reinforce the repair.

MUSCULAR DYSTROPHY

Muscular dystrophy is a term that describes a group of diseases, usually inherited, in which a progressive degeneration of skeletal muscles occurs. Symptoms are increasing wasting of muscles and loss of muscle function. There is no effective treatment. However, stretching exercises and surgery to release shortened muscles and tendons can benefit some sufferers.

DIAGNOSIS

The patient's symptoms and family history help in the diagnosis of muscular dystrophy. Tests include genetic screening and blood tests to identify an enzyme that is released by damaged muscle. A muscle biopsy involves the removal of a small piece of tissue, while electrical recording is utilized to demonstrate muscle activity.

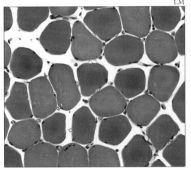
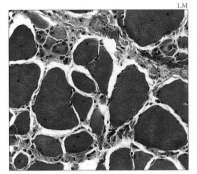

NORMAL MUSCLE FIBERS   ABNORMAL MUSCLE FIBERS

Muscle biopsy
A tissue sample is removed, using either a biopsy needle or scalpel incision, and is then examined under a microscope. The muscle fibers above right show the degeneration characteristic of muscular dystrophy.

MYASTHENIA GRAVIS

This autoimmune disorder is marked by severe muscle weakness and fatigue. This is caused by antibodies that gradually reduce the number of receptors in the fibers that stimulate muscle contractions. A thymus disorder may trigger the disease; the gland is removed and immunosuppressant drugs are often part of the treatment.

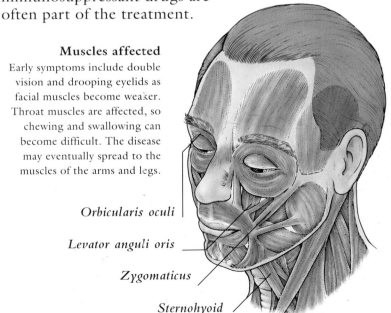

Muscles affected
Early symptoms include double vision and drooping eyelids as facial muscles become weaker. Throat muscles are affected, so chewing and swallowing can become difficult. The disease may eventually spread to the muscles of the arms and legs.

Orbicularis oculi

Levator anguli oris

Zygomaticus

Sternohyoid

CHAPTER 4

The NERVOUS SYSTEM

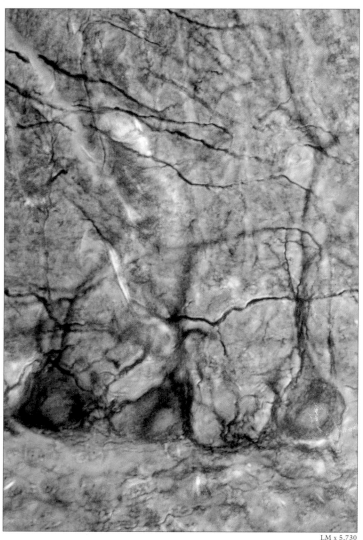

Two large nerve cells
from the cerebellum,
a part of the brain

LM x 5,730

INTRODUCTION

Our brains – not our hearts – are where we feel emotions such as love and anger. They are also where we think, decide, and initiate and control our actions. Nerve impulses are constantly flashing into the brain, around it, and away from it via the spinal cord and a network of cablelike nerves distributed throughout the body. The brain is often thought of as a computer, but this comparison is rather misleading. Communication between the billions of nerve cells in the brain is by means of chemical signals as well as electrical ones, which is why alcohol and drugs affect us.

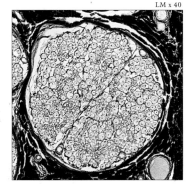

LM x 40

Cross-section of the sciatic nerve

More significantly, the complex connections between nerve cells are capable of growing and developing, and therefore of reacting to events for which the brain has not been programmed. The brain is capable of creativity in a way no computer has yet achieved. But it is delicate: nerve tracts within the brain or spinal cord, if they are damaged by injury or disease, are unable to repair themselves. The search for an understanding of the brain's cognitive functions poses a real challenge to science. On the other hand, recent advances in biochemistry and imaging techniques have shed new light on its physical workings, and on a variety of conditions such as stroke, tumors, and Alzheimer's disease.

A section of the spinal cord within the spinal column

THE NERVOUS SYSTEM

ORGANIZATION of the NERVOUS SYSTEM

THE BODY AND BRAIN ARE CONSTANTLY ALIVE with billions of electrical and chemical signals. This incessant signaling is the result of activity from neurons, or nerve cells, and their filamentous but far-reaching fibers. Along with support cells called neuroglia, neurons are found in the brain and spinal cord, also called the central nervous system or CNS. Extensions of the neurons also form the peripheral nerves that connect the CNS with the rest of the body. Most nerve signaling occurs unconsciously and keeps the body safe and functioning properly.

NERVE NETWORKS

The long, slender nerve fibers of individual neurons band together in groups outside the CNS to form the cablelike peripheral nerves. This body-wide network reports to the CNS on the state of events outside and inside the body. Most of the peripheral nerves divide and branch in order that nerve fibers make contact with as many parts and tissues as possible. Some form groups, called plexuses, so that important areas, such as the hand and fingers, are under finely tuned control.

Median nerve

Ulnar nerve

Lateral cutaneous branches of intercostal nerves

Medial cutaneous branches of intercostal nerves

Iliohypogastric nerve

Ilioinguinal nerve

Radial nerve

Musculo-cutaneous nerve

Vagus nerve

Phrenic nerve

Lateral pectoral nerve

Intercostal nerves

Dorsal branches of intercostal nerves

Subcostal nerve

Auriculotemporal nerve

Brain

Optic nerve

Facial nerve

Brachial plexus

Supraclavicular nerve

Axillary nerve

Deltoid nerve

Spinal ganglion

Spinal cord

Radial nerve

Muscular branches of median nerve

Filum terminale

Femoral nerve

Muscular branches of sciatic nerve

Obturator nerve

Muscular branches of femoral nerve

Anterior cutaneous branches of femoral nerve

Sciatic nerve

Muscular branch of tibial nerve

Common peroneal nerve

Infrapatellar branch of saphenous nerve

Deep peroneal nerve

Saphenous nerve

Interosseous nerve

Superficial peroneal nerve

Medial dorsal cutaneous nerve

Dorsal digital nerves

Medial plantar nerve

Pudendal nerve

Sciatic nerve

Tibial nerve

Infrapatellar branch of saphenous nerve

Common peroneal nerve

Cutaneous branches of saphenous nerve

Interosseous nerve

Deep peroneal nerve

Saphenous nerve

Superficial peroneal nerve

Dorsal cutaneous nerves

Intermediate dorsal cutaneous nerve

Lateral plantar nerve

Tibial nerve

Gluteal nerve

Median nerve

Ulnar nerve

Deep branch of ulnar nerve

Common palmar digital nerve

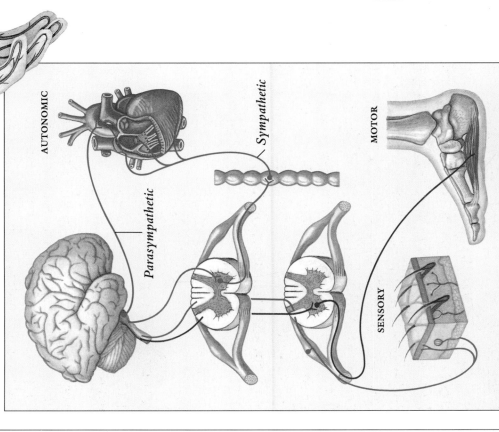

AUTONOMIC

Parasympathetic

Sympathetic

MOTOR

SENSORY

PERIPHERAL NERVOUS SYSTEM DIVISIONS

The peripheral nervous system is composed of three divisions: autonomic, sensory, and motor. Autonomic nerve fibers (blue), which may be parasympathetic or sympathetic, carry instructions from the CNS to the body's organs and glands. Sensory nerve fibers (red) relay information about inner bodily sensations and events occurring in the outside world. Motor nerve fibers (purple) serve voluntary skeletal muscles.

NERVE CELLS *and* NERVES

THE BASIC UNIT OF THE NERVOUS SYSTEM is the neuron. The body of this specialized nerve cell has projections that either receive electrical messages from or transmit messages to other neurons, muscles, or glands. The billions of interconnecting neurons that make up the nervous system are protected by other supporting nerve cells known as glial cells. These cells, located between neuron bodies and axons, are nonexcitable and compose over half the nerve cells in the nervous system.

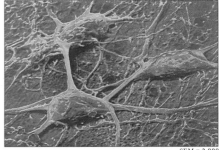

SEM x 2,880

Three neurons in the cerebral cortex

NEURON STRUCTURE

Like other body cells, neurons have a cell body with a central nucleus and a number of other structures that are important for the maintenance of cell life. Extending from the cell body are a variable number of projections, called processes or neurites. Axons are the neurites that carry impulses away from the cell body, while dendrites are the processes that receive impulses.

TYPES OF NEURON

The shape and size of neuron cell bodies vary greatly, as do the type, number, and length of their processes. Some general types of neuron are shown in the illustrations below.

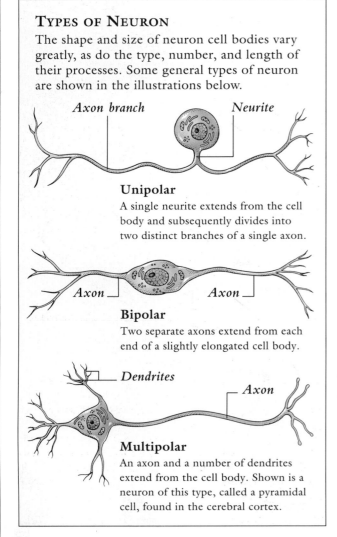

Axon branch   Neurite

Unipolar
A single neurite extends from the cell body and subsequently divides into two distinct branches of a single axon.

Axon   Axon

Bipolar
Two separate axons extend from each end of a slightly elongated cell body.

Dendrites

Axon

Multipolar
An axon and a number of dendrites extend from the cell body. Shown is a neuron of this type, called a pyramidal cell, found in the cerebral cortex.

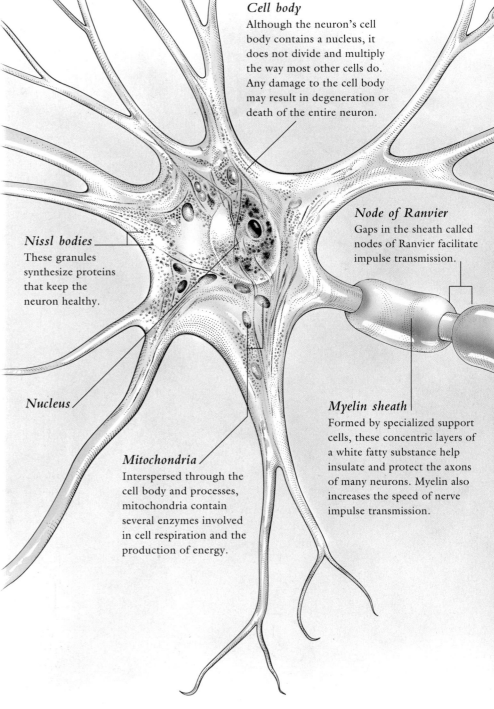

Cell body
Although the neuron's cell body contains a nucleus, it does not divide and multiply the way most other cells do. Any damage to the cell body may result in degeneration or death of the entire neuron.

Nissl bodies
These granules synthesize proteins that keep the neuron healthy.

Node of Ranvier
Gaps in the sheath called nodes of Ranvier facilitate impulse transmission.

Nucleus

Mitochondria
Interspersed through the cell body and processes, mitochondria contain several enzymes involved in cell respiration and the production of energy.

Myelin sheath
Formed by specialized support cells, these concentric layers of a white fatty substance help insulate and protect the axons of many neurons. Myelin also increases the speed of nerve impulse transmission.

SUPPORT CELLS

Unlike neurons, supporting nerve cells, known as glial cells, are not involved with the transmission of nerve impulses. Instead, they act to protect and nourish the neurons. Several types of these specialized cells exist. The smallest cells are called microglia; they engulf and destroy microorganisms. Other types of cells help insulate axons or circulate the cerebrospinal fluid.

Oligodendrocytes

These cells wrap their plasma membranes around neurons of both the brain and spinal cord to form myelin sheaths.

Astrocytes

Delicate projections of cytoplasm give these star-shaped cells their name ("astro" means star). Some cell processes connect with capillaries and help regulate the flow of substances from the blood to the brain and spinal cord.

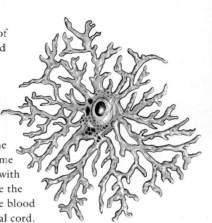

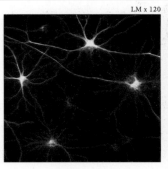

LM x 120

"Star-cell" constellation
Astrocytes, which are the most numerous type of glial cell, form complex networks in the gray matter of the brain.

Dendrite
Branching from the cell body are tapering filaments called dendrites, which receive electrical signals from other neurons.

Synaptic knobs
These bulges at the ends of the axon's terminal fibers contain vesicles, or sacs, of chemicals known as neurotransmitters that help transmit information from one cell to another.

Axon terminal fiber

Axon
The longest process that extends from a cell body is the axon, also known as a nerve fiber. Axons conduct impulses in one direction only. Some of them reach a length of more than 3ft (1m), and axons with a wide diameter conduct impulses very quickly.

NERVE STRUCTURE

Cordlike nerves are formed from bundles, called fascicles, of axons that project from several neurons. Most nerves travel to a specific site in the body and carry two types of fiber: sensory (afferent) fibers, which convey impulses from receptors in the skin, sense organs, and internal organs back to the brain and spinal cord; and motor (efferent) fibers, which transmit signals from the brain and spinal cord to a muscle or gland.

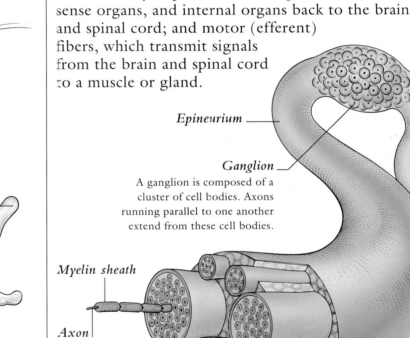

Epineurium

Ganglion
A ganglion is composed of a cluster of cell bodies. Axons running parallel to one another extend from these cell bodies.

Myelin sheath

Axon

Fascicle
The word "fascicle" means bundle; a nerve fascicle is composed of a bundle of neuronal axons.

Perineurium

A major nerve
This cross-section shows the sciatic nerve with its many fascicles. It is the thickest nerve in the body, measuring 0.8in (2cm) in diameter at the point where it emerges from the spinal cord. Branches of the sciatic nerve innervate the muscles and skin of the leg.

LM x 40

NEURON BEHAVIOR

IN ORDER TO PRODUCE ELECTRICAL NERVE IMPULSES, neurons must be triggered by a stimulus, which is anything inside or outside the body that evokes a physical or a psychological response. A neuron's capacity to respond to a stimulus is known as excitability. Electrical nerve impulses may also be blocked, or inhibited, by some neurotransmitters or by drugs.

TEM x 14,000

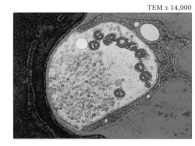

The synaptic cleft
Tiny sacs, called vesicles, containing neurotransmitters cluster near the synaptic cleft between an axon (seen in yellow) and a muscle fiber (red).

THE SYNAPSE

To "fire" a neuron, a stimulus must convert the electrical charge on the inside of the cell membrane from negative to positive. The nerve impulse travels down the axon to a synaptic knob, and triggers the release of chemicals that may stimulate a response in the target cell.

Axon terminal fiber

Neurofilaments
These act as scaffolding to help give the nerve cell its shape.

Cell membrane
This conveys electrical impulses away from the cell body.

Microtubules
These structures are thought to help in the transport of neurotransmitter molecules to the synaptic membrane.

Synaptic vesicles
These sacs contain molecules of the neurotransmitter and are drawn toward the synaptic cleft by calcium ions.

Receptor sites
The neurotransmitter combines with protein receptors on the target cell membrane, which then becomes permeable to specific ions.

Synaptic knob
Each knob at the end of an axon terminal fiber lies close to the neuron cell body, its axons or dendrites, another synaptic knob, or a muscle fiber.

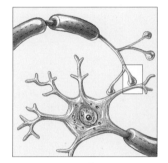

The synapse
This communication point between neurons (above and enlarged at left) comprises the synaptic knob, the synaptic cleft, and the target site.

Mitochondrion

Neurotransmitter molecules
These chemical molecules are released from the vesicles into the synaptic cleft, where they influence impulse transmission.

Synaptic cleft

Na$^+$

Na$^+$

Na$^+$

Membrane channels
Excitation occurs when enough positive sodium ions (Na$^+$) have passed through channels in the membrane to change the charge on the inside of the cell membrane from negative to positive.

Target cell membrane

SENDING A NERVE IMPULSE

The level at which a stimulus begins to transmit an electrical impulse is called a threshold. If a stimulus is too weak, or below the threshold, there is only a very brief local response in the membrane. If, however, the threshold is reached, the impulse travels along the entire length of the fiber. The speed of transmission can vary: fibers that are cold (as when ice is applied in order to dull pain), those with small diameters, and those without myelin sheaths conduct impulses more slowly.

REGENERATION

Peripheral nerve fibers that are crushed or only partially cut may slowly regenerate if the cell body and the segments of myelin sheath remain continuous. Regeneration does not occur in nerves in the brain or spinal cord; instead, injured nerve fibers are wrapped in scar tissue and inactivated.

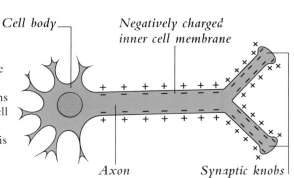

1 In a neuron's "resting" state when no impulse is being transmitted, positive sodium ions diffuse from the inside of the cell membrane at a continuous rate; the inner membrane of the cell is thus negatively charged.

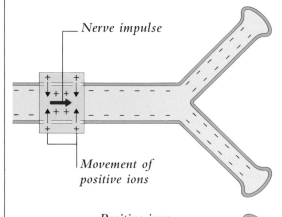

2 When stimulated by a nerve impulse, positively charged ions in the fluid outside the membrane cross the cell membrane. At these local sites the electrical charge on the inside of the cell membrane changes from negative to positive.

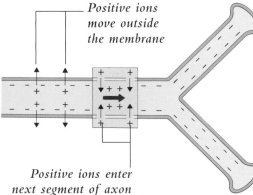

3 This localized reversal of charge across the membrane stimulates similar changes in the subsequent membrane segments. The electrical impulse continues down the axon; as it passes along, previous segments of the membrane revert back to the "inside-negative" state.

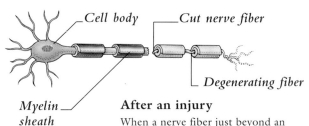

After an injury
When a nerve fiber just beyond an injury and farthest from the cell body no longer receives vital proteins and enzymes, it begins to degenerate and the myelin sheath becomes hollow.

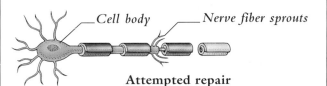

Attempted repair
The undamaged neuron cell body stimulates the growth of several nerve sprouts in the remaining portion of the fiber. One of these sprouts may eventually find its way through the empty but intact myelin sheath.

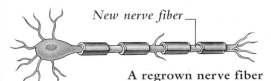

A regrown nerve fiber
Growing at a rate of about 1.5mm a day, the new nerve fiber reaches its previous connection. Function and sensation are slowly restored, and unused nerve sprouts degenerate.

4 The impulse reaches the synaptic cleft. At right is a microscopic image of the branching terminal fibers of a motor neuron axon (pink); these fibers end in synaptic knobs that lie very close to skeletal muscle fibers (red). When a neurotransmitter is released from vesicles in the synaptic knob, it crosses the synaptic cleft and stimulates the muscle fibers to contract.

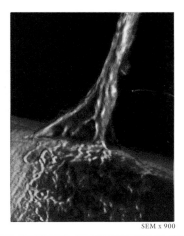

SEM x 900

INHIBITION

During inhibition, or blockage of electrical impulses, channels that are sensitive to chloride or potassium ions may open rather than channels that are sensitive to sodium. Positive potassium ions (shown as K^+) escape from the target cell, or else negative chloride ions (shown as Cl^-) permeate the cell membrane. In both cases, the electrical charge inside the target cell membrane stays negative, the neuron cannot be "fired," and the nerve impulse is inhibited.

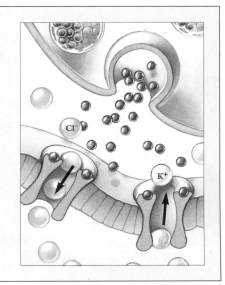

The BRAIN I

LOCATED IN THE SKULL, THE BRAIN is composed of more than 12 billion neurons and 50 billion supporting glial cells, but it weighs less than 3lb (1.4kg). With the spinal cord, the brain monitors and regulates many unconscious bodily processes, such as heart rate, and coordinates most voluntary movement. Most important, it is the site of consciousness and of all the intellectual functions that allow humans to think and create.

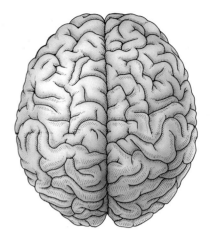

Halves of the brain
A longitudinal fissure, seen from above, divides the brain into two halves, called hemispheres, that communicate with each other.

OUTER BRAIN STRUCTURE

The most obvious feature about the cerebrum – the largest part of the brain – is its heavily folded surface, the pattern of which is different in each human being. The grooves are called sulci when shallow and fissures when deep. Fissures and some large sulci outline specific functional areas that are called lobes. A ridge on the brain's surface is called a gyrus.

Central sulcus

Cerebral cortex
The entire cerebrum is covered by a layer of gray matter about 2 to 6mm thick. Underneath is the brain's white matter as well as islands of gray matter.

Gyrus

Parietal lobe
A variety of bodily sensations such as touch, pressure, pain, and temperature are both perceived and interpreted here.

Occipital lobe
This area detects and interprets visual images.

Frontal lobe
Speech production, the elaboration of thought and emotion, and skilled movements are controlled by neurons found in this part of the brain.

Temporal lobe
The recognition of specific tones and loudness takes place in the temporal lobe. This area also plays a role in memory storage.

Sylvian or lateral fissure

Cerebellum
Neurons of the cerebellum link with other regions of the brain and the spinal cord, facilitating smooth, precise movement, and controlling balance and posture. It also plays a role in speech.

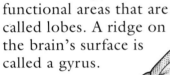

LM x 360

Cells of the cerebral cortex
Higher intellectual functions such as memory and the interpretation of sensory impulses are assisted by the complex network of neurons located in the cerebral cortex.

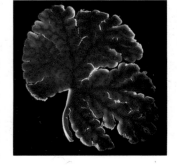

The tree of life
This image shows a section through the cerebellum, which is made up of many myelinated nerve fibers. The distinctive pattern resembles the veins of a leaf and is described as the arbor vitae ("tree of life").

INNER BRAIN STRUCTURES

Within the center of the brain lies the thalamus, which is the brain's information relay station. Surrounding the thalamus is a group of structures, the limbic system, which is involved in survival behavior and emotions such as rage and fright. Closely linked with the limbic system is the hypothalamus, which has overall control of the body's automatic processes.

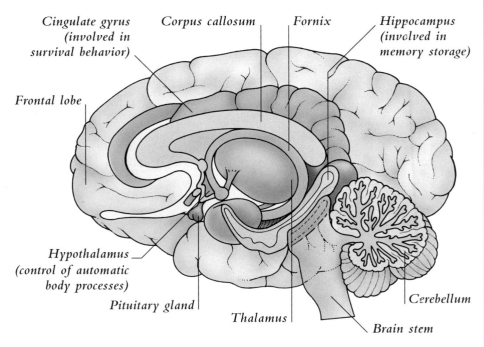

Cingulate gyrus (involved in survival behavior)

Corpus callosum

Fornix

Hippocampus (involved in memory storage)

Frontal lobe

Hypothalamus (control of automatic body processes)

Pituitary gland

Thalamus

Cerebellum

Brain stem

GRAY AND WHITE MATTER

The brain's gray matter is made up of groups of neuron cell bodies. White matter, by contrast, is composed mainly of the myelin-covered axons, or nerve fibers, that extend from the neuron cell bodies. Fatty myelin sheaths insulate the axons and increase the transmission speed of the nerve impulses.

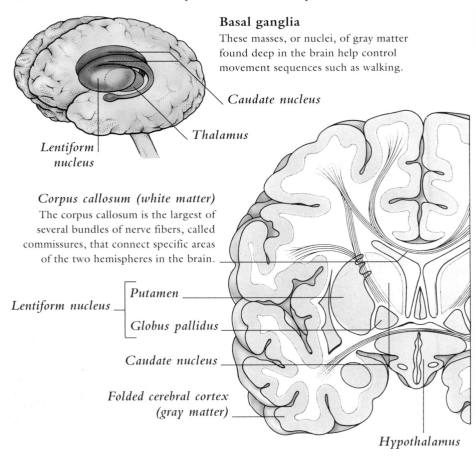

Basal ganglia
These masses, or nuclei, of gray matter found deep in the brain help control movement sequences such as walking.

Caudate nucleus

Thalamus

Lentiform nucleus

Corpus callosum (white matter)
The corpus callosum is the largest of several bundles of nerve fibers, called commissures, that connect specific areas of the two hemispheres in the brain.

Lentiform nucleus

Putamen

Globus pallidus

Caudate nucleus

Folded cerebral cortex (gray matter)

Hypothalamus

VERTICAL LINKS

Myelinated fibers organized into so-called projection tracts transmit impulses to and from the cerebral cortex and the lower brain and spinal cord. These nerve tracts pass through a communication link called the internal capsule, a compact band of fibers, and intersect the corpus callosum.

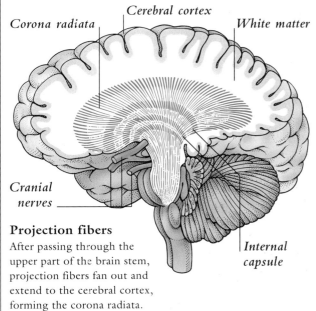

Corona radiata

Cerebral cortex

White matter

Cranial nerves

Internal capsule

Projection fibers
After passing through the upper part of the brain stem, projection fibers fan out and extend to the cerebral cortex, forming the corona radiata.

THE THALAMUS AND BRAIN STEM

The thalamus is a relay station that sorts, interprets, and directs sensory signals received from both the spinal cord and midbrain to the cerebral cortex and appropriate sites of the cerebrum. The brain stem contains centers that regulate several functions that are vital for survival; these include blood pressure, heartbeat, respiration, digestion, and certain reflex actions such as swallowing and vomiting.

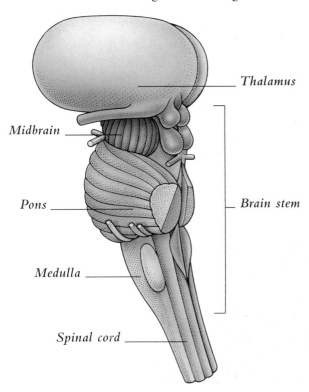

Thalamus

Midbrain

Pons

Brain stem

Medulla

Spinal cord

The BRAIN II

THE SOFT TISSUE OF THE BRAIN FLOATS within the bony casing of the skull in a watery medium, which also flows around the spinal cord. Called cerebrospinal fluid, this colorless liquid, which is renewed on a continuous basis, is produced inside the ventricles (chambers) of the brain. The fluid contains glucose, necessary to provide energy for cell function in the brain and spinal cord, as well as proteins and lymphocytes to guard against infection.

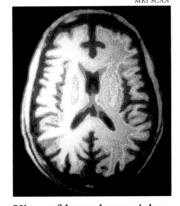

MRI SCAN

View of lateral ventricles
Each of the two lateral ventricles, one in each hemisphere, has a front and a longer back horn, which extend from the center of the brain. When seen from above, these ventricles have the appearance of a large X.

Site of fluid production (choroid plexuses)
Cerebrospinal fluid is produced in clusters of thin-walled capillaries, called choroid plexuses, that line the walls of the ventricles.

Direction of flow
Cerebrospinal fluid moves from the lateral ventricles into the third and fourth ventricles. It then flows up the back of the brain, down around the spinal cord, and up the front of the brain (arrows).

Site of reabsorption (arachnoid granulations)
After circulating, cerebrospinal fluid is reabsorbed into blood through arachnoid granulations, projections from the arachnoid layer of the meninges that connect with veins via the venous sinus.

PROTECTION FOR THE BRAIN

The solid bones of the skull may fracture if struck sufficiently hard. However, the cerebrospinal fluid within the skull absorbs and disperses excessive mechanical forces that might otherwise cause serious injury to the brain. An analysis of its chemical constituents and the pressure of its flow offers vital clues in the diagnosis of many disorders of the brain, such as meningitis.

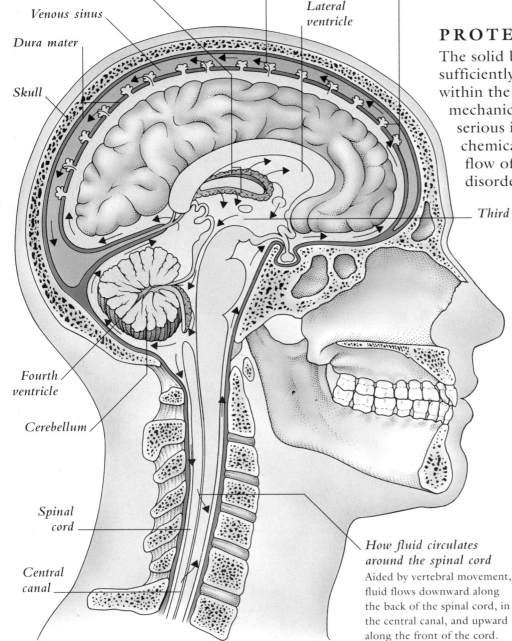

Venous sinus

Dura mater

Skull

Lateral ventricle

Third ventricle

Fourth ventricle

Cerebellum

Spinal cord

Central canal

How fluid circulates around the spinal cord
Aided by vertebral movement, fluid flows downward along the back of the spinal cord, in the central canal, and upward along the front of the cord.

FLUID-FILLED CHAMBERS

Fluid produced in the lateral ventricles drains via the interventricular foramen into the third ventricle close to the thalamus. It then flows through the cerebral aqueduct and into the fourth ventricle, which is located in front of the cerebellum. Circulation is aided by the pulsations of the cerebral arteries.

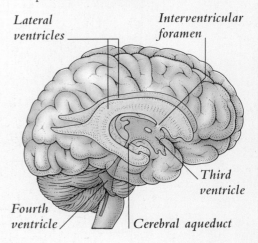

Lateral ventricles

Interventricular foramen

Third ventricle

Fourth ventricle

Cerebral aqueduct

THE MENINGES

Three membranes known as the meninges cover the brain. Lining the inside of the skull is the outermost membrane, the dura mater, which contains veins and arteries that nourish the cranial bones. The middle layer is known as the arachnoid ("spiderlike"); this consists of a type of webbed and elastic connective tissue. Next to the surface of the cerebral cortex is the pia mater; between this delicate, innermost layer and the arachnoid is the subarachnoid space, which contains cerebrospinal fluid as well as blood vessels.

Area shown enlarged

Dura mater

Arachnoid granulations

Venous sinus

Arachnoid

Bone of skull

Cerebrum

Pia mater

Artery

Subarachnoid space

BLOOD SUPPLY TO THE BRAIN

Although the brain accounts for only about 2 percent of the total weight of the body, it requires 20 percent of the body's blood. Both oxygen and glucose are transported by blood; without these essential elements, brain function quickly deteriorates, and dizziness, confusion, and loss of consciousness may occur. Within only 4 to 8 minutes of oxygen deprivation, brain damage or death results.

Arteries supplying blood to the brain

Two front and two back arteries join up at the base of the brain to form an arterial ring called the circle of Willis. From this point, branching blood vessels provide the brain with oxygenated blood.

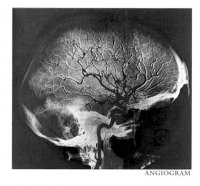

ANGIOGRAM

Blood-brain barrier

A controlled molecule flow to the brain is essential for stable brain function. Endothelial cells that are located in the capillary walls create an almost impassible layer. In addition, capillaries are also wrapped within fibers of the protective neurons (astrocytes). Oxygen, water, and glucose, relatively small molecules, pass through this two-layered barrier easily, but many drugs and chemicals cannot pass through at all.

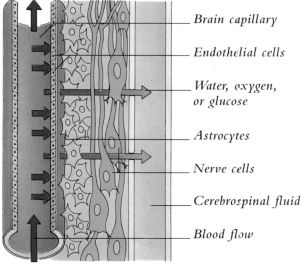

Brain capillary

Endothelial cells

Water, oxygen, or glucose

Astrocytes

Nerve cells

Cerebrospinal fluid

Blood flow

BRAIN DEVELOPMENT

Brain growth, the most important part of embryonic life, occurs much more quickly than the development and growth of the limbs or the internal organs. From small clusters of tissue, highly specialized areas of brain function emerge.

At 3 weeks

A tube of neural tissue develops along the back of the embryo. Three bulges, called the primary vesicles, develop into the main divisions of the brain.

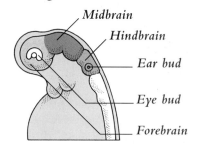

Midbrain

Hindbrain

Ear bud

Eye bud

Forebrain

At 7 weeks

The neural tube flexes and cranial nerves sprout from the hind-brain. Bulges form on the forebrain, one of which will develop into the cerebrum.

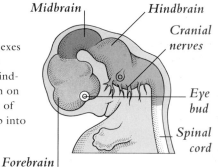

Midbrain

Hindbrain

Cranial nerves

Eye bud

Spinal cord

Forebrain

At 11 weeks

The hindbrain separates into the cerebellum, the pons, and the medulla. The forebrain develops further and the cerebrum starts to grow back over the hindbrain.

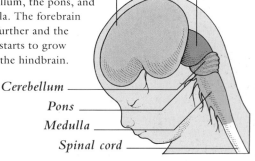

Cerebrum

Midbrain

Cerebellum

Pons

Medulla

Spinal cord

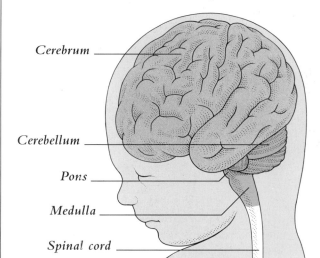

Cerebrum

Cerebellum

Pons

Medulla

Spinal cord

At birth

As the cerebrum enlarges to become the largest part of the brain, folding of the cerebral cortex occurs. Every individual has a unique folding pattern.

The SPINAL CORD

THE SPINAL CORD IS A CABLE about 17in (43cm) in length that descends from the brain stem to the lumbar part of the back. A slightly flattened cylinder, it is about as wide as a finger for most of its length, tapering to a threadlike tail. Through 31 pairs of spinal nerves, the spinal cord is connected to the rest of the body and relays information received via these nerves about its internal and external environment to and from the brain.

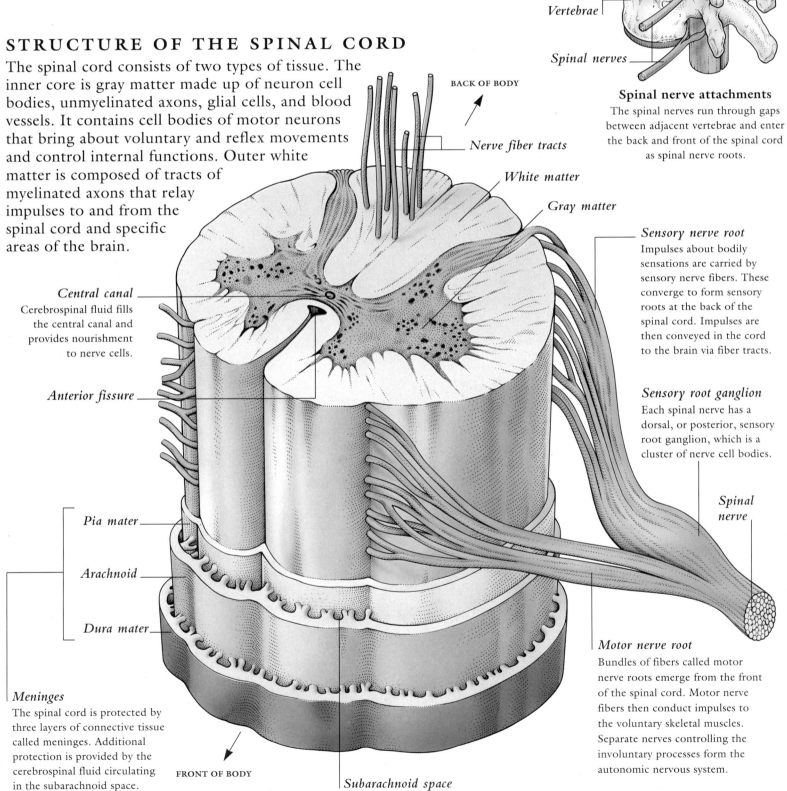

Spinal nerve attachments
The spinal nerves run through gaps between adjacent vertebrae and enter the back and front of the spinal cord as spinal nerve roots.

STRUCTURE OF THE SPINAL CORD

The spinal cord consists of two types of tissue. The inner core is gray matter made up of neuron cell bodies, unmyelinated axons, glial cells, and blood vessels. It contains cell bodies of motor neurons that bring about voluntary and reflex movements and control internal functions. Outer white matter is composed of tracts of myelinated axons that relay impulses to and from the spinal cord and specific areas of the brain.

Spinal cord

Vertebrae

Spinal nerves

BACK OF BODY

Nerve fiber tracts

White matter

Gray matter

Sensory nerve root
Impulses about bodily sensations are carried by sensory nerve fibers. These converge to form sensory roots at the back of the spinal cord. Impulses are then conveyed in the cord to the brain via fiber tracts.

Central canal
Cerebrospinal fluid fills the central canal and provides nourishment to nerve cells.

Anterior fissure

Sensory root ganglion
Each spinal nerve has a dorsal, or posterior, sensory root ganglion, which is a cluster of nerve cell bodies.

Spinal nerve

Pia mater

Arachnoid

Dura mater

Motor nerve root
Bundles of fibers called motor nerve roots emerge from the front of the spinal cord. Motor nerve fibers then conduct impulses to the voluntary skeletal muscles. Separate nerves controlling the involuntary processes form the autonomic nervous system.

Meninges
The spinal cord is protected by three layers of connective tissue called meninges. Additional protection is provided by the cerebrospinal fluid circulating in the subarachnoid space.

FRONT OF BODY

Subarachnoid space

PROTECTION OF THE SPINAL CORD

The spinal cord is protected primarily by the bony segments of the vertebral column and its supporting ligaments. Also protective are the circulating cerebrospinal fluid, which acts as a shock absorber, and the epidural space, a cushioning layer of fat and connective tissue that lies in between the periosteum (the membrane that covers the vertebral bone) and the dura mater, the outer layer of the meninges.

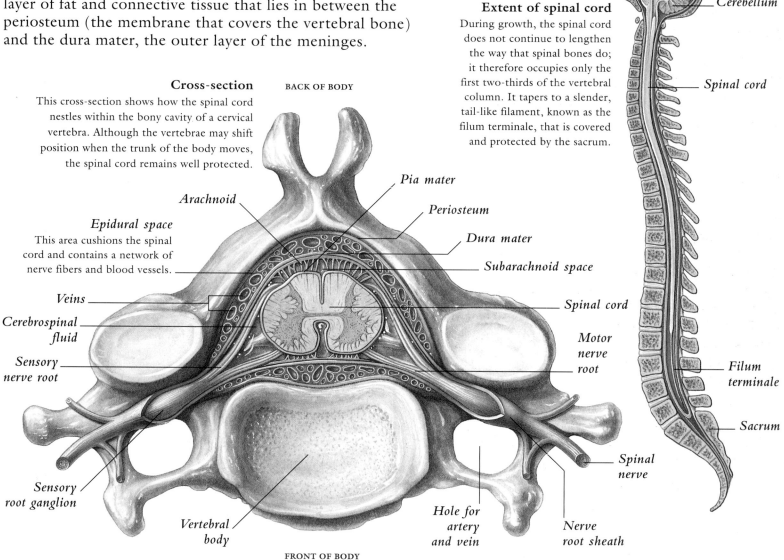

Cross-section
BACK OF BODY

This cross-section shows how the spinal cord nestles within the bony cavity of a cervical vertebra. Although the vertebrae may shift position when the trunk of the body moves, the spinal cord remains well protected.

Extent of spinal cord
During growth, the spinal cord does not continue to lengthen the way that spinal bones do; it therefore occupies only the first two-thirds of the vertebral column. It tapers to a slender, tail-like filament, known as the filum terminale, that is covered and protected by the sacrum.

Cerebrum

Skull

Cerebellum

Spinal cord

Pia mater

Arachnoid

Periosteum

Epidural space
This area cushions the spinal cord and contains a network of nerve fibers and blood vessels.

Dura mater

Subarachnoid space

Veins

Cerebrospinal fluid

Spinal cord

Sensory nerve root

Motor nerve root

Filum terminale

Sensory root ganglion

Vertebral body

Hole for artery and vein

Spinal nerve

Nerve root sheath

Sacrum

FRONT OF BODY

ORGANIZATION OF GRAY AND WHITE MATTER

Myelinated nerve fibers are grouped together into pathways according to the direction – whether to or from the brain – and the type of impulse they transmit and respond to, such as pain or temperature. Some of these tracts connect and relay impulses from one spinal cord level to another. Gray matter is organized into horns, also called columns.

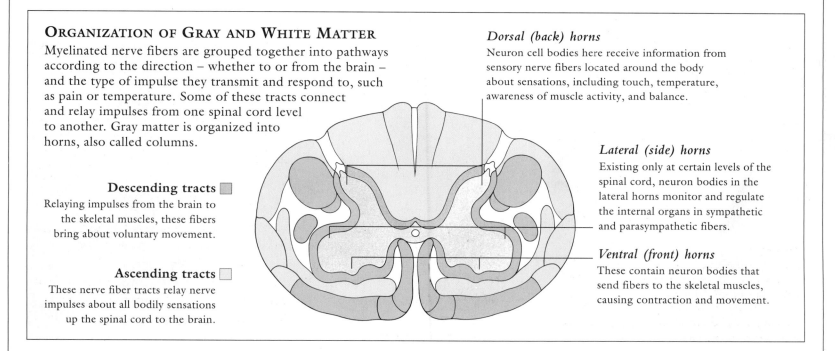

Descending tracts ▢
Relaying impulses from the brain to the skeletal muscles, these fibers bring about voluntary movement.

Ascending tracts ▢
These nerve fiber tracts relay nerve impulses about all bodily sensations up the spinal cord to the brain.

Dorsal (back) horns
Neuron cell bodies here receive information from sensory nerve fibers located around the body about sensations, including touch, temperature, awareness of muscle activity, and balance.

Lateral (side) horns
Existing only at certain levels of the spinal cord, neuron bodies in the lateral horns monitor and regulate the internal organs in sympathetic and parasympathetic fibers.

Ventral (front) horns
These contain neuron bodies that send fibers to the skeletal muscles, causing contraction and movement.

The PERIPHERAL NERVES

THE PERIPHERAL NERVES TRANSMIT INFORMATION both to and from the brain. Sensory fibers receive stimuli from the skin, the internal organs, and the outside world, while motor fibers initiate the contraction of skeletal muscle. Autonomic nerve fibers regulate the internal organs and glands, and ensure that balance is maintained in our internal environment (homeostasis).

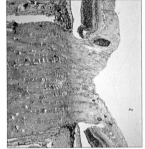

The optic nerve
The image shows the optic nerve as it enters the back of the eyeball. The white area is the vitreous humor, a gel that maintains eye shape; the deep red layer is the choroid, which is filled with blood vessels.

LM x 10

Olfactory nerve (I)
Relaying information about smells, the olfactory nerve connects the inside of the nose with the olfactory centers in the brain.

Optic nerve (II)
Each optic nerve is a bundle of approximately a million fibers that send visual signals from the retina to the brain.

Trigeminal nerve (V)
Branches of this tripartite nerve all contain sensory fibers that relay signals from the head, face, and teeth; the motor fibers innervate the chewing muscles. The branches are known as ophthalmic, maxillary, and mandibular.

Oculomotor (III), trochlear (IV), and abducent nerves (VI)
These nerves carry stimuli for voluntary movements of the eye muscles and eyelids. They also control pupil dilation and changes in the lens during focusing.

Facial nerve (VII)
Branches of this nerve innervate the taste buds, the skin of the external ear, and the salivary and lacrimal glands. They also control the muscles used in facial expressions.

Vestibulocochlear nerve (VIII)
Sensory fibers in the vestibular and cochlear branches of this nerve transmit information about sound, balance, and the orientation of the head.

Spinal accessory nerve (XI)
This nerve brings about movement in the head and shoulders. It also innervates muscles in the pharynx and larynx, and is involved in the production of voice sounds.

Glossopharyngeal (IX) and hypoglossal nerves (XII)
Motor fibers of these nerves are involved in swallowing, while the sensory fibers relay information about pain, taste, touch, and heat from the tongue and pharynx.

Vagus nerve (X)
The name vagus means "wanderer," and this nerve's sensory, motor, and autonomic fibers are involved in many vital bodily functions, including gland function, digestion, and heartbeat.

I   II   III   IV   VI   V   VII   VIII   IX   XII   XI   X

THE CRANIAL NERVES
Emanating from the undersurface of the brain, the 12 pairs of cranial nerves perform motor and/or sensory functions mainly in the head and neck region. The nine nerves with predominantly motor fibers also contain proprioceptive sensory fibers that convey information about the tension of the muscles they serve to the central nervous system.

THE SPINAL NERVES

The 31 pairs of peripheral spinal nerves emerge from the spinal cord and extend through spaces between the vertebrae. Each nerve divides and subdivides into a number of branches; two main subdivisions serve the front and the back of the body in the region innervated by that particular nerve. The branches of one spinal nerve may join up with other nerves to form plexuses ("braids"); these innervate certain areas of complex function or movement, such as the shoulder and neck.

Cervical region (C1 to C8)
The eight pairs of cervical spinal nerves interconnect, forming two networks, the cervical plexus (C1 to C4) and the brachial plexus (C5 to C8, and T1). These innervate the back of the head, the neck, shoulders, arms, and hands, as well as the diaphragm.

Thoracic region (T1 to T12)
Apart from T1, which is considered part of the brachial plexus, thoracic spinal nerves are directly connected with muscles lying between the ribs, the deep back muscles, and regions of the front of the abdomen.

Lumbar region (L1 to L5)
Four of the five pairs of lumbar spinal nerves (L1 to L4) form the lumbar plexus, which supplies the lower back as well as parts of the thighs and legs. L4 and L5 also interconnect with the first four sacral nerves (S1 to S4).

Sacral region (S1 to S5)
Two nerve networks, the sacral plexus (L5 to S3) and the coccygeal plexus (S4, S5, and the coccygeal nerve, Co1), innervate the thighs, the buttocks, the muscles and skin of the legs and feet, and the anal and genital area.

AREAS OF SENSATION

It is possible to create a map that delineates the surface skin into zones, called dermatomes, that are served by specific spinal nerves. Neurologists use either pinpricks or cotton swabs to identify sites of neural damage; absence of sensation in an area may reveal damage far removed from the area being investigated.

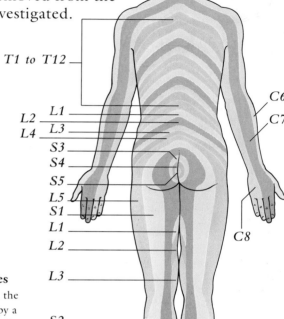

The dermatomes
The dermatomes in the trunk, each served by a corresponding pair of spinal nerves, are roughly horizontal, while those in the limbs are longitudinal. Since they overlap slightly, nerve root distribution can only be approximated on a dermatome map.

SPINAL REFLEX
A reflex is an involuntary and predictable response to a stimulus. The patellar spinal reflex measures the function of the spine's neural pathways. Tapping the patellar tendon stretches the front thigh muscle, stimulating a sensory neuron that transmits a nerve signal to the spinal cord. Motor nerve fibres then relay the signal to the muscle, which contracts and causes a slight kick.

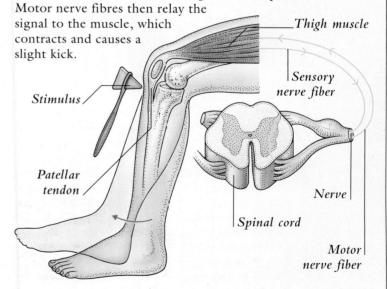

Thigh muscle

Sensory nerve fiber

Stimulus

Patellar tendon

Spinal cord

Nerve

Motor nerve fiber

The AUTONOMIC NERVOUS SYSTEM

STRUCTURED TO ENABLE IMMEDIATE, INVOLUNTARY RESPONSES, the autonomic nervous system (ANS) regulates internal body functions in order to maintain homeostasis, or inner balance. Nerve fibers monitor the organs and internal activities, such as heart rate and digestion. This information is integrated in the hypothalamus, brain stem, or spinal cord. Two divisions of the ANS – the sympathetic and the parasympathetic – then issue commands to the involuntary smooth muscles, to the heart muscle, and to the glands.

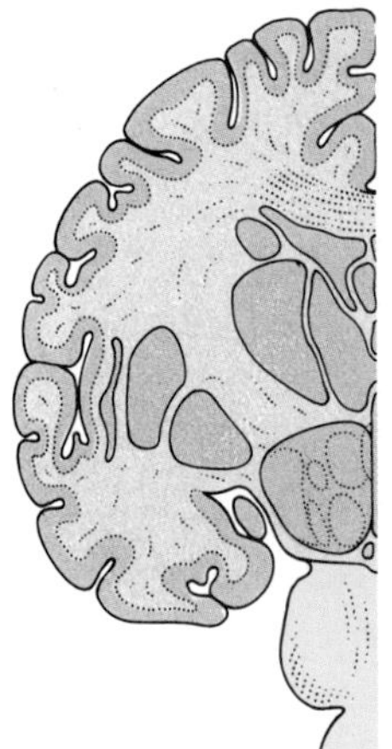

THE TWO DIVISIONS

The sympathetic division is principally an excitatory system that prepares the body for stress. The parasympathetic division maintains or restores energy. Although both divisions innervate many organs and structures, the number and position of ganglia – clusters of nerve cells where axons communicate in a synapse – are different. The activating chemicals, called neurotransmitters, and their effects are also different.

GUIDE TO ILLUSTRATION

The two divisions of the autonomic nervous system connect to both sides of the spinal cord; for clarity, the illustration shows one division on each side. The skin and blood vessels are innervated at all levels. More details of neural organization are explained in the box, Pathway Structures, on the facing page.

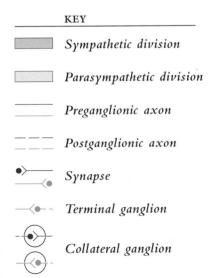

KEY

- *Sympathetic division*
- *Parasympathetic division*
- *Preganglionic axon*
- *Postganglionic axon*
- *Synapse*
- *Terminal ganglion*
- *Collateral ganglion*

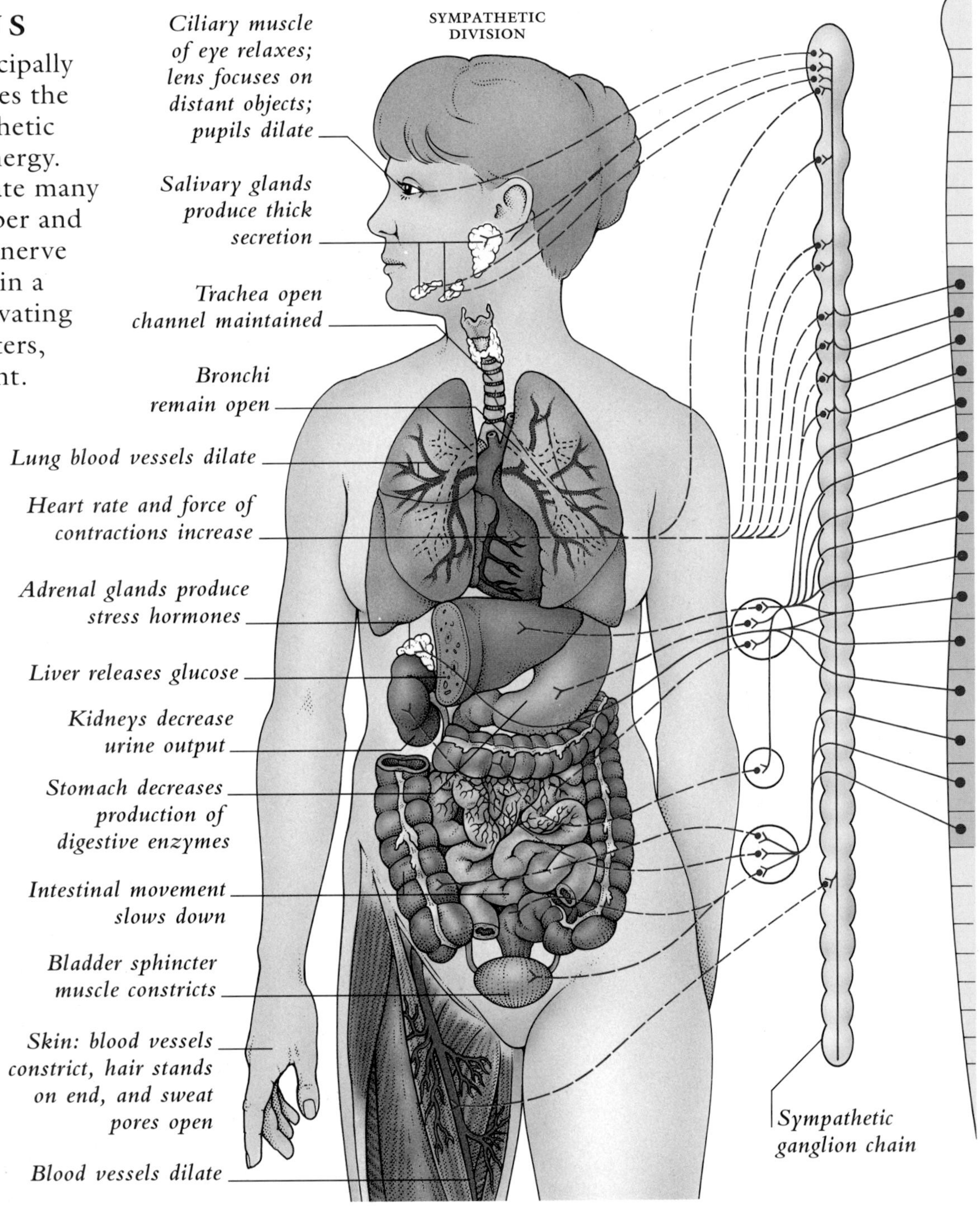

SYMPATHETIC DIVISION
Ciliary muscle of eye relaxes; lens focuses on distant objects; pupils dilate
Salivary glands produce thick secretion
Trachea open channel maintained
Bronchi remain open
Lung blood vessels dilate
Heart rate and force of contractions increase
Adrenal glands produce stress hormones
Liver releases glucose
Kidneys decrease urine output
Stomach decreases production of digestive enzymes
Intestinal movement slows down
Bladder sphincter muscle constricts
Skin: blood vessels constrict, hair stands on end, and sweat pores open
Blood vessels dilate
Sympathetic ganglion chain

PATHWAY STRUCTURES

In the sympathetic division, ganglia are located some distance from their target organs. Many are linked in a chain next to the spinal cord. In the parasympathetic division, ganglia lie close to or within the organs.

—— *Preganglionic axon* —— *Postganglionic axon*

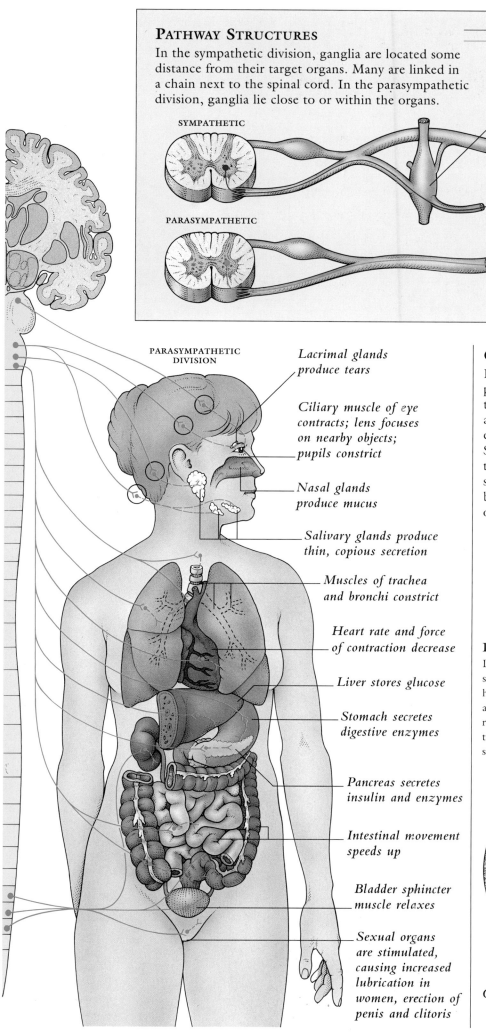

SYMPATHETIC

PARASYMPATHETIC

Sympathetic ganglion chain

Visceral organ (urinary bladder)

Collateral ganglion

Terminal ganglion

Smooth muscle cell

PARASYMPATHETIC DIVISION

Lacrimal glands produce tears

Ciliary muscle of eye contracts; lens focuses on nearby objects; pupils constrict

Nasal glands produce mucus

Salivary glands produce thin, copious secretion

Muscles of trachea and bronchi constrict

Heart rate and force of contraction decrease

Liver stores glucose

Stomach secretes digestive enzymes

Pancreas secretes insulin and enzymes

Intestinal movement speeds up

Bladder sphincter muscle relaxes

Sexual organs are stimulated, causing increased lubrication in women, erection of penis and clitoris

COORDINATION OF RESPONSE

In the eyes, involuntary changes in the size of the pupils occur constantly. Smooth muscle fibers in the irises are arranged concentrically in one band and radially in another, and each is innervated by either sympathetic or parasympathetic nerve fibers. Sensory receptors in the eyes respond to light and to the proximity or distance of objects. Nerve signals travel to the brain. A response is relayed back from the brain, and one or the other set of muscles constricts to adjust pupil size.

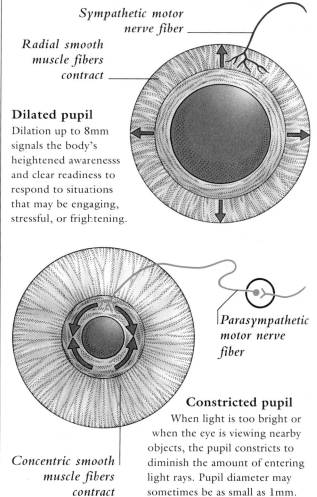

Sympathetic motor nerve fiber

Radial smooth muscle fibers contract

Dilated pupil
Dilation up to 8mm signals the body's heightened awarenesss and clear readiness to respond to situations that may be engaging, stressful, or frightening.

Parasympathetic motor nerve fiber

Concentric smooth muscle fibers contract

Constricted pupil
When light is too bright or when the eye is viewing nearby objects, the pupil constricts to diminish the amount of entering light rays. Pupil diameter may sometimes be as small as 1mm.

The PRIMITIVE BRAIN

THE LIMBIC SYSTEM EVOLVED EARLY IN HUMAN ANCESTRY, and influences unconscious, instinctive behavior similar to animal responses that relate to survival, such as the "fight-or-flight" response and reproduction. In humans, many of these innate, "primitive" behaviors are modified by the cerebral cortex. Humans plan for the future, feel hope, joy, and remorse, and their behavior is influenced by conscious moral, social, and cultural codes.

Cingulate gyrus *Corpus callosum*

Thalamus

Location of the limbic system
The limbic system encircles the top of the brain stem and forms a border (the meaning of "limbic") linking cortical and midbrain areas with lower centers that control automatic, internal body functions.

PARTS OF THE LIMBIC SYSTEM

The components of this ring-shaped system play a complex and important role in the expression of instincts, drives, and emotions. They mediate the effects of moods on external behavior and influence internal changes in bodily function and their appropriate expression. The association of feelings with sensations, such as smell and sight, and the formation of memories are also influenced by the limbic system.

Cingulate gyrus
This area, together with the parahippocampal gyrus and the olfactory bulbs, comprises the limbic cortex, which modifies behavior and emotions.

Septum pellucidum
A thin sheet of nervous tissue connects the fornix to the corpus callosum.

Fornix
The fornix is a pathway of nerve fibers that transmits information from the hippocampus and other limbic areas to the mamillary body.

Column of fornix

Mamillary body
This tiny nucleus acts as a relay station, transmitting information to and from the fornix and thalamus.

Midbrain
The limbic areas influence physical activity via the basal ganglia, the large clusters of nerve cell bodies below the cortex. Limbic midbrain areas also connect to the cortex and the thalamus.

Olfactory bulbs
The connection of these structures with the limbic system helps explain why the sense of smell evokes long-forgotten memories and emotions.

Amygdala
This structure influences behavior and activities so that they are appropriate for meeting the body's internal needs. These include feeding, sexual interest, and emotional reactions such as anger.

Parahippocampal gyrus
With other structures, this area helps modify the expression of emotions such as rage and fright.

Pons

Hippocampus
This curved band of gray matter is involved with learning and memory, the recognition of novelty, and the recollection of spatial relationships.

THE HYPOTHALAMUS

Composed of many tiny clusters of nerve cells called nuclei, the brain's hypothalamus is able to adjust consciousness, behavior, and internal functions. By direct nerve stimuli and neurohormonal secretions, these nuclei have links with the autonomic nervous, limbic, and endocrine systems. Although the various functions of the hypothalamus are well understood, the roles played by every nucleus are not yet clear.

FUNCTIONS

The hypothalamic nuclei monitor and regulate body temperature, food intake, water-salt balance, blood flow, the sleep-wake cycle, and the activity of the hormones secreted by the pituitary gland. The nuclei also mediate the responses to emotions such as anger and fear.

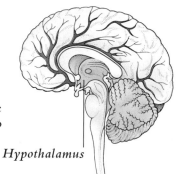

Hypothalamus

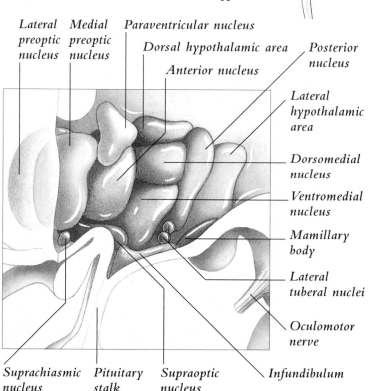

Lateral preoptic nucleus
Medial preoptic nucleus
Paraventricular nucleus
Dorsal hypothalamic area
Anterior nucleus
Posterior nucleus
Lateral hypothalamic area
Dorsomedial nucleus
Ventromedial nucleus
Mamillary body
Lateral tuberal nuclei
Oculomotor nerve
Infundibulum
Suprachiasmic nucleus
Pituitary stalk
Supraoptic nucleus

BRAIN STEM FUNCTIONS

The reticular formation, located in the brain stem, comprises at least four distinct neural systems, each with its own neurotransmitter. One of its functions is to operate the "reticular activating system," or RAS (an arousal system) that keeps the brain awake and alert. The brain stem also controls sleep, modulates spinal reflexes, maintains muscle tone and posture, and sustains breathing and heart rate.

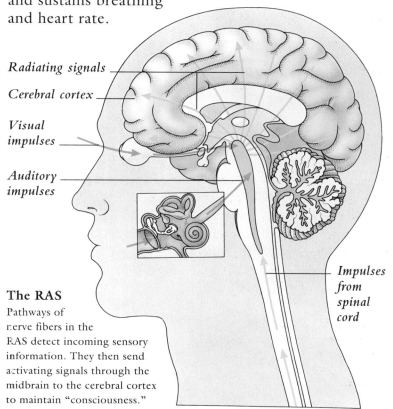

Radiating signals
Cerebral cortex
Visual impulses
Auditory impulses
Impulses from spinal cord

The RAS

Pathways of nerve fibers in the RAS detect incoming sensory information. They then send activating signals through the midbrain to the cerebral cortex to maintain "consciousness."

LM x 90

Nerve fibers in the pons

Fibers of the reticular formation in the pons are shown at left. Much of the function of this section of brain stem is unknown, but it does help regulate breathing and is involved in reflexes mediated by the fifth, sixth, seventh, and eighth cranial nerves, such as pupil constriction.

SLEEP

Nerve cells in the brain do not rest but, in a typical 7- or 8-hour sleep period, carry out different activities than those that occur when "awake." The patterns of both non-rapid eye movement (NREM) and rapid eye movement (REM) sleep, when most dreams occur, can be detected by recording the electrical activity of the brain. As sleep deepens, breathing and blood pressure are reduced, and body temperature drops.

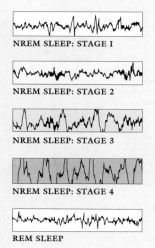

NREM SLEEP: STAGE I
NREM SLEEP: STAGE 2
NREM SLEEP: STAGE 3
NREM SLEEP: STAGE 4
REM SLEEP

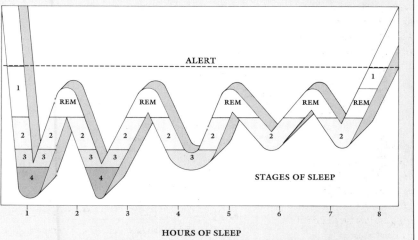

ALERT

REM

STAGES OF SLEEP

HOURS OF SLEEP

INFORMATION PROCESSING

THE INFORMATION RECEIVED FROM THE SENSES or generated by thought is processed in many different parts of the brain. Some areas process sensory data, such as light or sound, while others issue commands that initiate or coordinate voluntary movements. Other areas of the brain file away important data for future use. All of these areas are connected by bundles of nerve fibers. While the functions of some areas have been clearly defined, exact details of their intercommunication and integration are still not completely understood.

Motor nerve cell

Motor cortex

Caudate nucleus (body)

Basal ganglia
These structures plan and initiate complex movement. Links to other parts of the brain (not shown) update the "motor program."

Putamen

Globus pallidus

Thalamus

White matter

Gray matter

Cerebellum

Brain stem

VOLUNTARY MOVEMENT

Stimulated by sensory neurons or by conscious thought and intention, the premotor cortex area (see The Brain Map on facing page) formulates a central motor program. This plan is sent to the motor cortex, which then sends instructions to voluntary muscles. As movement progresses, it is coordinated and continually updated by corrective nerve signals sent from the cerebellum, which also controls balance and the body's position in space.

PATHWAY	SPECIALIZED FUNCTION
——	*Cortex sends motor message to muscle, resulting in movement*
——	*Sensory cell in muscle (which monitors movement) sends signal to cerebellum*
——	*Cerebellum sends corrective signals via thalamus to cortex to keep movement on course*
——	*Cerebellum also sends messages via spinal cord to muscle to correct muscle directly*

Pathways used in precise movements of the hand

THE BRAIN MAP

Scientists mapped the cortex into specific functional areas with PET scanning, by observing the effects of damage to or removal of specific parts of the brain, or by direct stimulation using electrodes. Large parts of the cortex are taken up by "association areas." These areas analyze and interpret neural information received from the primary sensory areas, helping coordinate voluntary movements.

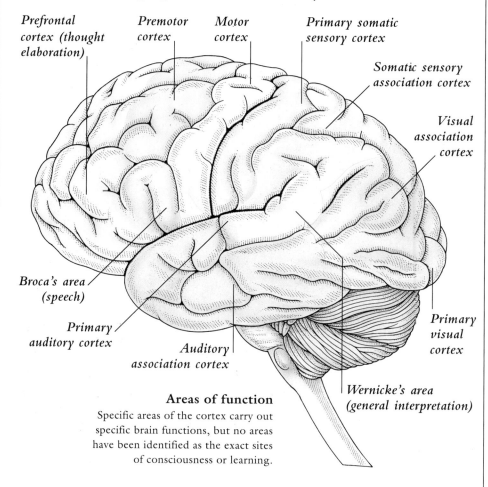

Prefrontal cortex (thought elaboration)

Premotor cortex

Motor cortex

Primary somatic sensory cortex

Somatic sensory association cortex

Visual association cortex

Broca's area (speech)

Primary auditory cortex

Auditory association cortex

Primary visual cortex

Wernicke's area (general interpretation)

Areas of function
Specific areas of the cortex carry out specific brain functions, but no areas have been identified as the exact sites of consciousness or learning.

GLUCOSE AND BRAIN ACTIVITY WITH PET

An increase in glucose metabolism is a reliable measure of intense brain activity. A modified glucose molecule is tagged with a radioisotope and injected into volunteers and patients; scans are taken as they perform certain tasks, or when they are exposed to stimuli such as music. Specific areas of activity are revealed (shown in red in the images below).

Visual stimulation
When the eyes are closed (left), or open (center), or when a complex scene is observed (right), the level of activity in the brain differs markedly.

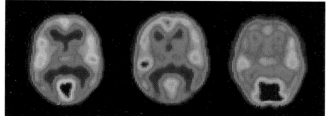

PET SCANS

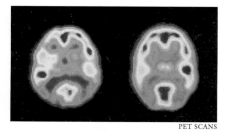

PET SCANS

Musical appreciation
When listening to music, trained musicians use the dominant, left cerebral hemisphere (left) associated with logical reasoning and sequential analysis. Untrained people use the right intuitive hemisphere (right), which "grasps things whole" without analysis.

MEMORY

Memories are the brain's storehouses of information, whether these are learned items or emotionally significant events. In order to create memories, nerve cells are thought to form new protein molecules and new interconnections. No one region of the brain stores all memories because the storage site depends on the type of memory: how to type or ride a bike are memories held in motor areas, while those about music are held in the auditory areas.

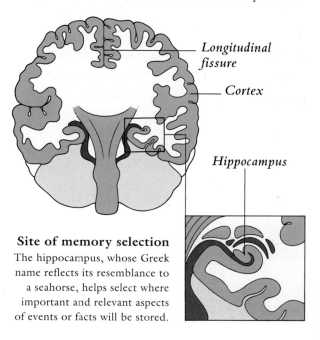

Longitudinal fissure

Cortex

Hippocampus

Site of memory selection
The hippocampus, whose Greek name reflects its resemblance to a seahorse, helps select where important and relevant aspects of events or facts will be stored.

THREE DEGREES OF MEMORY

Sensory memory, such as the brief recognition of a sound, is stored only for milliseconds. If retained and interpreted, this sensory input may become short-term memory for a few minutes. The transfer of short-term to long-term memory is known as consolidation, and requires attention, repetition, and associative ideas. How easily information is recalled depends upon how it was consolidated.

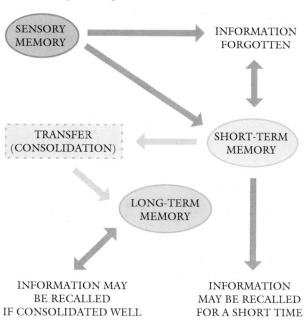

SENSORY MEMORY

INFORMATION FORGOTTEN

TRANSFER (CONSOLIDATION)

SHORT-TERM MEMORY

LONG-TERM MEMORY

INFORMATION MAY BE RECALLED IF CONSOLIDATED WELL

INFORMATION MAY BE RECALLED FOR A SHORT TIME

NEUROLOGICAL DISORDERS

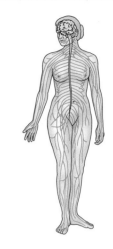

STRUCTURAL, BIOCHEMICAL, OR ELECTRICAL CHANGES in the brain and spinal cord, or the nerves leading to or from them, may cause disorders that result in paralysis, weakness, poor coordination, seizures, or loss of sensation. The introduction of scanners has brought rapid advances in diagnosis; increased understanding of brain function has generated improvements in treatment. However, some common disorders are due to conditions that are difficult to reverse. All that can be offered is some relief of symptoms.

EPILEPSY

One person in 200 is affected by the repeated seizures of epilepsy. These episodes of uncontrolled, chaotic electrical activity in the brain alter consciousness and may induce involuntary movements. Often the cause is unknown, but epilepsy that first appears in adults may be due to a brain condition such as a tumor or abscess, head injury, stroke, or chemical imbalance.

NORMAL EEG

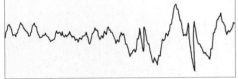

EEG DURING A SIMPLE PARTIAL SEIZURE

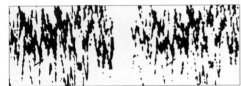

EEG DURING GRAND MAL SEIZURE

Types of seizure

In a grand mal attack, the victim falls to the ground unconscious and makes twitching movements for as long as several minutes. In a petit mal seizure, also called an absence seizure, the victim may be unaware of the outside world for as much as half a minute but does not fall. In a partial seizure there is usually no full loss of consciousness.

TEMPORAL LOBE EPILEPSY

This type of partial seizure affects one of the temporal lobes. Attacks may be preceded by an aura in which the victim experiences smells or sounds that others cannot detect. There may be involuntary movements during the attack, especially chewing and sucking, and a partial loss of consciousness. The attack may also cause the victim to have irrational feelings of fear or anger.

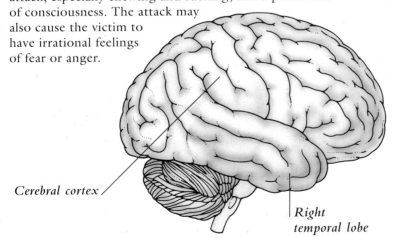

Cerebral cortex

Right temporal lobe

MULTIPLE SCLEROSIS

Multiple sclerosis (MS) is the most common disabling disorder of the nervous sytem affecting the young: one in every 1,000 people is affected. MS can cause episodes of blurred or double vision, partial paralysis, clumsiness, and problems with walking. There may also be interference with sensation. These episodes can last a few weeks and may sometimes be followed by months or years of relief from symptoms.

Sheath damage

MS is due to immune system damage to the myelin sheaths that protect nerve fibers. Macrophages, which are a type of scavenger cell, remove damaged sections of myelin, so that fibers are exposed and conduct impulses poorly or not all.

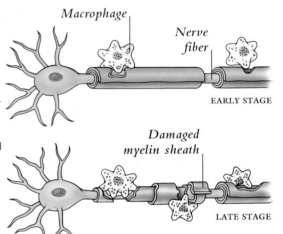

Macrophage

Nerve fiber

EARLY STAGE

Damaged myelin sheath

LATE STAGE

MRI SCAN

MRI SCAN

TREATMENT

Currently, there is no cure for MS. Injections of corticosteroid drugs can hasten recovery during exacerbations, and physical therapy relieves spastic muscle contractions. Treatment with a substance called beta interferon has been shown to prolong the intervals between relapses; other research has suggested that this treatment may help slow the progress of the disease. Side effects have limited the use of beta interferon among some patients.

Disease progression

A scan (top) shows MS lesions as three white spots, which usually grow as the disease progresses. A scan 6 weeks later (bottom) shows increased activity. This progression may be slowed in patients who are given beta interferon.

PARKINSON'S DISEASE

Parkinson's disease is a degenerative condition of the brain that occurs in about one in 200 people over the age of 60. More men than women are affected. The disease causes weakness and stiffness of the muscles and interferes with speech, walking, and performance of daily tasks. Emotional stimuli usually prompt little change of facial expression, and there is often a tremor of the person's hands when they are at rest.

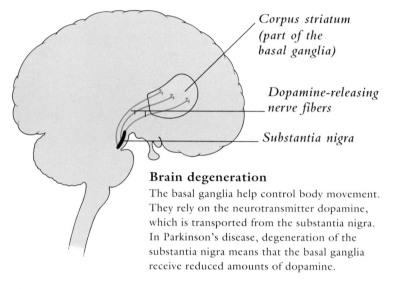

Corpus striatum (part of the basal ganglia)

Dopamine-releasing nerve fibers

Substantia nigra

Brain degeneration
The basal ganglia help control body movement. They rely on the neurotransmitter dopamine, which is transported from the substantia nigra. In Parkinson's disease, degeneration of the substantia nigra means that the basal ganglia receive reduced amounts of dopamine.

TREATMENT
The aim of treatment is to restore missing dopamine in the brain or to damp down the action of the dopamine antagonist acetylcholine. Drugs that increase dopamine levels include levodopa, selegilene, and bromocriptine. Anticholinergic drugs lower acetylcholine levels.

Normal chemical balance
In a normal brain, the levels of dopamine and acetylcholine are evenly balanced.

Acetylcholine

Dopamine

Chemical imbalance
In Parkinson's disease, the levels of dopamine are reduced and acetylcholine is relatively overactive.

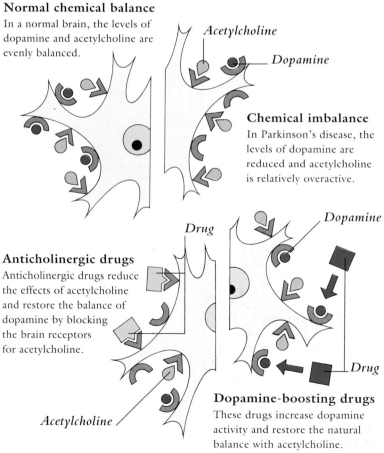

Drug

Dopamine

Anticholinergic drugs
Anticholinergic drugs reduce the effects of acetylcholine and restore the balance of dopamine by blocking the brain receptors for acetylcholine.

Acetylcholine

Drug

Dopamine-boosting drugs
These drugs increase dopamine activity and restore the natural balance with acetylcholine.

DEMENTIA

About one-third of the population over age 80 show symptoms of dementia, which may include a loss of memory for recent events, neglect of appearance, and repeated questions while ignoring replies or answers. In later stages, victims may become both bedridden and incontinent. Symptoms of dementia may appear as early as age 60 in one type of Alzheimer's.

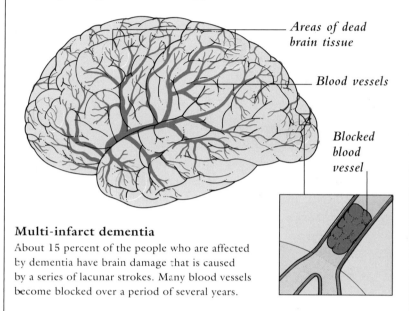

Areas of dead brain tissue

Blood vessels

Blocked blood vessel

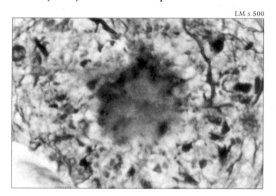

Multi-infarct dementia
About 15 percent of the people who are affected by dementia have brain damage that is caused by a series of lacunar strokes. Many blood vessels become blocked over a period of several years.

ALZHEIMER'S DISEASE
About 55 percent of dementia cases are the result of early- or late-onset Alzheimer's disease. Each type has a different genetic cause, but in both cases brain damage occurs due to the abnormal production of the protein amyloid. No cure has been found and some drugs may slow the progression for only very limited time periods.

Senile plaque
Brain tissue taken from a person afflicted with Alzheimer's disease reveals a deposit of a protein called amyloid (at center), which is a typical feature of the disease. Another main feature is a tangle of filaments within the brain's gray matter.

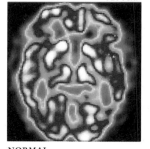

LM x 500

PET SCAN

PET SCAN

NORMAL

ALZHEIMER'S

Decreased brain activity
PET scanning shows how much energy is being used by brain cells. The brain scan of an Alzheimer's patient shows substantially lower activity than a normal brain. Yellow represents highest activity and blue the lowest.

CEREBROVASCULAR DISORDERS

THE TERM CEREBROVASCULAR DISORDERS covers any problem that affects the blood vessels supplying the brain. Stroke is the most serious consequence of such disorders: about one-third of victims die, one-third are left with some disability, and one-third make a good recovery. Migraine is another disorder of the vessels but it usually does not cause any permanent loss of function.

THE CAUSES OF A STROKE

A stroke may be caused by an interruption of the brain's blood supply or by the loss of blood onto the brain's surface or deep within its tissue. Any disruption of blood supply to the brain starves some of the nerve cells of oxygen and nutrients. These affected cells are unable to communicate with the parts of the body they serve, which results in a temporary or permanent loss of function. Hemorrhage impairs normal functioning of the brain by compressing and irritating tissue.

Blockage of tiny vessels
Prolonged high blood pressure or diabetes may damage some of the tiny blood vessels that penetrate deep within the brain. This may lead to localized blockages known as lacunar strokes that sometimes result in a form of dementia.

Thrombus
A buildup of fatty deposits within artery walls, called atherosclerosis, narrows the vessel and may encourage formation of a blood clot, or thrombus. If a thrombus blocks off an artery to the brain, a stroke follows as oxygen-starved brain tissue is damaged or even dies.

Embolus
The blockage of a cerebral artery, resulting in stroke, can be caused by a fragment of material that has traveled through the bloodstream and lodged in the vessel. Such a fragment, called an embolus, may be a piece of a clot from atherosclerotic neck arteries or from the heart lining.

Branches of the anterior cerebral artery

Posterior cerebral artery

Basilar artery

External carotid artery

Internal carotid artery

Vertebral artery

Common carotid artery

BLEEDING WITHIN BRAIN TISSUE

Bleeding within the brain, an intracerebral hemorrhage, is a main cause of stroke in older people who have hypertension. High blood pressure may put extra strain on small arteries in the brain, which causes them to balloon out and rupture.

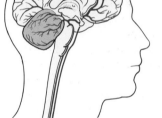

Blood vessels of the brain   *Hemorrhage*

Sudden bleeding

An intracerebral hemorrhage occurs suddenly. A severe headache and vomiting are common initial signs, followed by progressive paralysis and a loss of consciousness.

CT SCAN

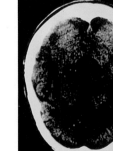

Making a diagnosis

A CT or MRI brain scan is essential to distinguish between a stroke caused by a thrombosis and one due to a hemorrhage, since symptoms may be similar. Bleeding appears as a yellow patch in the image above.

STROKES IN YOUNGER PEOPLE

Whereas strokes in older people are usually associated with advanced atherosclerosis or prolonged high blood pressure, in young people they are more likely to result from the leakage of blood due to arterial defects present from birth. In the majority of such cases, leakage occurs into the subarachnoid space, the area between the pia mater and the arachnoid layers of the meninges, the protective membranes covering the brain.

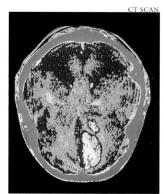

CT SCAN

Subarachnoid hemorrhage

The image seen at right reveals a subarachnoid hemorrhage (yellow) in the right frontal lobe. Bleeding was caused by the rupturing of a berry aneurysm (see below).

CONGENITAL CAUSES OF SUBARACHNOID HEMORRHAGE

The most common congenital condition leading to subarachnoid hemorrhage is the presence of berry aneurysms. These berrylike swellings of cerebral arteries are weak points and can rupture spontaneously. Malformed connections between cerebral blood vessels, from which blood can leak, are the other important congenital cause of subarachnoid hemorrhage. Such malformations are twice as common in men as in women.

Berry aneurysm

A berry aneurysm usually forms at arterial branches, often on the circle of Willis, the blood vessels at the base of the brain. Bleeding from a ruptured aneurysm can be halted by placing a clip around the neck of the aneurysm to seal it.

Neck of aneurysm

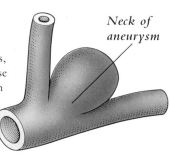

Capillaries

Arteriole

Venule

NORMAL

ABNORMAL

Arteriovenous malformation

This congenital defect is a tangle of blood vessels. Fewer capillary connections than normal exist between arterioles and venules. An increase in pressure results, which may cause blood to leak from the vessels into the subarachnoid space.

TRANSIENT ISCHEMIC ATTACK

A transient ischemic attack (TIA) temporarily interrupts the blood supply to the brain, resulting in strokelike symptoms usually lasting 2 to 30 minutes, but not more than 24 hours. The most common cause is an embolus, a tiny blood clot or a lipid fragment from elsewhere in the body. Up to one-third of TIA sufferers have a stroke within 5 years if untreated.

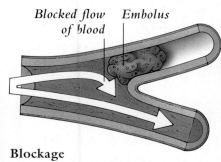

Blocked flow of blood   *Embolus*   *Dispersed particles*   *Blood flow resumes*

Blockage

A tiny embolus lodging in a cerebral artery may briefly deprive part of the brain of oxygenated blood. The symptoms depend on the part served by those arteries.

Dispersal

As normal blood flow breaks up and disperses the blood clot, oxygenated blood again reaches the starved section of the brain and symptoms disappear.

MIGRAINE

Migraine headaches are a recurring problem in over 10 percent of the population. The condition takes several different forms, with symptoms such as pain, dizziness, and visual disturbances often accompanied by nausea and sometimes vomiting. Complicated migraine attacks can cause disturbance of brain function. The symptoms are linked to changes in the diameter of blood vessels.

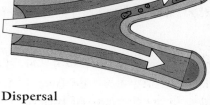

Onset of migraine attack

Some foods, red wine, stress, and drugs such as the contraceptive pill can trigger the blood vessels of the scalp and brain to narrow. This change in diameter may cause the victim to see flashing lights and experience temporary areas of blindness.

Constricted blood vessels

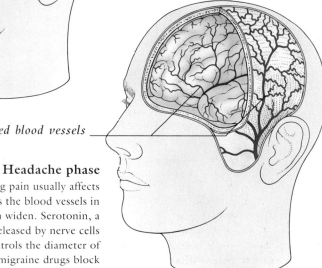

Dilated blood vessels

Headache phase

Severe, throbbing pain usually affects half of the head as the blood vessels in the scalp and brain widen. Serotonin, a neurotransmitter released by nerve cells in the brain, controls the diameter of blood vessels. Antimigraine drugs block the effects of serotonin in the brain.

NEUROLOGICAL INFECTIONS, TUMORS, *and* INJURIES

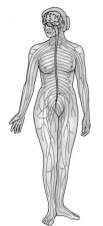

INJURIES AND DISORDERS AFFECTING THE BRAIN and nervous system can result in both physical and mental disabilities. The skull is a closed container, and any swelling of the brain raises pressure. This can cause damage to vital nerve tissue, with some loss of bodily control and function. Spinal injuries can damage nerve tracts, which results in varying degrees of either paralysis or sensory loss.

BRAIN INFECTIONS

A wide variety of viruses, bacteria, and parasites can infect the brain. Some viral and parasitic brain infections result from mosquito or other insect bites, while others develop from general viral diseases such as mumps and measles. In many countries, immunization on a wide scale has helped reduce the threat of viral infection affecting the brain.

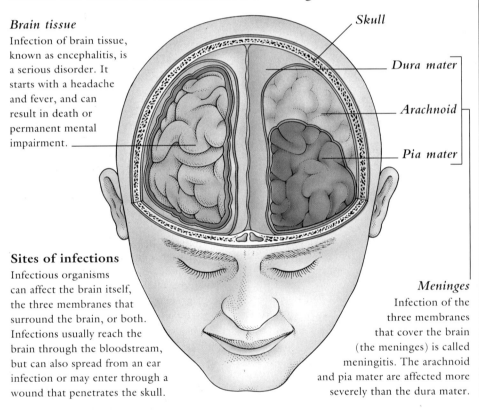

Brain tissue
Infection of brain tissue, known as encephalitis, is a serious disorder. It starts with a headache and fever, and can result in death or permanent mental impairment.

Skull

Dura mater

Arachnoid

Pia mater

Sites of infections
Infectious organisms can affect the brain itself, the three membranes that surround the brain, or both. Infections usually reach the brain through the bloodstream, but can also spread from an ear infection or may enter through a wound that penetrates the skull.

Meninges
Infection of the three membranes that cover the brain (the meninges) is called meningitis. The arachnoid and pia mater are affected more severely than the dura mater.

BRAIN ABSCESSES AND TUMORS

Both abscesses and tumors can develop inside the skull, either on the surface of the brain or within its tissue. Techniques such as CT and MRI scanning are used to identify the site of the abnormality and determine its size. Some types of tumors can be treated surgically; abscesses are drained or cut out, and then the person is treated with appropriate antibiotics.

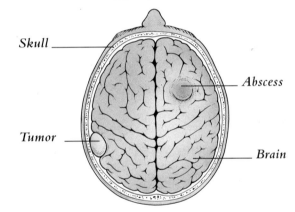

Skull

Abscess

Tumor

Brain

Similar symptoms
Abscesses and tumors both cause increased pressure inside the skull and produce similar symptoms, such as headaches, vomiting, muscle weakness, visual defects, and speech disorders.

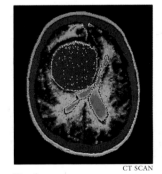

CT SCAN

Brain tumor
A tumor may be malignant (cancerous) or benign (noncancerous). The large, round area in the image is a glioma, a malignant tumor that arises from glial cells. This particular tumor grew slowly over the course of a number of years.

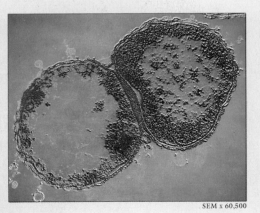

SEM x 60,500

MENINGITIS
The viral form of meningitis tends to occur in winter epidemics; its flulike symptoms usually clear up within a few weeks. The bacterial form of the disease is much more serious and may even be fatal in children. Another bacterial form, tuberculous meningitis, is found in regions of the world where tuberculosis is still prevalent.

Bacterial cause
Neisseria meningitidis organisms (shown left) are one cause of bacterial meningitis.

HEAD INJURIES

Accidents and assaults causing blows or wounds to the head can have extremely serious consequences. If both the scalp and skull are penetrated, there may be damage to the brain and a high risk of infection. Such an injury must be treated immediately by a neurosurgeon to remove foreign matter, clean the opening thoroughly, and repair the wound.

CLOSED HEAD INJURIES

Closed head injuries that do not open the skull may result from a fall or a blow. These injuries often cause a loss of consciousness and impaired brain function that may last only a few minutes or a whole lifetime. Results of treatments now available are inconclusive and unpredictable.

KEY

➡ *Direction of movement*

◼ *Stationary*

✦ *Blow to the head*

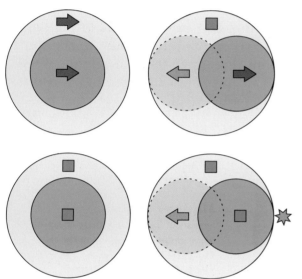

Deceleration

When a person moving rapidly is stopped suddenly, as in a fall, the brain may be injured as it smashes against the hard inner surface of the skull and then rebounds against the opposite surface of the inner skull.

Acceleration

If a stationary head is suddenly struck, as in boxing, the brain may be compressed against the inside of the skull close to the place of impact. The brain may then bounce off the opposite inner surface of the skull.

BLEEDING WITHIN THE SKULL

Closed injuries can be fatal because of the bleeding that occurs within the skull. There may not be any immediate symptoms, but a headache, confusion, drowsiness, and a noticeable change in personality may appear gradually as blood collects and clots. Immediate neurosurgical treatment is necessary. Removal of the blood clot by surgery brings dramatic relief of the symptoms.

Extradural hemorrhage
Bleeding that occurs between the inner surface of the skull and the dura mater is called an extradural hemorrhage.

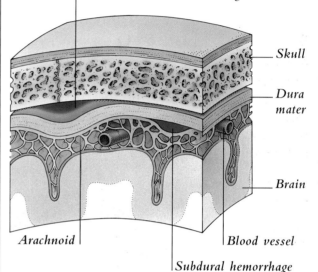

Skull

Dura mater

Brain

Arachnoid

Blood vessel

Subdural hemorrhage
Bleeding that occurs between the dura mater and the arachnoid is known as a subdural hemorrhage. Such bleeding may occur slowly over a long period.

PARALYSIS

Paralysis or weakness of various body areas results from damage to motor areas of the brain or neural pathways of the spinal cord. Voluntary muscle activity, including functions such as breathing, may be affected, and there may be a loss of sensation. Neither consciousness nor intellectual function is affected by paralysis.

Paraplegia

Damage to the middle or lower area of the spinal cord can cause paralysis of both legs and possibly part of the trunk, called paraplegia. Control of the bladder and bowel may also be affected, causing incontinence.

Quadriplegia

Damage to the spinal cord in the lower neck area can cause paralysis of the whole trunk plus arms and legs, called quadriplegia. If damage is between C1 and C2 or higher, the person is unlikely to survive.

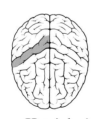

Hemiplegia

Damage to the motor areas on one side of the brain can lead to paralysis of the opposite side of the body. This one-sided type of paralysis is known as hemiplegia.

KEY

◼ *Area of body affected*

◼ *Site of damage*

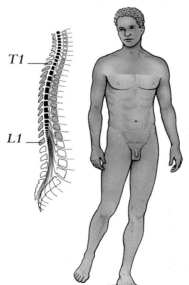

T1

L1

C4
C7

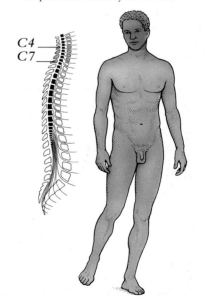

TOUCH, TASTE, *and* SMELL

SENSATION IS NOT ONLY OUR LINK with the outside world but also provides the body with important information about its internal environment. General sensory receptors respond to stimuli such as touch, pressure, pain, and temperature and are widespread throughout the body. Taste and smell, along with vision, hearing, and balance, are called special senses because their receptors are complex and respond to specific stimuli at very localized sites.

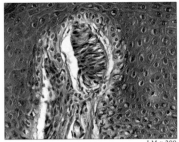

Meissner's corpuscle
LM x 200

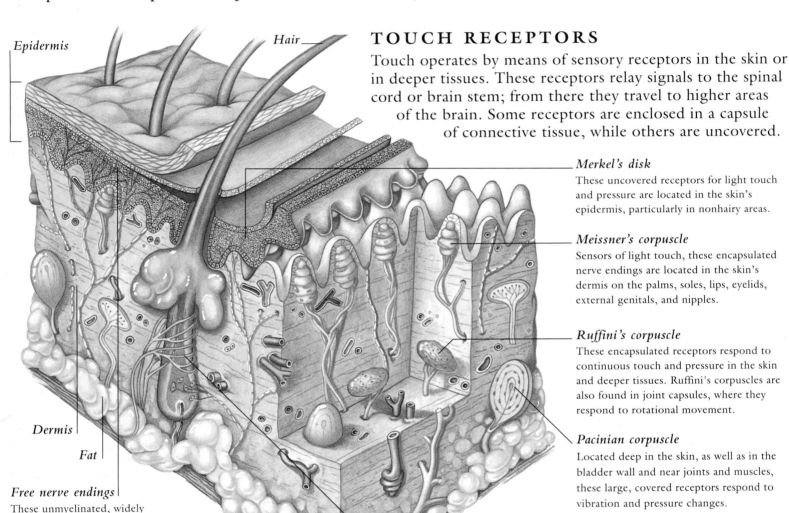

Epidermis

Hair

Dermis

Fat

Free nerve endings
These unmyelinated, widely distributed sensors of light touch, pressure, pain, and temperature are found everywhere in the skin, and in all types of connective tissue.

TOUCH RECEPTORS

Touch operates by means of sensory receptors in the skin or in deeper tissues. These receptors relay signals to the spinal cord or brain stem; from there they travel to higher areas of the brain. Some receptors are enclosed in a capsule of connective tissue, while others are uncovered.

Merkel's disk
These uncovered receptors for light touch and pressure are located in the skin's epidermis, particularly in nonhairy areas.

Meissner's corpuscle
Sensors of light touch, these encapsulated nerve endings are located in the skin's dermis on the palms, soles, lips, eyelids, external genitals, and nipples.

Ruffini's corpuscle
These encapsulated receptors respond to continuous touch and pressure in the skin and deeper tissues. Ruffini's corpuscles are also found in joint capsules, where they respond to rotational movement.

Pacinian corpuscle
Located deep in the skin, as well as in the bladder wall and near joints and muscles, these large, covered receptors respond to vibration and pressure changes.

Sensory receptor of hair shaft
Free nerve endings around hair follicles respond to touch and slight movement.

PAIN

Pain receptors are widespread, specialized free nerve endings that respond to extremes in temperature, pressure, and the chemical prostaglandin released from damaged cells. They transmit the location and intensity of the pain to the brain, and may stimulate the release of pain-blocking endorphins. Pain-relieving drugs function either by blocking prostaglandins or inhibiting pain impulses.

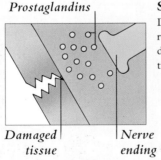

Prostaglandins

Damaged tissue

Nerve ending

Simple analgesics
Drugs such as aspirin prevent release of prostaglandins from damaged tissue, forestalling the response leading to pain.

Narcotic drugs

Imitating brain endorphins, narcotic drugs like morphine stop the transmission of pain signals between nerve cells.

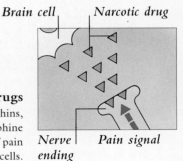

Brain cell

Narcotic drug

Nerve ending

Pain signal

RECEPTORS FOR TASTE

Taste receptor cells, or taste buds, are located mainly within protuberances called papillae on the surface of the tongue. Some buds are also on the palate, throat, and epiglottis. Taste buds on different parts of the tongue respond more strongly to one or another of four basic tastes – sweet, bitter, sour, and salty. Many other, more subtle taste sensations are made possible by the combination of these tastes with other stimuli that are associated with taste, such as odors.

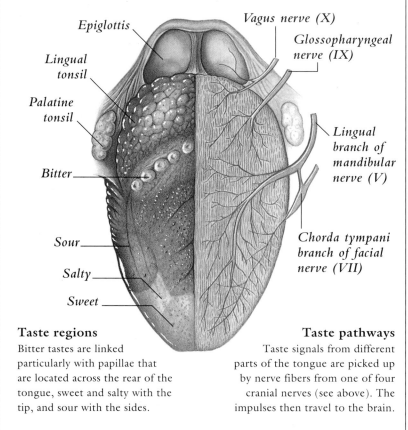

Epiglottis

Vagus nerve (X)

Lingual tonsil

Glossopharyngeal nerve (IX)

Palatine tonsil

Bitter

Lingual branch of mandibular nerve (V)

Sour

Salty

Chorda tympani branch of facial nerve (VII)

Sweet

Taste regions
Bitter tastes are linked particularly with papillae that are located across the rear of the tongue, sweet and salty with the tip, and sour with the sides.

Taste pathways
Taste signals from different parts of the tongue are picked up by nerve fibers from one of four cranial nerves (see above). The impulses then travel to the brain.

TASTE BUDS

Taste buds consist of a cluster of receptor, or "taste," cells and supporting cells. Projecting from the top of a receptor cell are tiny taste hairs, which are exposed to saliva that enters through taste pores. A substance taken into the mouth and dissolved in saliva interacts with receptor sites on the taste hairs, thus generating a nerve impulse that is transmitted to the brain.

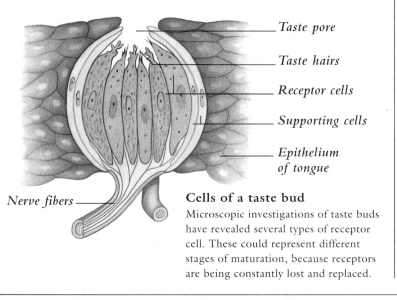

Taste pore

Taste hairs

Receptor cells

Supporting cells

Epithelium of tongue

Nerve fibers

Cells of a taste bud
Microscopic investigations of taste buds have revealed several types of receptor cell. These could represent different stages of maturation, because receptors are being constantly lost and replaced.

SMELL

The human sense of smell is much more sensitive than taste, and more than 10,000 odors can be detected. Since the olfactory structures tend to deteriorate with age, children can usually distinguish more odors than adults. Most other mammals possess an even sharper sense of smell than humans. In addition to warning of dangers such as smoke or noxious gases, smell also makes an important contribution to the sense of taste.

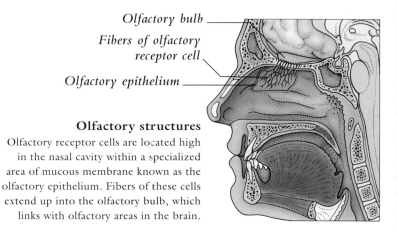

Olfactory bulb

Fibers of olfactory receptor cell

Olfactory epithelium

Olfactory structures
Olfactory receptor cells are located high in the nasal cavity within a specialized area of mucous membrane known as the olfactory epithelium. Fibers of these cells extend up into the olfactory bulb, which links with olfactory areas in the brain.

MECHANISM OF SMELL

Odor molecules dissolve in nasal mucus and stimulate the hairlike endings – cilia – of the receptor cells, generating a nerve impulse. The impulse travels along the fibers of the cells; these pass through holes in the cribriform plate of the ethmoid bone into the olfactory bulb, where they synapse with fibers of the olfactory nerves.

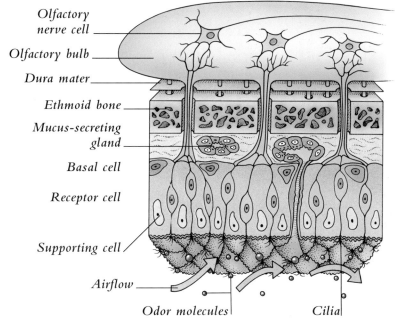

Olfactory nerve cell

Olfactory bulb

Dura mater

Ethmoid bone

Mucus-secreting gland

Basal cell

Receptor cell

Supporting cell

Airflow

Odor molecules

Cilia

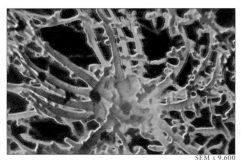

SEM x 9,600

Cilia
Each olfactory receptor cell ends in a bulbous swelling called an olfactory vesicle, from which six to 20 cilia project. The image at left shows a single vesicle with numerous cilia radiating from its surface.

EAR STRUCTURE, HEARING, and BALANCE

THE EARS ARE ORGANS OF BOTH HEARING AND BALANCE. The structures for these sensory functions are located in separate areas of the inner ear. However, both depend on stimulation of specialized receptors called hair cells that respond to sound waves or to movement. Nerve fibers leaving the auditory and balancing structures form the vestibulocochlear nerve, which carries nerve impulses to the brain for interpretation.

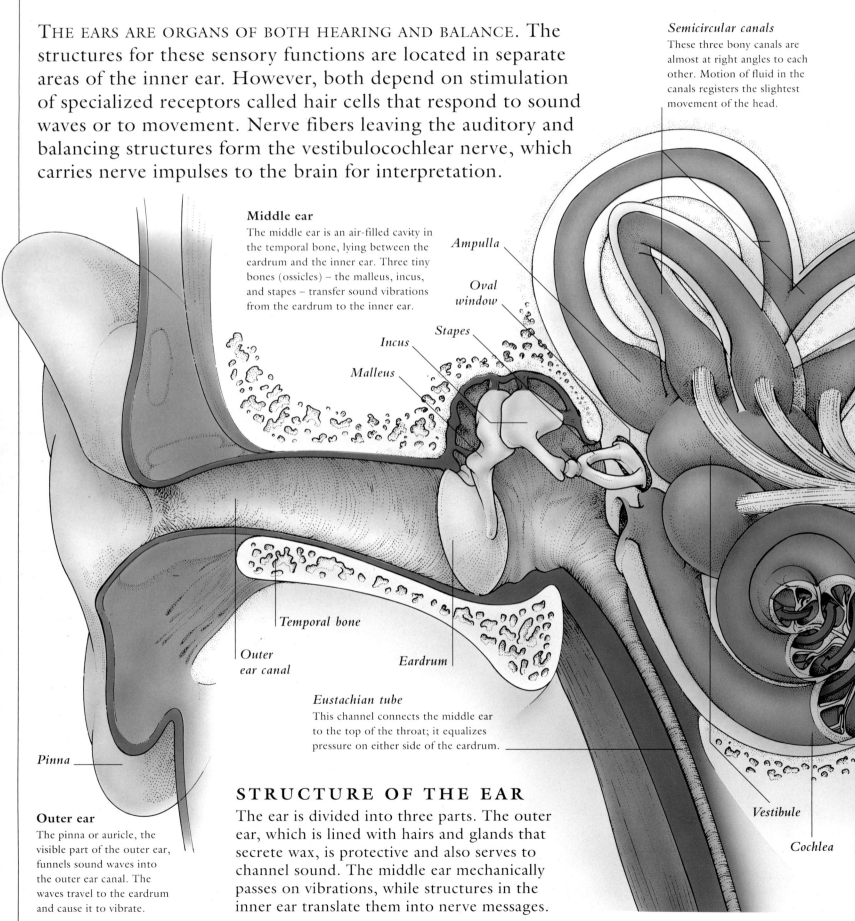

Semicircular canals
These three bony canals are almost at right angles to each other. Motion of fluid in the canals registers the slightest movement of the head.

Middle ear
The middle ear is an air-filled cavity in the temporal bone, lying between the eardrum and the inner ear. Three tiny bones (ossicles) – the malleus, incus, and stapes – transfer sound vibrations from the eardrum to the inner ear.

Ampulla

Oval window

Stapes

Incus

Malleus

Temporal bone

Outer ear canal

Eardrum

Eustachian tube
This channel connects the middle ear to the top of the throat; it equalizes pressure on either side of the eardrum.

Pinna

Vestibule

Cochlea

STRUCTURE OF THE EAR
The ear is divided into three parts. The outer ear, which is lined with hairs and glands that secrete wax, is protective and also serves to channel sound. The middle ear mechanically passes on vibrations, while structures in the inner ear translate them into nerve messages.

Outer ear
The pinna or auricle, the visible part of the outer ear, funnels sound waves into the outer ear canal. The waves travel to the eardrum and cause it to vibrate.

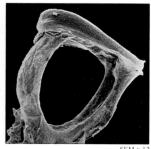

Stapes
The innermost ossicle, the stapes, is the smallest bone in the body. It closely resembles a stirrup for which it is named. It is attached to the incus by a ball-and-socket joint and, like the other two earbones, is held in place by ligaments.

Vestibular nerve (branch of vestibulocochlear nerve)

Cochlear nerve (branch of vestibulocochlear nerve)

Inner ear
Also called the labyrinth, the inner ear consists of a complex system of membranous canals with a bony casing. The organ of hearing is located in the snail-like cochlea. The sense of balance operates from structures in the vestibule and semicircular canals.

HEARING
Sound waves entering the ear canal make the eardrum vibrate. The ossicles pass the vibrations to the oval window, a membrane at the entrance to the inner ear. When this membrane vibrates, it sets off wavelike motions in the fluid that fills the cochlea, resulting in stimulation of hair cells.

THE COCHLEA
The cochlea is subdivided into three fluid-filled chambers, which spiral in parallel around a bony core. The central channel, the cochlear duct, contains the spiral organ of Corti, the organ of hearing. Located on the basilar membrane, the spiral organ consists of supporting cells and many thousands of sensory hair cells arranged in rows.

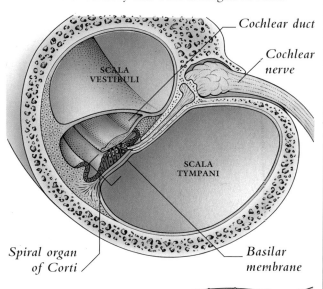

Cochlear duct

Cochlear nerve

SCALA VESTIBULI

SCALA TYMPANI

Spiral organ of Corti

Basilar membrane

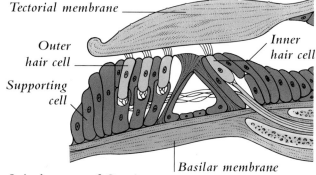

Tectorial membrane

Outer hair cell

Supporting cell

Inner hair cell

Basilar membrane

Spiral organ of Corti
From each hair cell of the spiral organ tiny sensory hairs project to make contact with the tectorial membrane above. When the basilar membrane vibrates, the hair cells are stimulated as they squeeze against the tectorial membrane.

Hair cells
Shown in yellow at the right are four rows of hair cells in the organ of Corti. Each hair cell has up to 100 bristle-like hairs that translate mechanical movement into electrical sensory impulses transmitted directly to the brain.

SEM x 640

BALANCE
The sense of balance relies not only on the sensory organ in the inner ear, but also on visual input and on information received from receptors in the body, especially those around joints. The information is processed by the cerebellum and cerebral cortex to enable the body to cope with changes in acceleration or direction of the head.

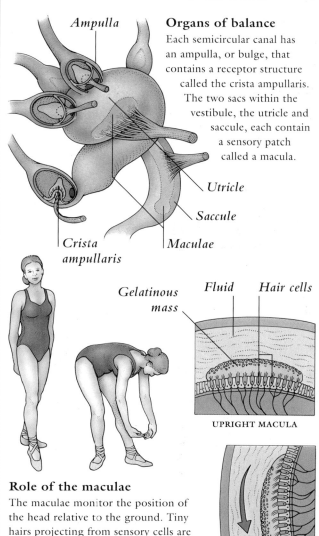

Ampulla

Organs of balance
Each semicircular canal has an ampulla, or bulge, that contains a receptor structure called the crista ampullaris. The two sacs within the vestibule, the utricle and saccule, each contain a sensory patch called a macula.

Utricle

Saccule

Crista ampullaris

Maculae

Gelatinous mass

Fluid   *Hair cells*

UPRIGHT MACULA

Role of the maculae
The maculae monitor the position of the head relative to the ground. Tiny hairs projecting from sensory cells are embedded in a gelatinous mass. If the head is tipped, gravity pulls the mass down, stimulating the hair cells.

DISPLACED MACULA

Role of the crista ampullaris
The crista ampullaris responds to rotational movements. The hair cells of each crista are embedded in a conical gelatinous mass, the cupula. When the fluid in the semicircular canals swirls during movement, it displaces the cupula, stimulating the hair cells.

Cupula   *Hair cells*   *Fluid*

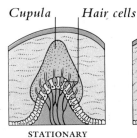

STATIONARY   ROTATING

EYE STRUCTURE *and* VISION

OF THE FIVE SENSES, VISION IS THE MOST SPECIALIZED and also the most complex, a result of sensory reception and intellectual judgement. Light rays that enter the pupils and register on the retinas at the back of the eyes create two-dimensional images. These images are converted into electrical impulses, which are transported through the optic nerve of each eye to parts of the brain, especially the occipital lobe, where they are interpreted.

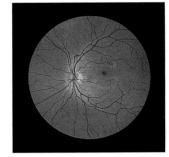

The eyeball's blood supply
The retinal artery enters the eyeball through the optic disk, known as the eye's white or "blind spot," and then branches over the retina's surface.

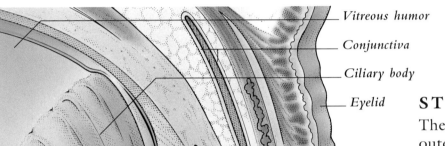

Vitreous humor

Conjunctiva

Ciliary body

Eyelid

Ligaments attached to ciliary muscle

Cornea

Lens

Iris

Choroid

Retina

Sclera

STRUCTURE OF THE EYE

The eyeball has three layers known as tunics. The outer fibrous tunic has two parts: the transparent, curved cornea, and the opaque white sclera, which helps maintain the eye shape. The middle vascular tunic contains the iris, ciliary body, and choroid, whose blood vessels supply all three tunics. The back part of the third, innermost layer, the retina, is where light converges and images form.

THE CAVITIES OF THE EYE

The anterior and posterior chambers of the front cavity of the eye are filled with aqueous humor, a fluid that provides oxgyen, glucose, and proteins. The back cavity of the eye is filled with a clear gel called vitreous humor. Produced by the ciliary body, both substances contribute to the constant internal pressure maintaining the shape of the eye.

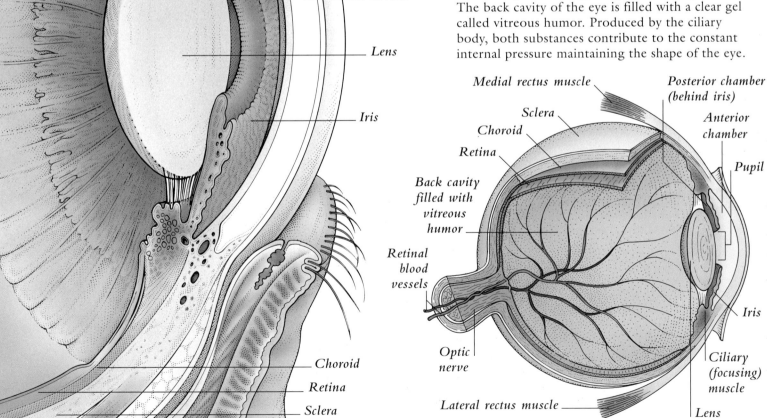

Medial rectus muscle

Posterior chamber (behind iris)

Sclera

Anterior chamber

Choroid

Retina

Pupil

Back cavity filled with vitreous humor

Retinal blood vessels

Iris

Optic nerve

Ciliary (focusing) muscle

Lateral rectus muscle

Lens

VISUAL PATHWAYS

Light passes through the cornea and lens to converge on the retina, creating an upside-down image. The medial (inner) and lateral (outer) parts of each retina transmit signals through the optic nerve; signals from the medial part of each retina intersect at the optic chiasm, which is located at the base of the brain, and cross to the opposite side of the brain. In the visual cortex, the image is turned upright and interpreted.

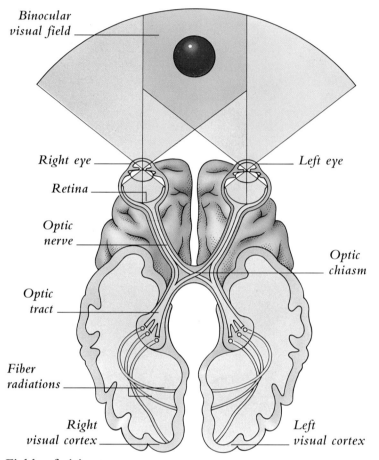

Fields of vision
Each eye sees a slightly different image, but the visual field of one eye partially overlaps the visual field of the other one. The area of binocular vision permits depth perception, the ability to judge an object's distance.

ACCOMMODATION

Ciliary muscles in the eye automatically respond to the proximity or the distance of an object by altering the shape of the lens; this changes the angle of incoming light rays and allows for sharper focus on the retina. Because the elasticity of the lens decreases as the body ages, so does the speed and power of accommodation.

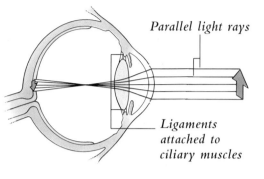

Distant objects
To focus on objects in the distance, the ciliary muscles relax and the lens flattens and thins. Light rays are slightly refracted (bent) by the lens.

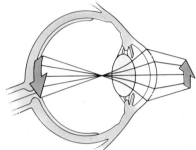

Nearby objects
To view objects that are nearby, the ciliary muscles contract and the lens becomes more rounded. The point at which the image of a close object becomes blurred is called the near point of vision; it occurs when the lens reaches its maximum curvature.

ACCESSORY STRUCTURES

The eyes depend on accessory structures that support, move, lubricate, and protect them. These include the orbital bones of the eye socket, muscles of the eyeball, eyebrows, eyelids, and eyelashes, as well as lacrimal (tear) glands and ducts. Vision may be impaired if any of these structures is irritated, infected, or misshapen.

Tear glands
These produce tears that help cleanse the eye.

Conjunctiva
A transparent mucous membrane covers and moistens the sclera and the inside of the eyelids.

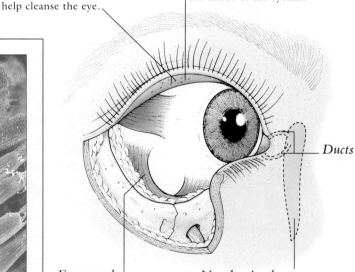

Ducts

Eye muscle
Six muscles (three are shown) attached to the sclera control movement.

Nasolacrimal sac
Excess tears evaporate or drain through ducts into a sac that is connected to the nasal cavity.

PHOTORECEPTIVE NEURONS

Two types of neurons can be found in the retina. Rods discern light and dark, shape, and movement, and contain only one light-sensitive pigment. Cones, which need more light than rods to be activated, are of three types; each contains a pigment that responds to a different light wavelength (green, red, or blue). The combination of these three wavelengths permits color discrimination.

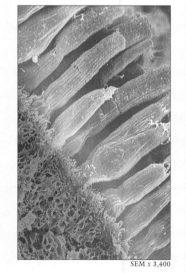

SEM x 3,400

Rods and cones
Each eye has about three million cones, mainly in the retina's macula, or yellow spot. Approximately 100 million rods (shown in blue) are in the periphery.

EAR *and* EYE DISORDERS

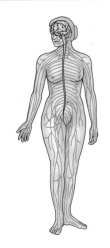

THE EARS AND THE EYES ARE VULNERABLE to many disorders, ranging from damage caused by an excess of sound and light to the degeneration of the senses that occurs with age. Hearing and vision are mutually supportive, so that when one of them suffers reduced performance, the other may become more acute to compensate. Some sensory disorders may be inherited. It is important to diagnose and correct ear and eye problems in young children as soon as possible because crucial learning is affected during the early years.

DEAFNESS

Conductive deafness results from impaired transmission of sound waves to the inner ear. In adults it is usually due to blockage by earwax. Other causes are damage to the eardrum or fusion of the stapes' footplate at the entrance to the inner ear, which is called otosclerosis. Sensorineural deafness occurs when nerve impulses are poorly transmitted as a result of damage to inner ear structures or to the acoustic nerve.

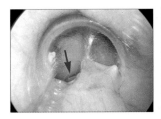

Perforated eardrum
The arrow indicates a small perforation in the eardrum. Punctures may be caused by infection, by poking objects into the ear, or by a blow.

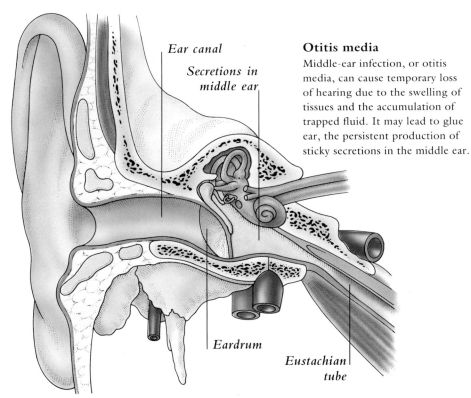

Ear canal

Secretions in middle ear

Eardrum

Eustachian tube

Otitis media
Middle-ear infection, or otitis media, can cause temporary loss of hearing due to the swelling of tissues and the accumulation of trapped fluid. It may lead to glue ear, the persistent production of sticky secretions in the middle ear.

TREATMENTS

Simple measures can be effective for treatment of conductive deafness, such as syringing of the ear for removing wax or administration of antibiotics for infections. Otosclerosis and glue ear sometimes require surgical treatment. Sensorineural deafness usually cannot be cured, but hearing aids are used by some to increase the volume of sound. Cochlear implants relay signals from an external transmitter to an electrode wired into the inner ear.

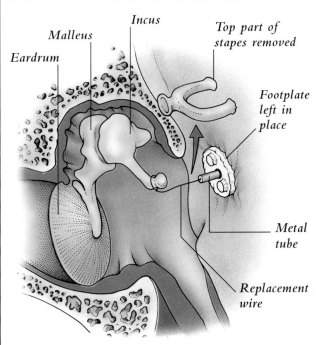

Malleus

Incus

Top part of stapes removed

Eardrum

Footplate left in place

Metal tube

Replacement wire

Stapedectomy for otosclerosis
In this operation, the top part of the stapes is removed. A laser beam is used to create a small hole in the footplate and a metal tube is inserted. A wire attached to the tube connects to the incus and transmits vibrations to the inner ear.

SENSORINEURAL DEAFNESS

Damage to inner ear structures may be present from birth. It may also be caused by some types of drugs, by prolonged exposure to loud noise, by increased fluid pressure in Meniere's disease, or by deterioration of ear structures with age.

Acoustic neuroma
This benign tumor (arrow) presses on the acoustic nerve, causing deafness. It can be removed surgically or by irradiation.

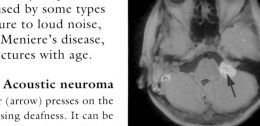

MRI SCAN OF BRAIN

Treating glue ear
Glue ear often clears up with antibiotics. If it does not, a small tube, called a grommet, can be placed into a hole made in the eardrum, so that excess fluid can drain out and air can flow in. In children, enlarged adenoids sometimes block the Eustachian tube and may need to be removed.

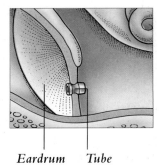

Eardrum   Tube

VISUAL PROBLEMS

Problems with focusing for near or far vision result from the eyeball being either too long or too short. Irregular curvature of the cornea causes astigmatism, in which parts of the visual field are blurred. Normal aging often brings on difficulty with near vision, as the lens gradually loses its elasticity and cannot easily adjust its shape; this condition is called presbyopia.

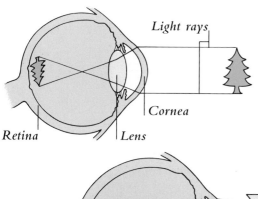

Nearsightedness

Nearsightedness, called myopia, occurs when the eyeball is too long. Instead of focusing on the retinal surface, the image focuses in front of it. Either concave eyeglasses or contact lenses can correct the focusing point.

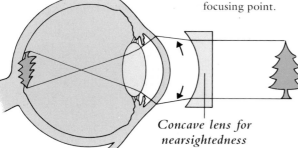

Concave lens for nearsightedness

Farsightedness

Farsightedness, called hyperopia, occurs when the eyeball is too short, causing the image to focus behind the retina instead of on its surface. Either convex eyeglasses or contact lenses correct the focusing point.

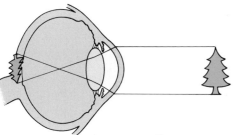

Convex lens for farsightedness

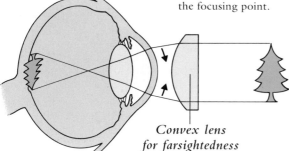

Laser surgery for nearsightedness

OPERATIONS FOR NEARSIGHTEDNESS

Certain forms of myopia and astigmatism can be treated by surgery. Refractive keratotomy involves making incisions in different parts of the cornea to change the way light rays are bent. A scalpel or laser is used to make the cuts. Sometimes the operation is done in two or three stages. Both types of operation carry risks, and long-term effects are uncertain.

CAUSES OF BLINDNESS

In most developed countries, blindness occurs mainly later in life. Glaucoma is rare before age 40. Disease of the retina (retinopathy) can result from diabetes mellitus or hypertension. The most common cause of blindness in people over 60 is macular degeneration, in which the central retina (macula) is covered with scar tissue. Cataracts occur in many people who are over age 60. When vision is impaired, some people may require an operation to remove the opaque lens.

CATARACTS

Cataracts are most often detected in older adults during routine examination. Many people are unaware of their cataracts, since they may not interfere with vision and cause no pain. The clouding of the lens is irreversible. Cataracts are sometimes congenital, caused when a pregnant woman has been infected with rubella (German measles) in early pregnancy. Diabetes mellitus and exposure to radiation are other possible factors.

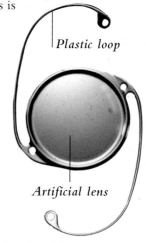

Plastic loop

Artificial lens

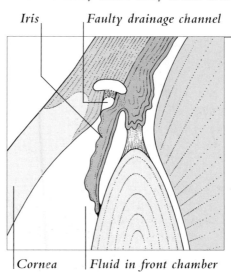

Cataract

This advanced cataract appears as a white mass behind the iris. Changes in protein fibers gradually made the lens opaque. An artificial replacement lens can restore normal vision.

Artificial lens

To treat a cataract, an incision is made in the cornea and the opaque lens is removed from its capsule. An artificial lens is then inserted and fixed in place by means of plastic loops.

GLAUCOMA

Glaucoma can cause blindness as a result of damage due to increasing pressure within the eye, which compresses the blood vessels that supply the optic nerve. This results in degeneration of nerve fibers. The pressure is created by buildup of aqueous humor, which normally drains away as fast as it is secreted.

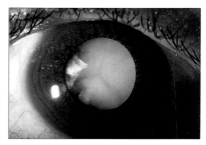

Iris   *Faulty drainage channel*

The cause

The build-up of fluid that occurs in glaucoma results from faults in the drainage channel between the back of the cornea and the iris. Drug treatment usually helps lower pressure. Surgery to open a blocked channel or create an artificial one may be needed.

Cornea   *Fluid in front chamber*

CHAPTER FIVE

C H A P T E R   5

The ENDOCRINE SYSTEM

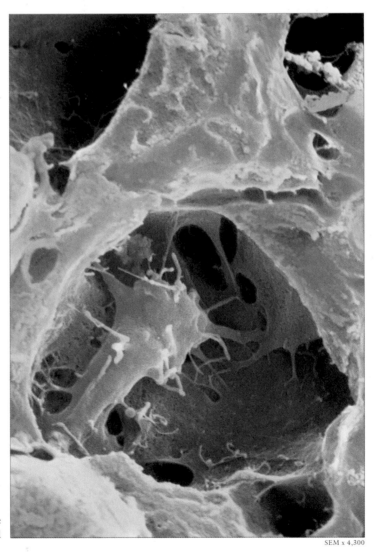

Tissue of the
pituitary gland

SEM x 4,300

INTRODUCTION

Hormones are the body's internal chemical messengers. They carry the information that controls the function of almost all of the body's cells and tissues. Most hormones are themselves controlled by a mechanism called feedback, which is like the thermostat in a central heating system. When a gland is working harder than the body needs it to, the hormone system switches it off; when the body needs the gland to speed up, the system turns on the switch again. Researchers have discovered how the main endocrine (hormone-producing) glands function; some, but not all, are controlled by the pituitary gland, which is controlled in turn by the hypothalamus. Both of these are located in the brain. The chemical makeup of virtually all hormones is now known, and if a gland fails, medications have been developed that will replace the missing hormones, and thus restore normal health. Another pharmaceutical innovation, the hormone blocker, is used to interfere with the activity of some hormones, or to counteract their oversecretion. Contraceptive drugs are hormone preparations that enable women to control their own fertility. Research workers are still discovering hormones and learning about the subtle controls by which the body functions. These findings should lead to better and less intrusive treatments across the whole range of human disease.

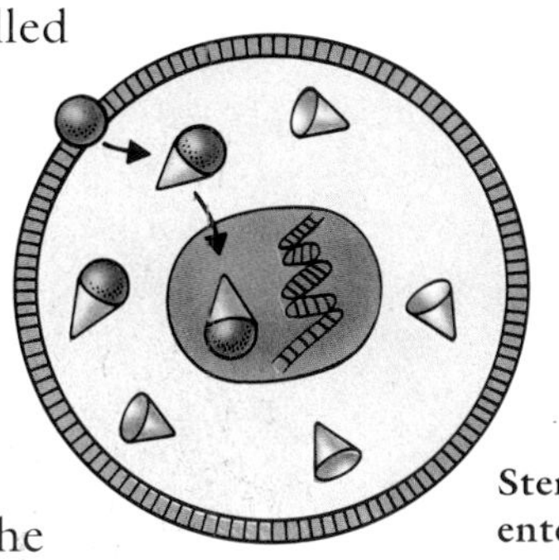

Steroid hormones entering a cell

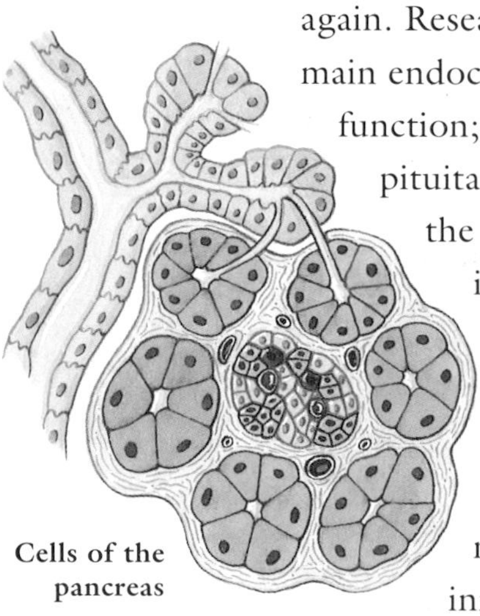

Cells of the pancreas

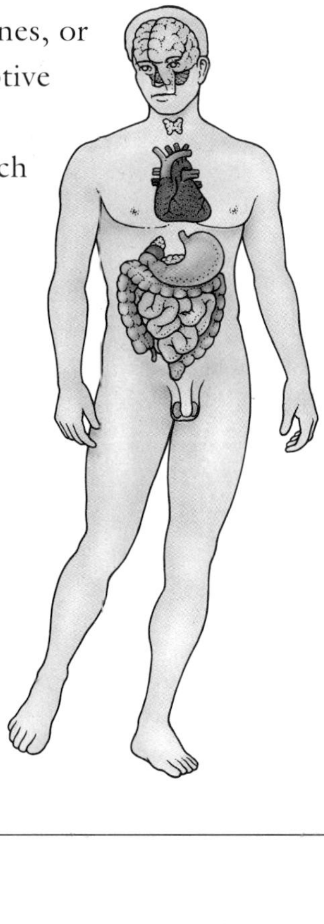

THE ENDOCRINE SYSTEM

HORMONE PRODUCERS

THE WORD HORMONE MEANS "TO SPUR ON." Each hormone is a complex chemical substance produced and secreted into the bloodstream by an endocrine gland, or secreted by specialized cells in other organs such as parts of the gastrointestinal tract or heart. Functions regulated by hormones include metabolism (the breaking down or building of chemical elements); water and mineral balance; growth and sexual development; and the body's reaction to stress in response to the nervous system.

HORMONE ACTION

Hormones integrate the activities of widely separated organs. While the effects of the nervous system appear rapidly but may be short-lived, the endocrine system's effects are slow to appear, longer-lasting, and usually occur at distant target sites. The prostaglandins are an example of a special class of hormones producing only local effects within the same tissue. They are known to be secreted within the brain and lungs.

Pineal gland
This tiny gland secretes melatonin, a hormone that controls body rhythms such as sleeping and waking and may influence sexual development.

Hypothalamus
Hormones from this cluster of nerve cells cause smooth muscle to contract, control water balance, and are responsible for breast milk.

Parathyroid glands

Pituitary gland
Once called the "master gland," it controls all other major endocrine glands.

Thyroid gland
This gland's major hormone controls metabolism, including heart rate and the rate of energy use. Another aids in controlling calcium metabolism.

Heart
The heart produces a hormone called atriopeptin, which reduces blood volume and blood pressure, and so helps regulate fluid balance.

PARATHYROID GLANDS

These four glands make parathormone, which increases blood calcium levels by the increase of intestinal calcium absorption, release of calcium stores from bones, and reduction of calcium excretion from the kidneys.

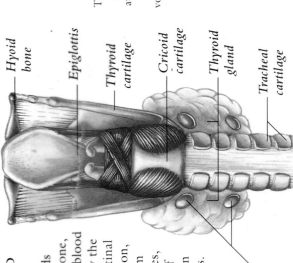

BACK VIEW

Hyoid bone

Epiglottis

Thyroid cartilage

Cricoid cartilage

Thyroid gland

Tracheal cartilage

Parathyroid glands

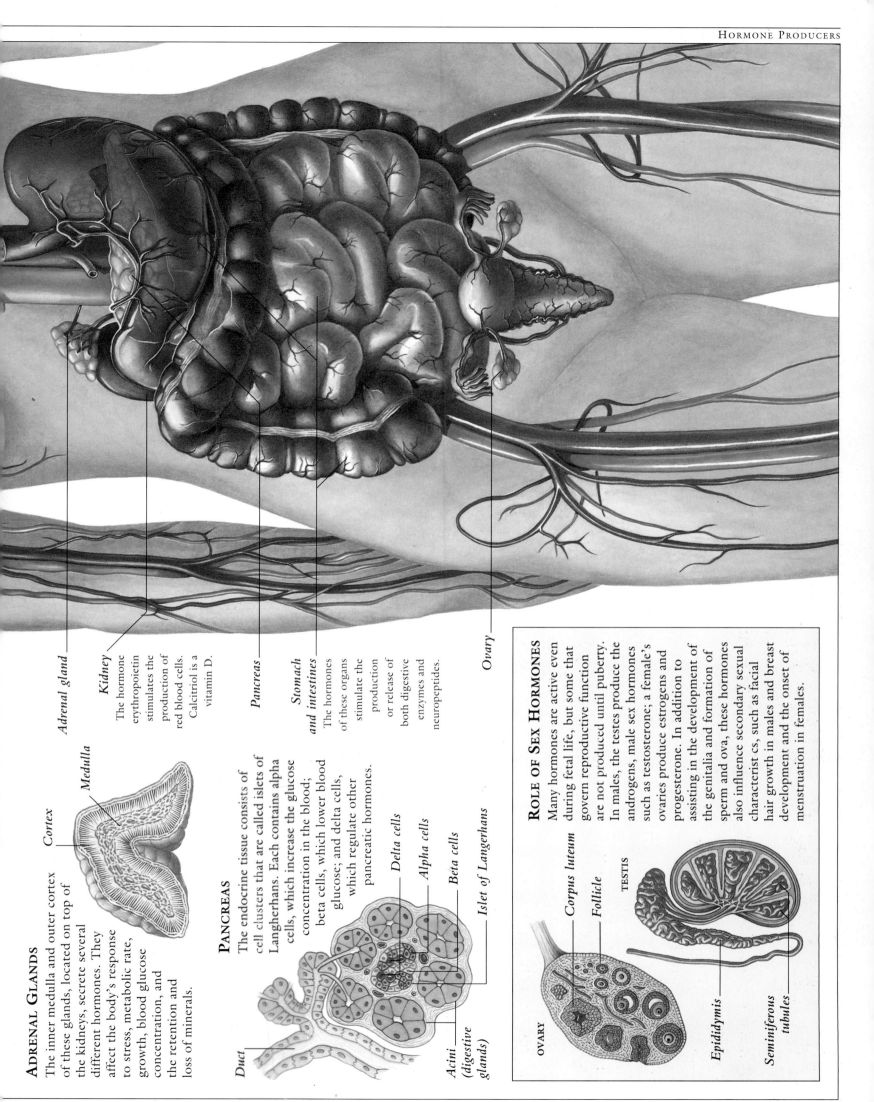

ADRENAL GLANDS

The inner medulla and outer cortex of these glands, located on top of the kidneys, secrete several different hormones. They affect the body's response to stress, metabolic rate, growth, blood glucose concentration, and the retention and loss of minerals.

Cortex

Medulla

Adrenal gland

Kidney
The hormone erythropoietin stimulates the production of red blood cells. Calcitriol is a vitamin D.

Pancreas

Stomach and intestines
The hormones of these organs stimulate the production or release of both digestive enzymes and neuropeptides.

Ovary

PANCREAS

The endocrine tissue consists of cell clusters that are called islets of Langerhans. Each contains alpha cells, which increase the glucose concentration in the blood; beta cells, which lower blood glucose; and delta cells, which regulate other pancreatic hormones.

Delta cells

Alpha cells

Beta cells

Islet of Langerhans

Duct

Acini (digestive glands)

ROLE OF SEX HORMONES

Many hormones are active even during fetal life, but some that govern reproductive function are not produced until puberty. In males, the testes produce the androgens, male sex hormones such as testosterone; a female's ovaries produce estrogens and progesterone. In addition to assisting in the development of the genitalia and formation of sperm and ova, these hormones also influence secondary sexual characterist cs, such as facial hair growth in males and breast development and the onset of menstruation in females.

Corpus luteum

Follicle

OVARY

TESTIS

Epididymis

Seminiferous tubules

97

HORMONAL CONTROL

HORMONES REACH EVERY PART OF THE BODY, and the membrane of every cell has receptors for one or more hormones that stimulate or retard a specific body function, such as sexual development. The hypothalamus, located at the base of the brain, acts as the mastermind that coordinates hormone production, producing regulatory, or releasing, hormones; these travel a short distance through special blood vessels and nerve endings to the pituitary gland, which was once called the "master gland."

THE MASTER GLAND

Attached to the hypothalamus by a short stalk, the pea-sized pituitary gland hangs from the base of the brain and is composed of two parts, an anterior and a posterior lobe. Some of its hormones act indirectly by stimulating target glands to release other hormones. Others have a direct effect on the function of target glands or tissue.

HORMONE CONTROL MECHANISMS

Made up of molecules derived from steroids or proteins, hormones produce their effect only when bound to a specific receptor on or in a target cell. Hormones derived from proteins bind to receptors on the outside of the cell membrane; steroid hormones pass into the cell before binding to receptors.

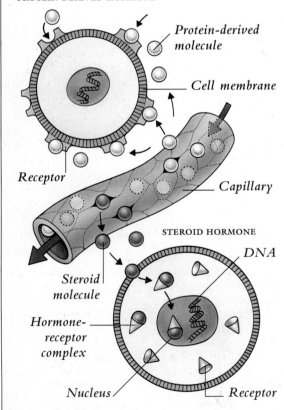

PROTEIN-DERIVED HORMONE

Protein-derived molecule

Cell membrane

Receptor

Capillary

STEROID HORMONE

DNA

Steroid molecule

Hormone-receptor complex

Nucleus

Receptor

Artery

Skin
Melanocytes in skin tissue are stimulated by MSH to produce more melanin, a pigment that darkens skin in response to sunlight.

MSH

Adrenal glands
The steroid hormones influence the use of fats, proteins, carbohydrates, and minerals. Along with other hormones, they also influence the response of the body to stress.

ACTH

TSH

Thyroid gland
Thyroid hormone affects body metabolism, growth, and heart rate. Calcitonin combats elevated calcium levels.

GH

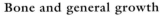

Bone and general growth
Growth hormone acts throughout the body to promote the protein synthesis that is essential for normal growth and development in children.

FSH, LH

Testis and ovary
Male and female hormones released by these glands control sexual development and reproductive function.

Anterior lobe
Influenced by trigger hormones from the hypothalamus, the anterior lobe of the pituitary gland makes at least six hormones.

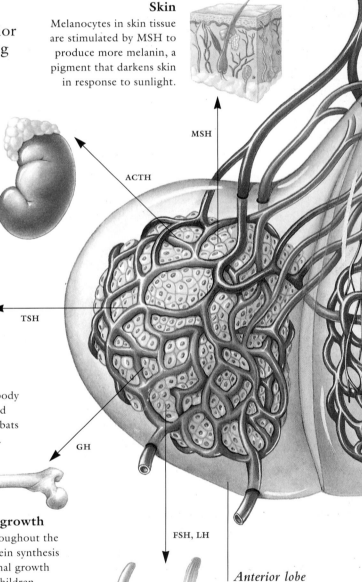

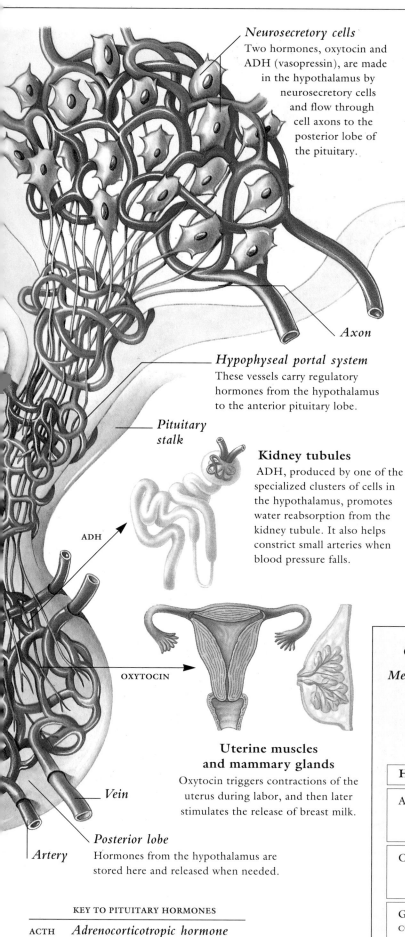

Neurosecretory cells
Two hormones, oxytocin and ADH (vasopressin), are made in the hypothalamus by neurosecretory cells and flow through cell axons to the posterior lobe of the pituitary.

Axon

Hypophyseal portal system
These vessels carry regulatory hormones from the hypothalamus to the anterior pituitary lobe.

Pituitary stalk

ADH

Kidney tubules
ADH, produced by one of the specialized clusters of cells in the hypothalamus, promotes water reabsorption from the kidney tubule. It also helps constrict small arteries when blood pressure falls.

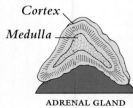

OXYTOCIN

Uterine muscles and mammary glands
Oxytocin triggers contractions of the uterus during labor, and then later stimulates the release of breast milk.

Vein

Artery

Posterior lobe
Hormones from the hypothalamus are stored here and released when needed.

KEY TO PITUITARY HORMONES

ACTH	*Adrenocorticotropic hormone*
TSH	*Thyroid-stimulating hormone*
GH	*Growth hormone*
FSH	*Follicle-stimulating hormone*
LH	*Luteinizing hormone*
MSH	*Melanocyte-stimulating hormone*
ADH	*Antidiuretic hormone*

FEEDBACK MECHANISMS

A feedback mechanism that involves the hypothalamus, pituitary gland, and target gland controls all hormone production. A feedback system can promote the release of another hormone (positive feedback) or, alternatively, inhibit its release (negative feedback). This mechanism helps maintain the balance in our internal environment.

1 Responding to levels of thyroid hormone, the hypothalamus makes TRH. This stimulates the anterior pituitary gland to release TSH. The thyroid gland is then triggered to produces its hormones.

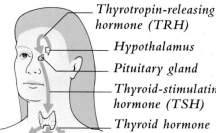

Thyrotropin-releasing hormone (TRH)
Hypothalamus
Pituitary gland
Thyroid-stimulating hormone (TSH)
Thyroid hormone

2 If thyroid hormone levels are too high, negative feedback alerts the hypothalamus so that it produces less TRH. A reduced TRH level results in a reduced level of TSH. The thyroid responds by producing less hormone.

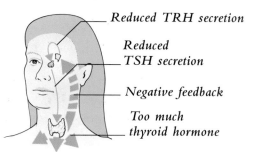

Reduced TRH secretion
Reduced TSH secretion
Negative feedback
Too much thyroid hormone

3 If thyroid hormone levels fall too low, the feedback mechanism is weakened. In response, the hypothalamus makes more TRH; TSH rises so that the levels of thyroid hormone also rise.

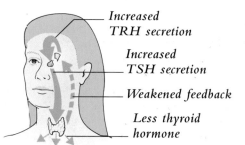

Increased TRH secretion
Increased TSH secretion
Weakened feedback
Less thyroid hormone

ADRENAL HORMONES

Cortex
Medulla

ADRENAL GLAND

The adrenal cortex has three zones, and each produces its own hormones. The medulla is almost a separate endocrine gland: it produces hormones, but its nerve fibers also link it up with the sympathetic nervous system and are involved in the "fight-or-flight" response.

HORMONES OF THE ADRENAL CORTEX AND ADRENAL MEDULLA

ALDOSTERONE	Secreted by the outermost zone of the adrenal cortex, this hormone inhibits the amount of sodium excreted in the urine and helps maintain blood volume and blood pressure.
CORTISOL	The middle layer of the cortex produces this steroid hormone. It controls how the body utilizes fat, protein, carbohydrates, and minerals and helps reduce inflammation.
GONADO-CORTICOIDS	Produced by the inner layer of the adrenal cortex, these sex hormones have only a slight effect on the sex organs. They influence sperm production in men, and the distribution of body hair and menstruation in women.
EPINEPHRINE AND NOREPINEPHRINE	These hormones from the adrenal medulla affect the body during stress. Norepinephrine raises heart rate and blood pressure; epinephrine stimulates carbohydrate metabolism.

CHAPTER SIX

C H A P T E R   6

The CARDIOVASCULAR SYSTEM

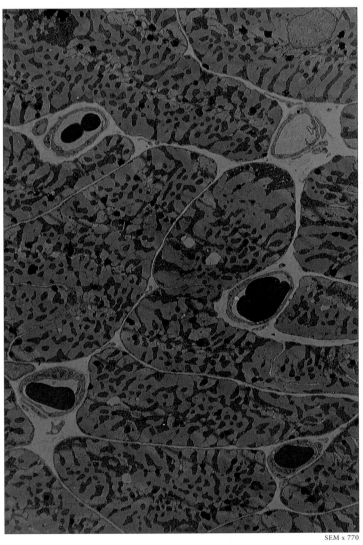

Myocardium, a special type of muscle tissue unique to the heart

SEM x 770

INTRODUCTION

For most of us the heart has a special symbolism, being linked with emotions and virtues such as love and courage. In fact, it is simply a pump. The assumed link between the heart and the emotions dates back to ancient times, when the pumping function of the heart was not well understood. The ancient Greeks and Romans believed that the arteries transported air rather than blood. Some two thousand years later, the English physician William Harvey discovered that the blood is pumped by the heart through two circuits, which carry it to and from the lungs and around the entire body. Heart disease has always been a major cause of death, but during the twentieth century the pattern has changed. In the early 1900s, the most common type of heart disease was valve damage caused by rheumatic fever in children and young adults; now, however, heart valve disease is rare in those under the age of 60. Today coronary heart disease is the leading cause of death in industrialized countries in people over age 35. Links between this disease and smoking, excessive cholesterol associated with high-fat diets, high blood pressure, and insufficient exercise are all well established. Available treatments include drugs, surgery, and even heart transplantation, but the key to preventing coronary heart disease is simply maintaining a healthy lifestyle.

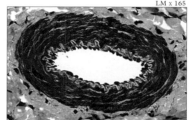

LM x 165

Cross-section of the aorta

Pathways of impulses from the heart's pacemaker

THE CARDIOVASCULAR SYSTEM

The HEART and CIRCULATION

BLOOD IS PUMPED OUT FROM THE HEART through strong, elastic tubes called arteries. Vessels leaving the right ventricle initiate the pulmonary circulatory system, which carries deoxygenated blood to the lungs for replenishment with fresh oxygen. The aorta, the single main artery from the left ventricle, branches to form the systemic circulation, which takes oxygen to all body tissues. Another network of veins returns deoxygenated blood to the heart. The smallest arteries (arterioles) and veins (venules) are linked by tiny capillaries. This system is 90,000 miles (150,000km) long.

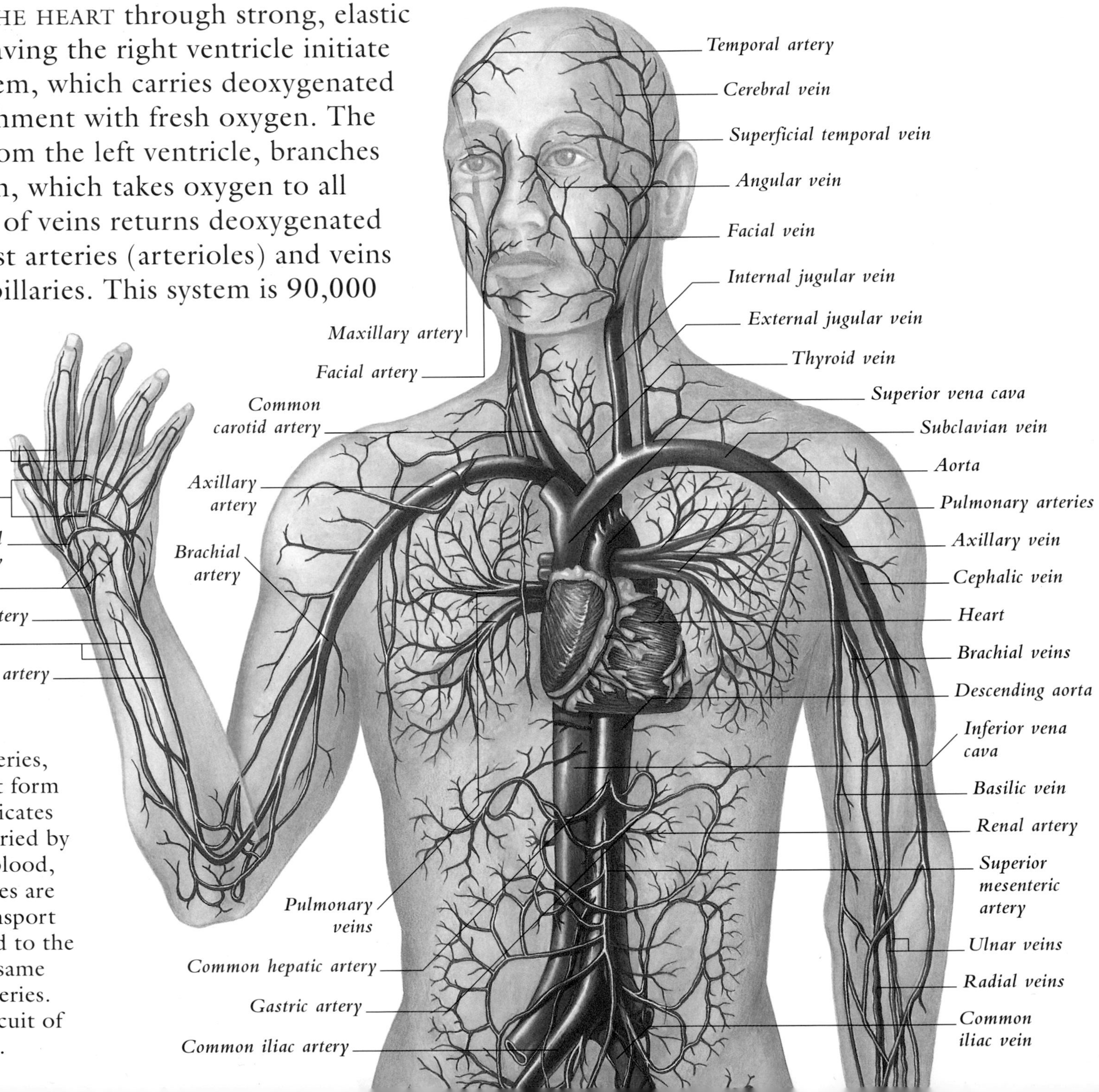

CHANNELS FOR BLOOD

The illustration shows many of the arteries, veins, and branching blood vessels that form the body's circulatory system. Red indicates oxygenated blood, which is usually carried by arteries; blue indicates deoxygenated blood, carried by veins. The pulmonary arteries are the only arterial blood vessels that transport deoxygenated blood. Blood is returned to the heart through the veins at exactly the same rate at which it is pumped out into arteries. On average, blood completes a full circuit of the body in approximately one minute.

Temporal artery
Cerebral vein
Superficial temporal vein
Angular vein
Facial vein
Internal jugular vein
External jugular vein
Thyroid vein
Superior vena cava
Subclavian vein
Aorta
Pulmonary arteries
Axillary vein
Cephalic vein
Heart
Brachial veins
Descending aorta
Inferior vena cava
Basilic vein
Renal artery
Superior mesenteric artery
Ulnar veins
Radial veins
Common iliac vein

Maxillary artery
Facial artery
Common carotid artery
Axillary artery
Brachial artery

Digital arteries
Palmar arches
Dorsal carpal artery
Radial artery
Ulnar artery
Palmar carpal arteries
Interosseous arteries

Pulmonary veins
Common hepatic artery
Gastric artery
Common iliac artery

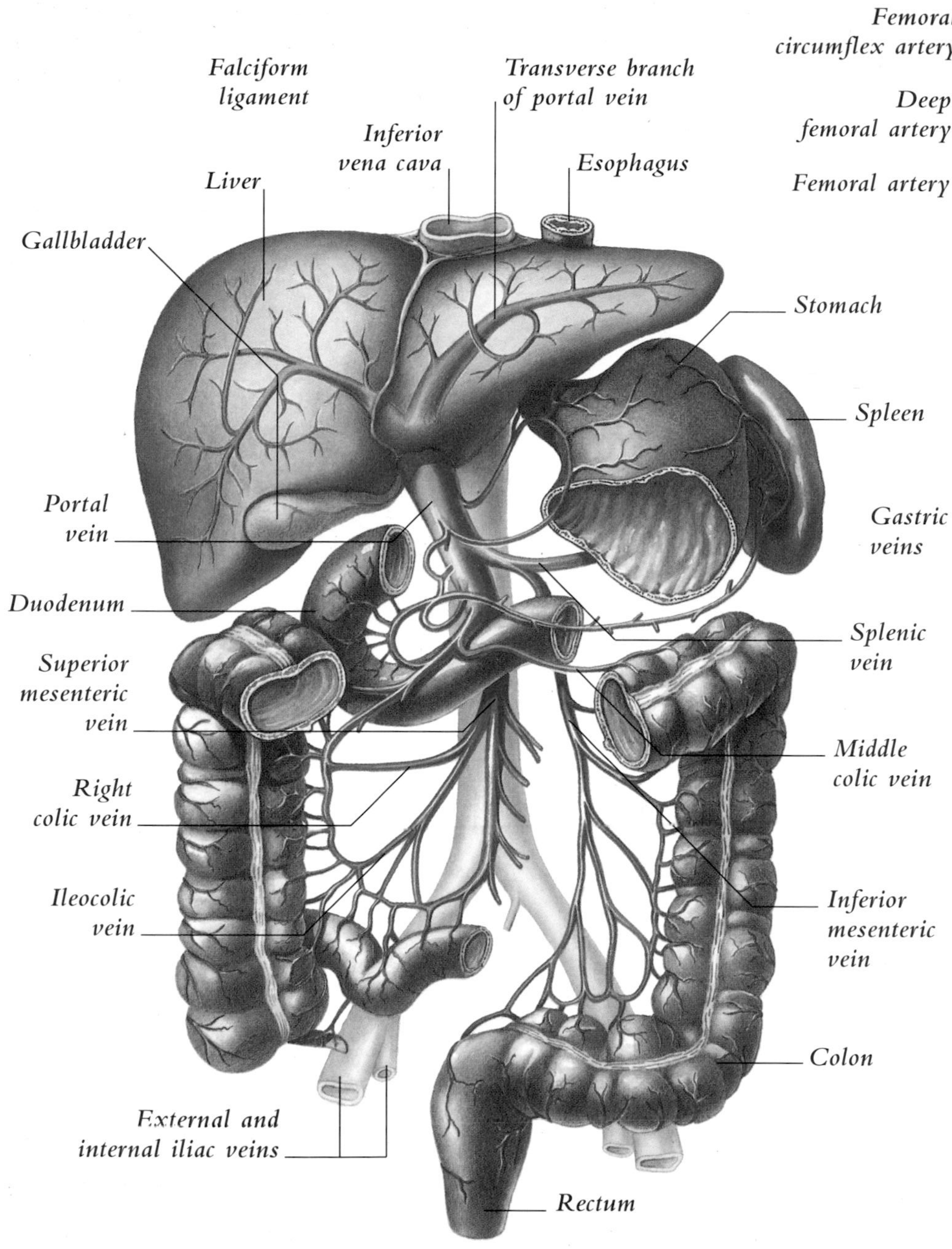

Venous network of the hand

Palmar venous arch

Digital veins

Great saphenous vein

Femoral vein

Accessory saphenous vein

Venous network of the knee

Popliteal vein

Perforating veins

Peroneal veins

Anterior tibial veins

Posterior tibial veins

Small saphenous vein

Plantar venous arch

Dorsal metatarsal veins

Dorsal venous arch

Dorsal digital veins

Perforating arteries

Arterial network of the knee

Descending genicular artery

Popliteal artery

Peroneal artery

Posterior tibial artery

Anterior tibial artery

Plantar arteries

Dorsal metatarsal arteries

Arcuate artery

Dorsal digital arteries

Femoral circumflex artery

Deep femoral artery

Femoral artery

Stomach

Spleen

Gastric veins

Splenic vein

Middle colic vein

Inferior mesenteric vein

Colon

Transverse branch of portal vein

Esophagus

Inferior vena cava

Falciform ligament

Liver

Gallbladder

Portal vein

Duodenum

Superior mesenteric vein

Right colic vein

Ileocolic vein

External and internal iliac veins

Rectum

THE PORTAL SYSTEM

The network of blood vessels between the liver and digestive organs is called the portal system. Blood from the stomach, spleen, intestines, and pancreas drains into a number of veins. These merge to become the portal vein, which carries gastrointestinal blood to the liver. This organ absorbs and stores nutrients, and also removes toxins (poisons) and pollutants. Detoxified blood returns to the heart and lungs in the inferior vena cava for oxygenation and redistribution.

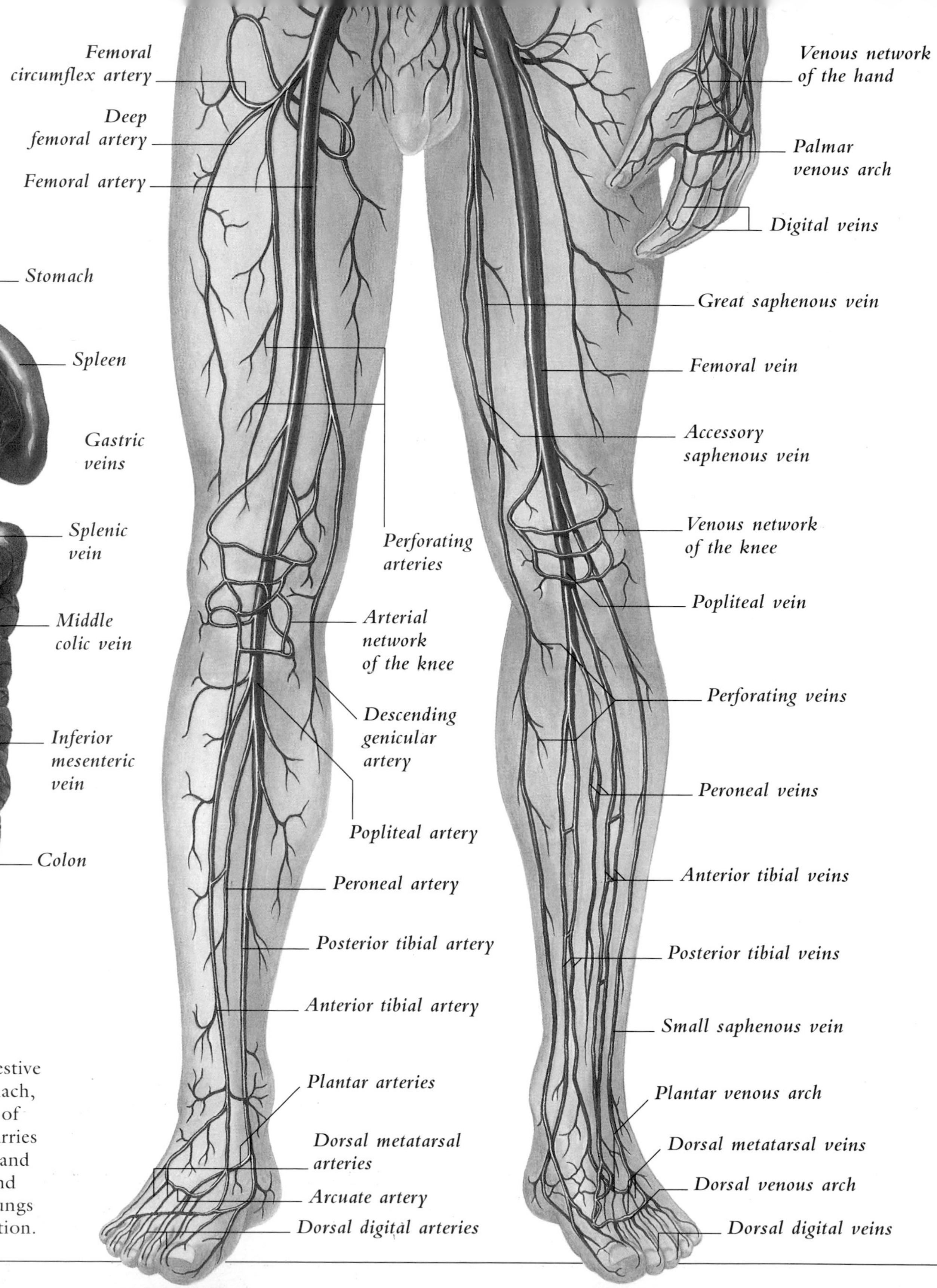

HEART STRUCTURE

THE HEART IS A POWERFUL MUSCLE about the size of a grapefruit. Located just left of the center of the chest, it operates as two coordinated pumps, continuously sending blood around the body. Oxygen and nutrients are carried to organs and tissues by this circulation, which also removes harmful wastes. A type of muscle that is called myocardium is unique to the heart.

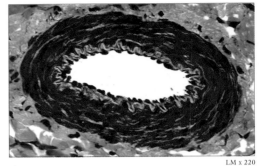

LM x 220

Aorta
This cross-section shows the aorta, the largest blood vessel in the body, with an inner diameter of about 1 inch (2.5cm). The aorta arches out from the left ventricle, and carries oxygenated blood to every other part of the body.

HEART STRUCTURE

The heart has four chambers: two upper chambers called atria and two thicker-walled lower chambers known as ventricles. A strong muscular wall, the septum, divides the two sides of the heart. The four heart valves are crucial for allowing blood to pass in and out of the heart chambers in only one direction.

Aorta

Pulmonary artery
The pulmonary artery branches after leaving the right ventricle, each branch carrying deoxygenated blood to one lung. It is the only artery that carries deoxygenated blood.

Superior vena cava
This large vein carries used blood from the head and arms into the right atrium.

Pulmonary veins
Newly oxygenated blood from the lungs returns to the left atrium via the four pulmonary veins; oxygenated blood is not carried by any other veins.

Pulmonary veins

Left atrium

Aortic valve

Mitral valve

Right atrium

Endocardium
This smooth membrane lines both the interior of the heart and its valves.

Septum
This thick muscle wall divides the heart into two halves.

Pericardium
The pericardium is a tough, fibrous sac surrounding the heart. It has an inner, fluid-covered membrane, cushioning the heart.

Pulmonary valve

Tricuspid valve

Inferior vena cava
Deoxygenated blood returns to the right atrium from the legs, pelvis, and abdomen along the inferior vena cava.

Right ventricle

Left ventricle

Myocardium
The heart's interconnected muscle fibers (cells) enable it to contract automatically.

TWO PUMPS IN ONE

Used blood from body tissues enters the right side of the heart and is pumped out to the lungs. Passage of blood through the lungs, the pulmonary circulation, enables blood to pick up oxygen. The reoxygenated blood returns to the left side of the heart and is then pumped out again to the body tissues in a route that is known as the systemic circulation. The full circuit around the lungs and body takes only approximately one minute when the body is at rest, and the heart pumps about 5 to 7 quarts of blood a day.

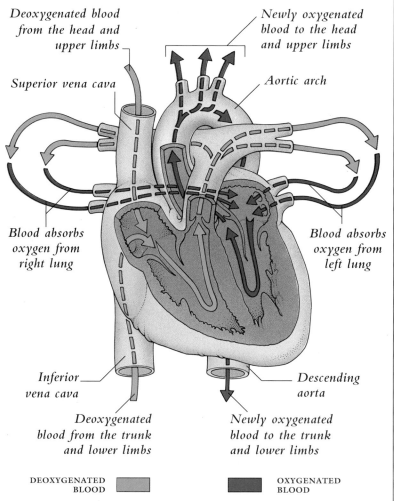

Deoxygenated blood from the head and upper limbs

Newly oxygenated blood to the head and upper limbs

Superior vena cava

Aortic arch

Blood absorbs oxygen from right lung

Blood absorbs oxygen from left lung

Inferior vena cava

Descending aorta

Deoxygenated blood from the trunk and lower limbs

Newly oxygenated blood to the trunk and lower limbs

DEOXYGENATED BLOOD

OXYGENATED BLOOD

BLOOD SUPPLY TO THE HEART

Because the heart needs a generous supply of oxygen, it needs a correspondingly large blood supply; only the brain requires more. Blood that flows through the chambers of the heart cannot seep through to reach the muscle cells, so the heart muscle has a separate network of blood vessels called the coronary system.

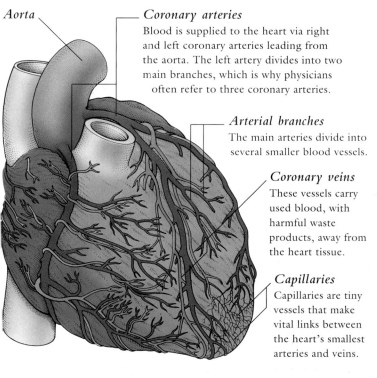

Aorta

Coronary arteries
Blood is supplied to the heart via right and left coronary arteries leading from the aorta. The left artery divides into two main branches, which is why physicians often refer to three coronary arteries.

Arterial branches
The main arteries divide into several smaller blood vessels.

Coronary veins
These vessels carry used blood, with harmful waste products, away from the heart tissue.

Capillaries
Capillaries are tiny vessels that make vital links between the heart's smallest arteries and veins.

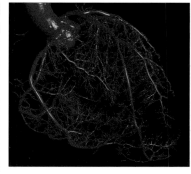

Resin cast of coronary arteries

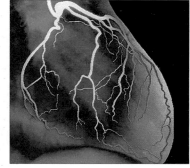

Angiogram of coronary arteries

CARDIAC SKELETON

A set of four tough fibrous rings, or cuffs, known as the cardiac skeleton, provides the points of attachment for the four heart valves and heart muscle. The muscle fibers of the left and right ventricles, which make up the bulk of the heart, are shown at right. The wraparound arrangement of the muscle fibers enables the ventricles to squirt blood out of the heart, similar to the way a closing fist squirts water out of a balloon.

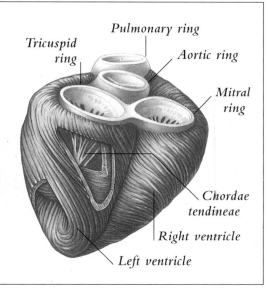

Pulmonary ring

Tricuspid ring

Aortic ring

Mitral ring

Chordae tendineae

Right ventricle

Left ventricle

BLOOD COLLECTION

The pattern formed by coronary veins closely mirrors that of the coronary arteries. Most of them drain into the coronary sinus, a large vein at the back of the heart. From this vein, blood flows into the right atrium. Some small veins empty directly into the right atrium.

Right pulmonary veins

Left pulmonary veins

Coronary sinus

Coronary veins

HEART FUNCTION

AS A DYNAMIC PUMP, the heart forces blood around an impressive network of blood vessels that would circle the Earth two-and-a-half times. The actual power comes from the ventricles, with their thick muscular walls that contract so that blood surges out into the arteries. This heart activity is repeated automatically, with the rate of beating and the amount of blood that is pushed out increasing or decreasing in response to the body's level of stress and exertion.

ECG recording
Electrocardiography (ECG) detects the flow of electrical impulses throughout the heart. Color-coding relates the tracing to various stages in the passage of impulses (see below).

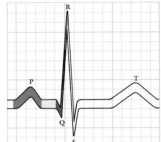

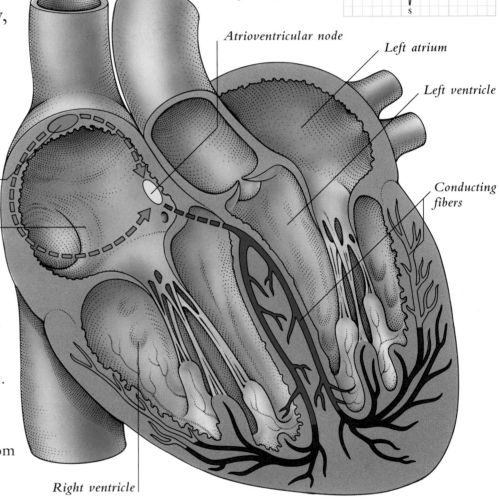

Atrioventricular node

Left atrium

Left ventricle

Sinoatrial node

Right atrium

Conducting fibers

Right ventricle

ELECTRICAL PATHWAYS

Regular, rhythmic beating of the heart is maintained by electrical impulses that originate in the sinoatrial node, which is the body's natural pacemaker. Impulses spread through the atria, stimulating a contraction, to the atrioventricular node. After a slight pause (0.2 seconds), the impulses pass along special conducting muscle fibers through the ventricles and cause them to contract. Any variation from this characteristic sequence indicates the possibility of a heart disorder.

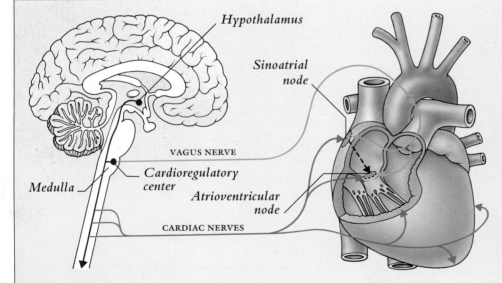

Hypothalamus

Sinoatrial node

VAGUS NERVE

Medulla

Cardioregulatory center

Atrioventricular node

CARDIAC NERVES

NERVOUS SYSTEM CONTROL

Without control by nerves, the heart would beat about 100 beats per minute. However, parasympathetic nerves, especially the vagus nerve, set a resting rate of 60 to 70 by means of impulses to and from the cardioregulatory center in the medulla. When the hypothalamus is stimulated by stress or exercise, it signals the sympathetic cardiac nerves to raise the heart rate. The release of hormones by the adrenal gland also increases the heart rate.

▢ PARASYMPATHETIC NERVES

■ SYMPATHETIC NERVES

THE HEART VALVES

Four valves allow blood to move through the heart chambers in only one direction. They consist of two or three half-moon flaps, or cusps, of fibrous tissue that attach to the walls of the heart. The cusps open when the blood is flowing correctly but close into a tight seal to prevent any backward movement. Opening and closing of the valves occurs in reaction to changing pressure on either side as blood surges through.

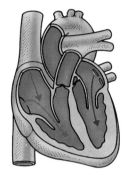

Pulmonary valve
Blood flows through the pulmonary valve from the right ventricle into the pulmonary artery. This three-cusped valve is essential to prevent backflow of blood when the ventricle, after its vigorous contraction, relaxes again.

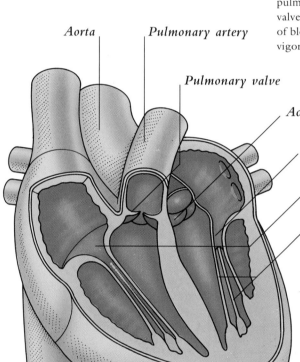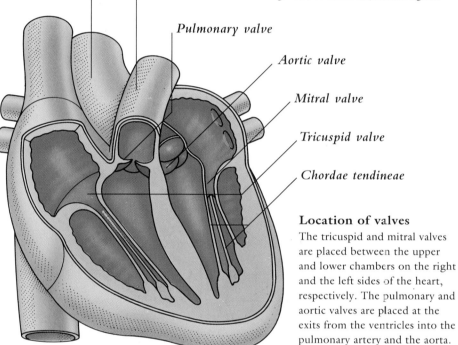

Aorta *Pulmonary artery*

Pulmonary valve

Aortic valve

Mitral valve

Tricuspid valve

Chordae tendineae

Location of valves
The tricuspid and mitral valves are placed between the upper and lower chambers on the right and the left sides of the heart, respectively. The pulmonary and aortic valves are placed at the exits from the ventricles into the pulmonary artery and the aorta.

Valve cusps
These thin, fibrous cusps of the valves are covered by a smooth membrane called endocardium and reinforced by dense connective tissue. The pulmonary, aortic, and tricuspid valves have three cusps; the mitral valve has two cusps.

TWO CUSPS   THREE CUSPS

Chordae tendineae
Fibrous strands known as chordae tendineae fasten the tricuspid and mitral valves to the ventricular walls. These cords prevent the valves from being pushed upward by the forceful pressure of flowing blood. The aortic and pulmonary valves are less likely to be pushed out by pressure and so do not need such fastenings.

HEARTBEAT CYCLE

Three separate and distinct phases make up the carefully timed, sequential beating of the heart. The relaxing of the heart and refilling with blood in the first stage is followed by stages of contracting and squeezing. The entire cycle usually takes only about four-fifths of a second, but its speed may more than double during vigorous exercise or in times of stress.

Diastole
During the first phase of the cycle, oxygenated blood enters the left atrium and deoxygenated blood simultaneously enters the right atrium. This blood then flows into the ventricles. By the end of this phase, the ventricles are filled to about 80 percent of capacity.

Atrial systole
Impulses from the sinoatrial node initiate the second phase of the cycle, during which both atria contract, squeezing any blood that remains in the atria into the ventricles.

Ventricular systole
The ventricles contract during the third phase of the heartbeat sequence. The valves at the exits of both ventricles open and the blood is forced into the aorta and pulmonary artery. As this phase ends, diastole starts again.

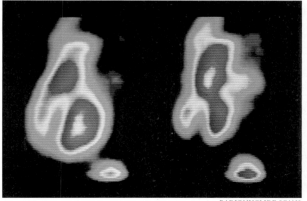

RADIONUCLIDE SCANS

Relaxation and contraction
The scans shown above were produced by using a gamma camera to detect radioactively tagged red blood cells. They show how the distribution of blood (red and yellow areas) in the heart varies at different stages of the pumping cycle. On the left, the heart is relaxed and filling with blood. On the right, the heart is squeezing blood out.

BLOOD VESSELS *and* BLOOD

A COMPLEX CIRCULATORY SYSTEM ENABLES BLOOD to perform its many different functions. Blood transports oxygen, nutrients, and waste products excreted from cells; it helps regulate the body's water content, temperature, and acid-alkali balance; and its specialized cells and proteins protect the body from infection or blood loss resulting from an injury.

BLOOD TRANSPORT

Without oxygen, all cells in the body would quickly die. The blood oxygenated in the lungs circulates through the muscular arteries into smaller vessels that are called arterioles. They end in capillaries, with walls a single cell layer thick, connected to small venules that feed into the larger veins.

VEIN STRUCTURE

Deoxygenated blood, which returns to the heart through the veins, is at low pressure. Its movement is helped by a succession of one-way valves that prevents backflow.

Valve flaps

Elastic layer

Elastic layer

Elastic layer

Tunica adventitia
Nerves, blood vessels, and lymph vessels are found in this connective tissue sheath.

Tunica media
This sheath consists of stacked muscle cells and fibrous tissue. It is the thickest sheath in large arteries.

Outer layer of tunica intima
The outer layer is composed of a connective tissue rich in elastic fibers.

Inner layer of tunica intima
In this layer, flattened epithelial cells are in direct contact with blood flowing in the central canal, or lumen.

Red blood cells
These pigmented cells give blood its color. Without them, the body tissues would not receive any oxygen.

White blood cell

Platelets

Plasma
More than half of blood is a straw-colored liquid that contains water and a variety of nutrients, minerals, and proteins.

White blood cells
The main function of these variously shaped cells, which are grouped under the name of leukocytes, is to defend the body against infection.

Platelets
The smallest blood cells, the platelets, play an essential role in the blood-clotting process that follows blood vessel injury.

BLOOD CLOTTING
Platelets trigger a series of complex chemical reactions that lead to the formation of fibrin strands. These create a mesh at the wound site, trapping red blood cells to form a blood clot.

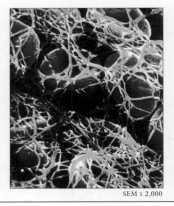

Red blood cells in fibrin strands

SEM x 2,000

THE CAPILLARY NETWORK

The two circulatory routes, arterial and venous, are linked by capillaries. These tiny blood vessels form a network, the extent of which varies with the tissue's metabolism. The heart has a very extensive capillary network, while that of the skin is less complex. The blood flows more slowly through capillaries than it does through arteries, thus permitting the exchange of oxygen, nutrients, and waste products.

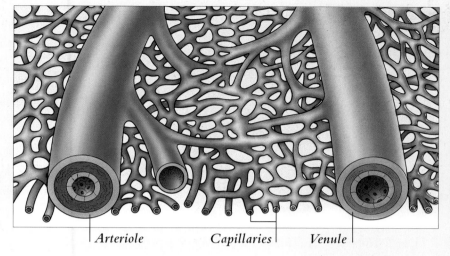

Site of exchange
The one-cell-thick capillary walls contain tiny spaces through which nutrient- and oxygen-bearing fluid flows out onto tissues and waste products such as carbon dioxide move back.

Arteriole　　*Capillaries*　　*Venule*

RED BLOOD CELL PRODUCTION

The life span of oxygen-bearing red blood cells, or erythrocytes, is only 80 to 120 days. Two million of them die every second, but these are replaced at the same rate by new ones generated in the body's red bone marrow in a process called erythropoiesis. This ensures that a constant and appropriate supply of essential oxygen is available for cell life and function.

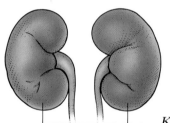

The role of the kidneys
If the kidneys are receiving only low amounts of oxygen, some of their cells stimulate the activation of a hormone known as erythropoietin.

— *Kidneys*

Site of production
Erythropoietin travels to the red bone marrow, and stimulates increased production of erythrocytes.

Erythropoietin

Bone marrow

Production of red blood cells

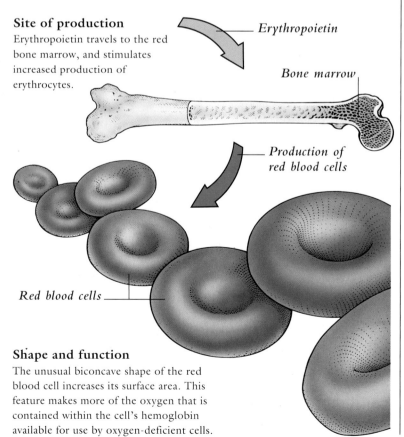

Red blood cells

Shape and function
The unusual biconcave shape of the red blood cell increases its surface area. This feature makes more of the oxygen that is contained within the cell's hemoglobin available for use by oxygen-deficient cells.

HEMOGLOBIN AND OXYGEN

Hemoglobin in the red blood cells is composed of heme, an iron-bearing red pigment, and of globin, ribbonlike protein chains. Oxygen from the lungs' tiny air sacs enters the red blood cells and combines with the iron to form oxyhemoglobin. In this form, oxygen is transported throughout the body. It is then released in the capillaries and passes into the fluid that bathes surrounding tissues. Cells take up this vital element and use it to produce energy.

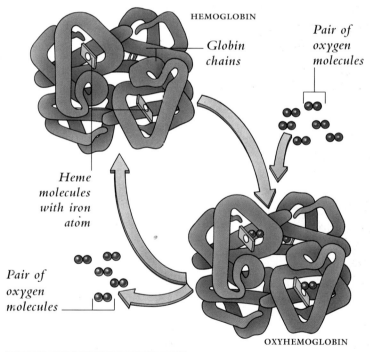

HEMOGLOBIN

Globin chains

Pair of oxygen molecules

Heme molecules with iron atom

Pair of oxygen molecules

OXYHEMOGLOBIN

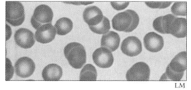

Normal blood
Molecules of hemoglobin, the oxygen-receptive pigment, give blood its red color, shown in this smear of normal red blood cells.

LM

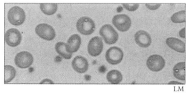

Iron-deficiency anemia
In anemia, red blood cells are few in number, pale due to the lack of hemoglobin, and have less oxygen than normal red blood cells.

LM

CORONARY HEART DISEASE

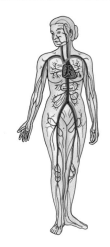

ANY HEART DISORDER DUE TO a restricted supply of blood to the heart muscle is called coronary heart disease. The most common manifestations are angina pectoris (gripping chest pains usually brought on by exertion) and myocardial infarction, also called a heart attack, which is the death of an area of heart muscle caused by a more complete deprivation of blood.

ATHEROSCLEROSIS

Coronary heart disease is usually caused by the narrowing of coronary arteries by atherosclerosis, the buildup of fatty deposits in the lining of arteries. The process that may lead to atherosclerosis begins with the accumulation of excess fats and cholesterol in the blood. These substances infiltrate the lining of arteries, gradually increasing in size to form a deposit known as an atheroma.

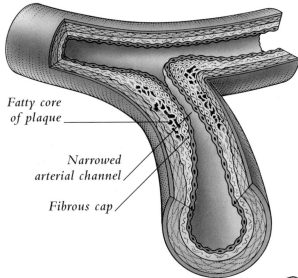

Fatty core of plaque

Narrowed arterial channel

Fibrous cap

Buildup of plaque
Deposits of atheromas gradually build up to form masses known as plaques. Consisting of a fatty core topped by a fibrous cap, plaques thicken arterial walls, narrowing the inner channel and impeding blood flow. If blood turbulence roughens the surface of the plaque, platelets and blood cells can collect, creating a blood clot that may block the artery completely.

Sites of atherosclerosis
Atherosclerosis can occur anywhere in the main coronary arteries or in smaller branches, but plaque usually builds up at stress points in the artery such as branch junctions.

Right coronary artery

Aorta

Left main coronary artery

Left circumflex artery

Left anterior descending artery

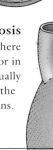

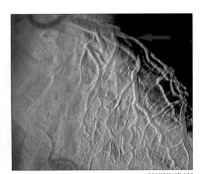

ANGIOGRAM

A narrowed artery
This image clearly shows the effects of atherosclerosis. Sections of the left coronary artery, seen at top right, are severely narrowed and only a minimal amount of blood flows through.

RISK FACTORS

Smoking, insufficient exercise, high blood pressure, diabetes, being overweight, and a diet high in saturated fats are the main risk factors for developing coronary heart disease. The stresses people face and how well they cope may also play a role.

NATIONAL PATTERNS

Rates of death from coronary heart disease vary markedly among different countries. The richer, more industrialized countries have higher rates, but these have been declining since the 1960s (except in Eastern Europe) due to improved medical treatments and increased consciousness of the influence of lifestyle on health. Generally men have rates that are somewhat higher than those of women, but the gap is narrowing.

CORONARY HEART DISEASE DEATHS* (MEN)			
COUNTRY	1960–64	1970–74	1985–89
US	333.6	316.1	165.7
UK	272.1	259.2	218.3
FINLAND	335.2	314.5	248.3
ITALY	172.0	123.9	91.5
JAPAN	72.4	48.7	32.6
AUSTRALIA	317.4	309.1	179.7
URUGUAY	167.4	187.8	111.6
COSTA RICA	79.3	76.1	133.8

*PER 100,000 POPULATION

CORONARY HEART DISEASE DEATHS* (WOMEN)			
COUNTRY	1960–64	1970–74	1985–89
US	168.5	156.3	83.2
UK	137.7	108.5	95.2
FINLAND	158.6	115.3	97.9
ITALY	119.1	65.9	38.3
JAPAN	50.6	27.6	17.8
AUSTRALIA	154.9	143.3	87.1
URUGUAY	93.0	105.9	56.6
COSTA RICA	65.6	57.2	75.8

*PER 100,000 POPULATION

ANGINA

Chest pains that come on with exertion are a warning sign that the cardiac muscle is not receiving enough blood for the effort being expended. An angina attack typically begins with a gripping or pressurelike pain behind the breastbone that sometimes radiates into the neck and jaw, and then down into the arms. The pain subsides rapidly with rest. Strong emotions, a heavy meal, or exposure to cold can mean that less exertion is needed to trigger an attack.

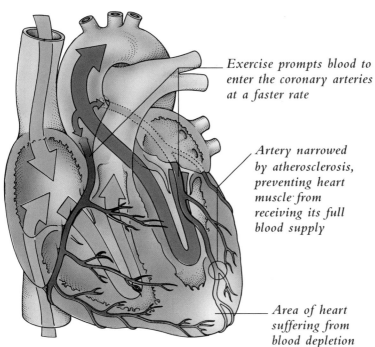

Exercise prompts blood to enter the coronary arteries at a faster rate

Artery narrowed by atherosclerosis, preventing heart muscle from receiving its full blood supply

Area of heart suffering from blood depletion

Effects on the heart
When blood supply is insufficient, the heart does not receive the oxygen and glucose it needs. The heart tries to create energy by alternative chemical processes, but produces waste products that the reduced blood supply cannot remove adequately. Pain may occur.

DRUG TREATMENT

Drugs used to treat angina act by widening the coronary arteries, thus improving blood flow. They also lower blood pressure and slow the heart, so that the work of the heart muscle can be reduced. Nitrate drugs, beta blockers, and calcium-channel blockers are frequently prescribed.

Narrowed blood vessels

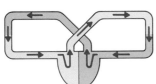

Resistance to blood flow makes heart work harder

Before drugs
Narrowed arteries deprive the heart muscle of the oxygen and glucose it needs to produce energy during exertion.

Widened blood vessels

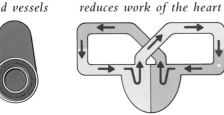

Less resistance to blood flow reduces work of the heart

After drugs
Nitrate drugs relax blood vessel walls, causing them to widen. This helps improve the blood flow and also the supply of oxygen.

HEART ATTACK

A heart attack, or myocardial infarction, usually occurs suddenly. Chest pain may be like that of angina, but is usually more severe and not necessarily brought on by exertion or relieved by rest. A victim may also sweat, feel weak, and even lose consciousness. If the attack leads to complete stoppage of the heart, also known as cardiac arrest, death usually follows.

Blocked blood supply
When a coronary artery becomes blocked, and remains blocked, the heart muscle it supplies dies. The severity of a heart attack depends on the amount of muscle affected and the health of other coronary arteries.

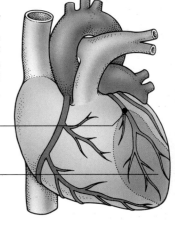

Blocked coronary artery

Damaged heart muscle

MEASURING ENZYME ACTIVITY

Proteins called enzymes regulate the body's chemical reactions. Muscle damage during a heart attack results in the release of certain enzymes into the bloodstream. Measuring enzyme activity helps reveal the extent of damage to the heart.

Site of blockage

Damaged area

Heart muscle fibers

Enzyme release
Enzymes from heart muscle fibers pass into capillaries and travel through coronary veins into the circulation.

Capillary

A number of different enzymes are released

DRUG TREATMENT

Doctors prescribe a number of different drugs to help keep blood flowing freely. Thrombolytic drugs perform by breaking down recent clots, and antiplatelet drugs and anticoagulants are useful to prevent clots from forming, thereby maintaining normal circulation of the blood.

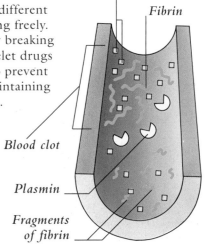

Thrombolytics
Blood clots form when strands of fibrin enmesh blood cells. Thrombolytic drugs cause the normally inactive substance plasminogen to change into plasmin, which breaks down fibrin and dissolves clots.

Platelets

Fibrin

Blood clot

Plasmin

Fragments of fibrin

SURGERY *for* CORONARY HEART DISEASE

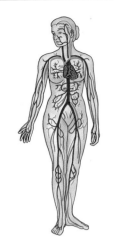

A NARROWED OR BLOCKED BLOOD VESSEL can be dealt with chemically by drugs, or surgically by clearing or bypassing the blockage. Coronary artery bypass requires the temporary use of a heart-lung machine, which permits stoppage of the heart. The more recent technique of balloon angioplasty uses an inflatable catheter, and a heart-lung machine is not needed.

BALLOON ANGIOPLASTY

Balloon angioplasty is particularly suited for patients in whom only one artery is critically narrowed. It is also preferable to avoid surgery in patients who have associated advanced lung disease. A catheter is used to reach the heart and inflate a balloon at the site of the blockage, cracking or flattening the plaque to widen the channel. The operation is brief and the recovery quick, but over a third of the obstructions recur.

1 An incision is made into the patient's arm (or leg), through which a guide wire is pushed into the brachial (or femoral) artery. Using X-ray or ultrasound guidance, the wire is then threaded into the affected coronary artery (via the aorta) to the blockage.

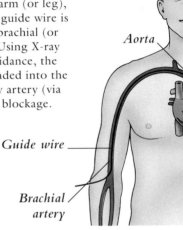

Aorta

Coronary artery

Guide wire

Brachial artery

Artery

Guide wire

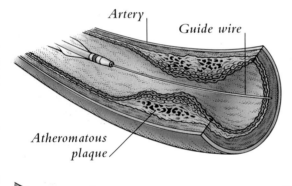

Atheromatous plaque

2 At the blockage, the wire is carefully maneuvered between the plaque accretions. Threaded on the guide wire, the catheter is pushed until the balloon at its tip is in place at the obstruction.

Supply tube   *Artery*   *Inflated balloon*

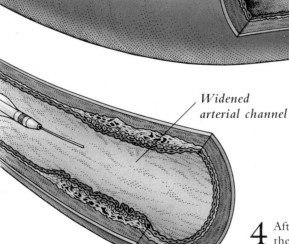

Widened arterial channel

3 A pump outside the body forces air or liquid along the catheter and into the balloon, inflating it to up to eight times atmospheric pressure. Pressure is exerted for up to 60 seconds on the blockage, then released. This procedure may be repeated several times.

Atheromatous plaque often splits when compressed by balloon

4 After inflations have squeezed the plaque against the artery wall, the blood pressure is checked to ensure that it is equal on either side of the site of the obstruction. The catheter is then withdrawn.

LASER CATHETERS

In a developing laser technique, the surgeon inserts an angioplasty balloon, which contains a fiberoptic channel through which the laser is passed. Obstructions are vaporized by the laser, then the balloon stretches. This process may cause damage to the vessel wall. Any fragments are then sucked out by a vacuum device.

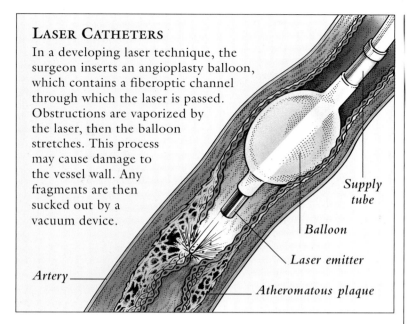

Supply tube

Balloon

Laser emitter

Artery

Atheromatous plaque

CORONARY ANGIOGRAMS

Angiography is a technique that shows the outline of arteries. Radiopaque dye, a contrast medium, is introduced via a catheter into the coronary arteries, and a series of X-ray pictures is then taken to record the dye's progress. The technique can be used to examine the coronary arteries in order to check on the success of balloon angioplasty. The angiograms below were taken before and after the operation.

NARROWED ARTERY

WIDENED ARTERY

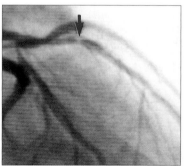

Balloon catheters

The balloons used in balloon angioplasty have to be tiny, strong, and flexible. They are uninflated when introduced into blood vessels in order to minimize friction and have to withstand great pressure when inflated. Liquid has now replaced air as an inflation medium, enabling a greater force to be exerted against the arterial wall.

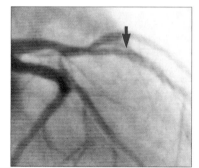

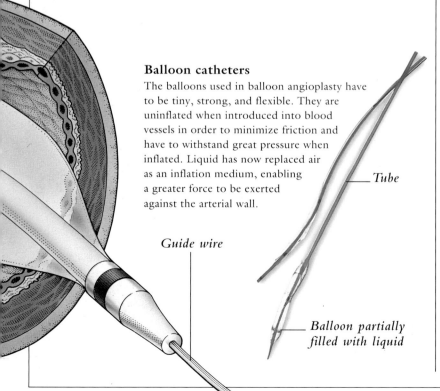

Tube

Guide wire

Balloon partially filled with liquid

CORONARY ARTERY BYPASS

Bypass surgery is the most common procedure used to treat critically narrowed or blocked coronary arteries, a cause of heart attack or uncontrollable angina. The operation involves the use of one or more of the patient's arteries or veins, often either the saphenous vein or the internal mammary artery, to bypass the blocked section. A heart-lung machine takes over temporarily, so the surgeon can operate on a nonbeating heart.

Site of chest incision

Saphenous vein

1 After general anesthesia, the chest is opened by a making a central incision down the breastbone. The pericardium is opened to expose the heart. Incisions have been made in the leg to allow a part of the saphenous vein to be removed.

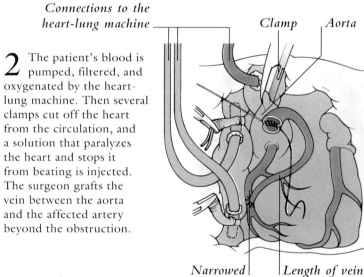

Connections to the heart-lung machine

Clamp *Aorta*

2 The patient's blood is pumped, filtered, and oxygenated by the heart-lung machine. Then several clamps cut off the heart from the circulation, and a solution that paralyzes the heart and stops it from beating is injected. The surgeon grafts the vein between the aorta and the affected artery beyond the obstruction.

Narrowed coronary artery *Length of vein*

Right coronary artery

Aorta

3 More than one coronary artery bypass can be carried out at the same time. A triple bypass is seen at left. After the surgeon has grafted the vein(s), the clamps are removed; if the heart does not start beating, electrical stimulation is used. The heart-lung machine is disconnected.

Left circumflex artery

Blockages

Blockage

Left anterior descending artery

HEART STRUCTURE DISORDERS

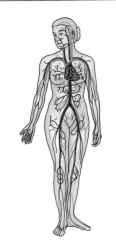

HEART STRUCTURE DISORDERS ARE COMMON and can affect people of any age. These disorders may be congenital or acquired through infections that affect heart muscle, such as rheumatic fever or endocarditis. They may even result from diseases, such as syphilis, or from a myocardial infarction (heart attack). Artificial valves and microsurgical techniques can repair many disorders.

HEART VALVE DISORDERS

Effective pumping by the heart depends on all four valves operating properly. There are two main types of disorder that may affect one or more of the valves. Stenosis, a condition in which a valve outlet is too narrow, may be congenital or due to rheumatic fever or even aging. In incompetence, or insufficiency, the cusps of the valve do not meet so that the valve fails to close properly. It may be caused by a prolapsed valve, coronary heart disease, or an infection.

HEART MURMURS

Normally, blood flow in the heart cannot be heard. A murmur commonly results from turbulent blood flow through a defective valve. So-called "innocent" heart murmurs may occur in childhood or can be associated with increased cardiac output that occurs in anemia or pregnancy. These are intermittent and fainter than those associated with structural abnormalities.

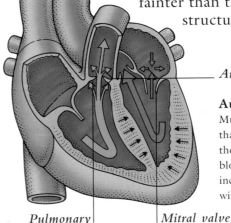

Areas of turbulence

Audible defects
Murmurs are produced by blood that rushes around and through the cusps of a stenosed valve, or blood that leaks back through an incompetent valve and collides with onrushing blood.

Pulmonary valve stenosis

Mitral valve incompetence

LOWER PRESSURE

Normal aortic valve
As the ventricles contract, high pressure forces the valve open, allowing blood through (left). When the ventricles relax and fill with blood, the pressure is higher on the other side of the valve. This causes the valve to close tightly, preventing blood from flowing backward (right).

HIGHER PRESSURE

HIGHER PRESSURE

LOWER PRESSURE

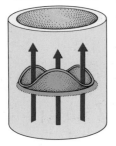

LOWER PRESSURE

Stenosis
A narrowed valve allows less blood through, so the heart must pump harder to maintain blood flow.

HIGHER PRESSURE

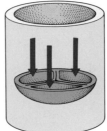

HIGHER PRESSURE

Incompetence
The leakage of blood back into the ventricles can occur when the cusps of a valve fail to close completely.

LOWER PRESSURE

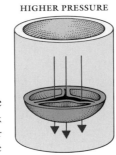

REPLACEMENT VALVES

Valves made from metal and plastic may use a caged-ball or tilting disk mechanism. Both are long-lasting but seem to cause blood clots, so patients may need anticoagulant drugs. Valves made from animal or human tissues are less durable but do not cause clots.

CAGED-BALL VALVE

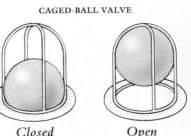

Closed   *Open*

TILTING DISK VALVE

Closed   *Open*

Modified tissue valve
Tissue valves may be taken from a pig (shown above), a human after death, or may be made of tendons from a patient's own body.

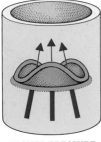

Incompetent aortic valve
Blood is pumped from the left ventricle into the aorta and from there to the rest of the body. If the aortic valve leaks (blue color), the blood flows back into the heart. The left ventricle strains to pump this blood pool and its muscle wall thickens.

CONGENITAL DEFECTS

If a woman contracts a viral infection (particularly rubella) during early pregnancy, the fetal heart may fail to develop normally. Congenital defects can also occur if a pregnant woman has diabetes that is not well controlled, or if the child has Down's syndrome (caused by a chromosomal abnormality). Ultrasound screening has made it possible to recognize and plan for the treatment of some heart defects before birth.

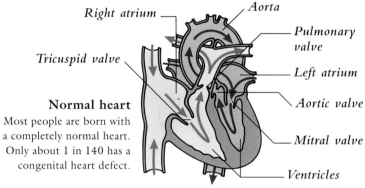

Normal heart
Most people are born with a completely normal heart. Only about 1 in 140 has a congenital heart defect.

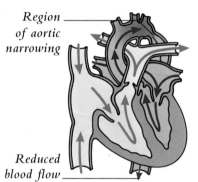

Coarctation of the aorta
In this defect a discrete area of the aorta is narrowed, which results in reduced blood flow to the lower body. An infant may be pale and find it difficult to breathe or eat. Corrective surgery is needed.

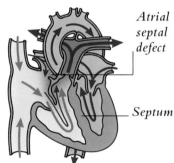

Atrial septal defect
A hole in the septum, or wall, that separates the atria may allow too much blood to flow to the lungs. Often seen in children with Down's syndrome, these defects may need surgery when a child is age 4 or 5.

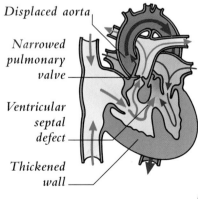

Tetralogy of Fallot
This is a combination of four structural defects: a hole in the septum between the ventricles, a thickened right ventricular wall, a displaced aorta, and a narrowed pulmonary valve. Symptoms are a bluish cast to the skin and a heart murmur. Surgery is beneficial.

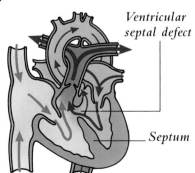

Ventricular septal defect
A hole in the septum separating the ventricles means that blood from the left ventricle pumps into the right. Too much blood is pumped at high pressure into the lungs. Although a small hole may often close as a child grows, larger holes require surgery.

OPERATION

SURGICAL REPAIR OF A VENTRICULAR SEPTAL DEFECT

Although 60 percent of ventricular septal defects heal on their own, others may be life-threatening and need surgical repair. The size of the septal hole will affect the child's symptoms, which may include a heart murmur, cyanosis (a bluish cast to the skin), and difficulty in breathing due to high blood pressure in the lungs. The repair is usually performed before a child is 2 years old.

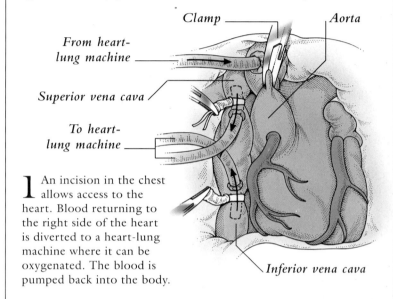

1 An incision in the chest allows access to the heart. Blood returning to the right side of the heart is diverted to a heart-lung machine where it can be oxygenated. The blood is pumped back into the body.

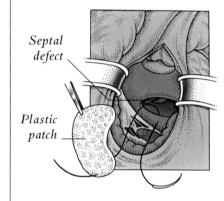

2 The pumping of the heart is stopped and its chambers emptied of blood. An incision is made in the wall of the right ventricle. The surgeon stitches a small plastic patch over the septal defect, and then stitches the incision in the ventricle wall tightly together. After the heart has been restarted, the patient is taken off the heart-lung machine.

3 Closing the hole restores normal circulation so that contractions of the left ventricle pump blood throughout the body and contractions of the right ventricle pump blood to the lungs. After the operation, the patient's symptoms and quality of life improve markedly.

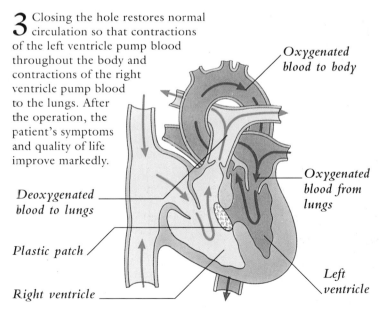

HEART RATE *and* RHYTHM DISORDERS

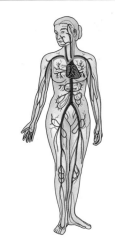

A NORMAL HEART BEATS STEADILY about 60 to 100 times per minute, although this rate rises during exercise or stress. If the rhythm becomes erratic, or the rate becomes unusually slow or fast, the condition is known as an arrhythmia. A common cause of such disorders is damage caused by coronary heart disease; however, almost all heart disorders, as well as hypothyroidism, can be factors.

DIAGNOSING THE CAUSE

Heartbeat occurs when electrical impulses are initiated by "pacemaker" cells that are located at the top of the heart. These impulses spread through the atria, and then travel along conducting fibers to the ventricles, rhythmically stimulating contractions. An irregular pattern or abnormal rate sometimes produces symptoms such as dizziness and fainting, palpitations, breathing difficulties, and chest pain.

Recording an ECG

An electrocardiogram (ECG) detects the site and type of arrhythmia. Electrodes placed on the chest, the wrists, and the ankles connect via wires to a machine. A tracing of the heartbeat is produced.

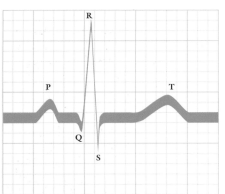

Reading an ECG

Each deflection represents a stage in the passage of impulses through the heart.

P: *Atria contract*
Q: *Impulses slow down through conducting fibers*
R: *Ventricles contract (positive charge)*
S: *Negative charge*
T: *Ventricles return to resting state*

|◄ 0.2 SEC ►|

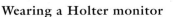

AMBULATORY ECG

Because an ECG is performed for a brief period of time only, a normal reading may be obtained even when an arrhythmia exists. Intermittent abnormalities in heartbeat may be detected by a Holter monitor worn over a 24-hour period. The patient also records when any symptoms, such as dizziness or pain, occur.

Wearing a Holter monitor

The patient continues with normal activities while wearing a monitor on a belt around the waist.

ABNORMAL PATTERNS

Arrhythmias are grouped into tachycardias, in which the heart beats faster than 100 times per minute, or bradycardias, in which the rate drops below 60 beats per minute. Patterns can also be classified by rhythm (regular or irregular), the part of the heart where the impulse originates, and the part of the heart affected. Arrhythmias are commonly caused by coronary heart disease, stress, caffeine, and some types of medication.

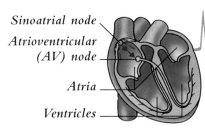

Sinoatrial node
Atrioventricular (AV) node
Atria
Ventricles

Sinus tachycardia

This regular but rapid pattern (100 beats or more per minute) can occur during exercise, stress, or a fever, or as a response to stimulants such as caffeine.

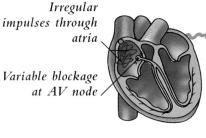

Irregular impulses through atria
Variable blockage at AV node

Atrial fibrillation

Random and extremely rapid atrial contractions (more than 300 to 500 beats per minute) trigger an irregular pattern in the ventricles.

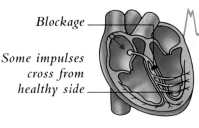

Blockage
Some impulses cross from healthy side

Bundle-branch block

Damage to a branch of the heart's bundle of conducting fibers impedes the passage of impulses. The rate slows if all of the branches are blocked.

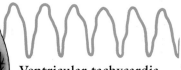

Damaged muscle
Slowed conduction through damaged area causes circular impulses

Ventricular tachycardia

Damaged heart muscle causes abnormal electrical activity. This triggers very rapid but ineffective ventricular contractions (more than 140 beats per minute).

REGULATING HEART RHYTHM

Arrhythmias are commonly caused by an inadequate blood flow to the cells that stimulate cardiac contractions. There are several types of pacemaker available that regulate heart action and correct these abnormal patterns. Occasionally, defibrillation or cardioversion (passing an electric shock through the heart) is used, or drugs may be prescribed.

PACEMAKERS

A pacemaker is a battery-operated device that can send timed electrical impulses to the heart to make it contract regularly. There are several types: some supply constant impulses at a predetermined rate, while others are activated only when the heart is not beating normally. Insertion of a pacemaker is usually carried out using local anesthesia.

Dual-chamber

With this device, the atria and the ventricles are served by separate wires that adjust heart rhythm automatically.

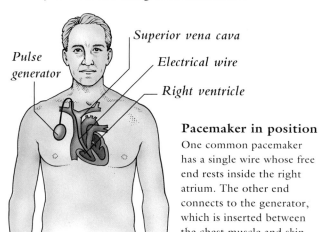

Superior vena cava

Pulse generator

Electrical wire

Right ventricle

Pacemaker in position

One common pacemaker has a single wire whose free end rests inside the right atrium. The other end connects to the generator, which is inserted between the chest muscle and skin.

Programmable

This type of pacemaker can be programmed by sending electromagnetic signals through the skin.

IMPLANTABLE DEFIBRILLATORS

To stabilize ventricular tachycardia, a potentially fatal arrhythmia, an implantable defibrillator may be used. The device is a small electric generator that has three wires. When it detects a racing heartbeat, an electric shock is produced. This stops the heart for a split second so that the sinoatrial node can restart normal heart rate.

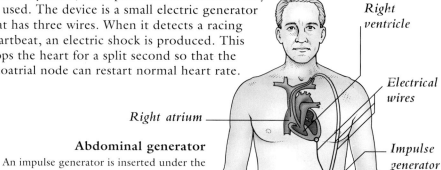

Right ventricle

Right atrium

Electrical wires

Impulse generator

Abdominal generator

An impulse generator is inserted under the skin of the abdomen. A wire is connected to the left lower heart surface and wires are fed into the right atrium and ventricle.

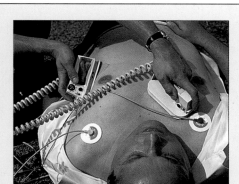

EMERGENCY DEFIBRILLATION

A heart attack sometimes brings on a ventricular fibrillation, which must be treated immediately. Two metal paddles are placed on the chest and then deliver an electric shock to the heart muscle. The two round electrodes shown here connect to an ECG machine in order to monitor heart activity.

CALCIUM-CHANNEL BLOCKERS

These drugs slow the passage of impulses through the heart muscle and so help correct some types of arrhythmia. Their effect results from their action in stopping the flow of calcium into the heart muscle fibers. Although calcium-channel blockers can have dramatic results in improving an arrhythmia, they cannot cure the underlying disorder.

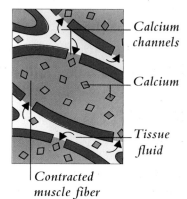

Before treatment

Calcium flows freely from the fluid that bathes cells through the membranes of cardiac muscle fibers. Calcium combines with a protein within the fiber, triggering muscle contractions.

Calcium channels

Calcium

Tissue fluid

Contracted muscle fiber

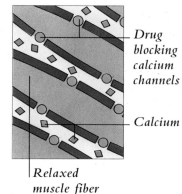

After treatment

The flow of calcium through the cardiac muscle membranes is blocked by the action of the drug. Muscle fibers relax, inhibiting passage of impulses through the heart and slowing the heart rate.

Drug blocking calcium channels

Calcium

Relaxed muscle fiber

CARDIAC GLYCOSIDES

Originally produced from the leaves of the foxglove plant and commonly known as digitalis drugs, the cardiac glycosides slow the heart rate by slowing the nerve impulses through the heart muscle; they also make contractions of the ventricles stronger.

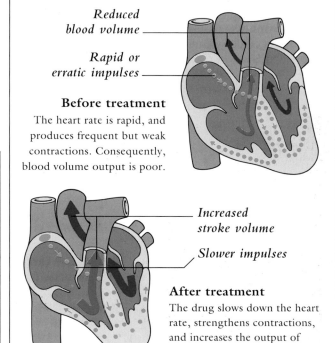

Reduced blood volume

Rapid or erratic impulses

Before treatment

The heart rate is rapid, and produces frequent but weak contractions. Consequently, blood volume output is poor.

Increased stroke volume

Slower impulses

After treatment

The drug slows down the heart rate, strengthens contractions, and increases the output of blood volume per heartbeat.

HEART MUSCLE DISEASE *and* HEART FAILURE

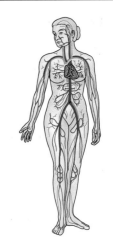

HEART PROBLEMS ARE MOST OFTEN the result of hypertension, a congenital defect, or coronary heart disease. Sometimes, however, they are caused by a disease of the heart muscle itself, or of the saclike pericardium surrounding the heart. All these disorders, if they are long-standing or particularly severe, can lead to heart failure and reduce the heart's pumping ability.

HEART MUSCLE DISEASE

Inflammation of the heart muscle, called myocarditis, is usually caused by a viral infection, but may also be the result of rheumatic fever or exposure to drugs, chemicals, or radiation. Many people recover without treatment. Noninflammatory heart muscle disease, or cardiomyopathy, may be due to a genetic disorder, a vitamin or mineral deficiency, or excessive alcohol. The three main types are shown below.

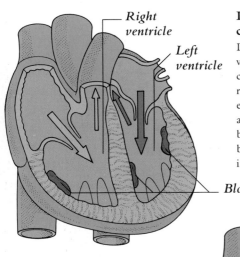

Right ventricle

Left ventricle

Blood clots

Dilated cardiomyopathy

Dilation (widening) of the ventricles causes the heart to contract less forcefully. As a result, not enough blood is ejected with each heartbeat and less oxygen reaches the body tissues. In some cases blood clots may form on the inner walls of the heart.

Hypertrophic cardiomyopathy

This type of cardiomyopathy is usually inherited, although its cause is still not known. Overgrowth of heart muscle fibers causes thickening, especially in the left ventricle and the septum.

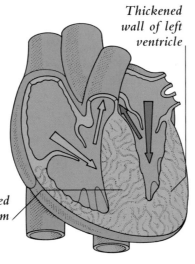

Thickened wall of left ventricle

Thickened septum

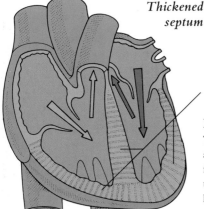

Rigid ventricular walls

Restrictive cardiomyopathy

The walls of the ventricles become abnormally rigid and do not allow for normal filling. The disease is usually caused by scarring of the lining of the heart, or by deposits of iron or protein.

PERICARDITIS

Inflammation of the pericardium, the membranous bag surrounding the heart, is usually due to a viral infection or a heart attack. It may also occur as a complication of rheumatic fever, cancer, tuberculosis, kidney failure, an autoimmune disease, or an injury from a penetrating wound. Symptoms such as fatigue, breathlessness, and fever may be partially relieved by anti-inflammatory drugs or, in some cases, surgery.

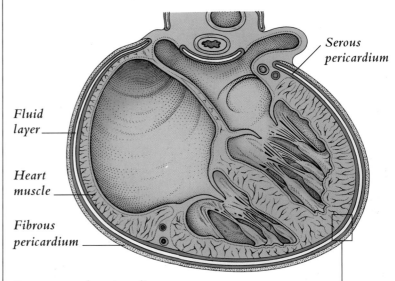

Serous pericardium

Fluid layer

Heart muscle

Fibrous pericardium

Structure of pericardium

The pericardium has two layers. The outer layer, the fibrous pericardium, is tough and inelastic. The inner layer is the serous pericardium. The layers are separated by a thin film of lubricating fluid secreted by the serous pericardium.

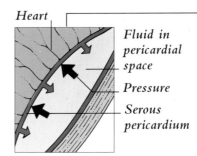

Heart

Fluid in pericardial space

Pressure

Serous pericardium

Pericardial effusion

An inflamed serous pericardium may produce too much fluid, which can compress the heart and interfere with pumping.

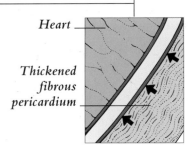

Heart

Thickened fibrous pericardium

Constrictive pericarditis

Inflammation sometimes scars the pericardium, causing it to thicken and shrink. The heart may not be able to fill between contractions.

HEART FAILURE

Heart failure is the inability of the heart to pump enough blood to meet the needs of the lungs and body tissues effectively. Symptoms of this disease include coughing, fatigue, edema (fluid in tissues), and breathlessness, and are related to whether one side of the heart is affected to a greater degree. A variety of drugs act to widen blood vessels, prevent fluid buildup, and strengthen heart contractions.

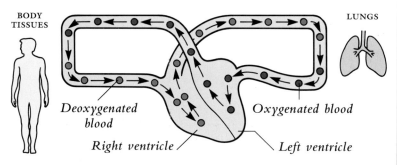

BODY TISSUES
LUNGS
Deoxygenated blood
Oxygenated blood
Right ventricle
Left ventricle

Normal circulation
Both sides of the heart normally pump out the same amount of blood after each beat, and take in the same amount as they pump out. There is no blood congestion anywhere in the circulation.

RIGHT-SIDED HEART FAILURE

LEFT-SIDED HEART FAILURE

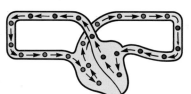
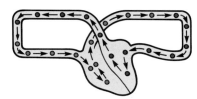

1 A diseased right side of the heart, possibly due to a valve defect or a respiratory disorder, is unable to pump blood to the lungs as fast as it returns from the body via the veins.

1 The left side of the heart may fail to pump blood out to the body as fast as it returns from the lungs through the pulmonary veins. This may be due to a heart structure defect or an arrhythmia.

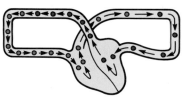
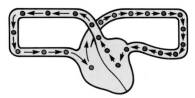

2 Blood begins to pool in the right side of the heart. The veins, continuing their attempt to return blood, become congested.

2 Blood that cannot reenter the circulation begins to back up into the pulmonary veins and the lungs, causing congestion.

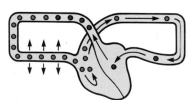
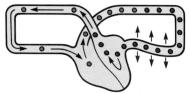

3 Increased congestion causes raised pressure in the veins, which forces fluid out through the capillary walls. The ankles swell and the liver becomes swollen and tender as fluid accumulates.

3 The pressure from increasing congestion causes fluid to collect in the lungs, preventing efficient transfer of oxygen to the blood. This leads to coughing, severe breathlessness, and fatigue.

OPERATION

HEART TRANSPLANTATION

This major operation is usually reserved for those people under 60 whose progressive heart failure has not been successfully treated by medications or previous surgery. The main risks are infection and the recipient's rejection of the donor heart. To prevent rejection from occurring, immunosuppressant drugs are commenced before the operation; they must be taken for the rest of the patient's life, and may have serious side effects.

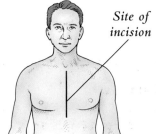

1 The patient is anesthetized, and the surgeon makes an incision in the patient's chest. The sternum (breastbone) is split apart, and the pericardial membranes are cut open to expose the defective heart.

Site of incision

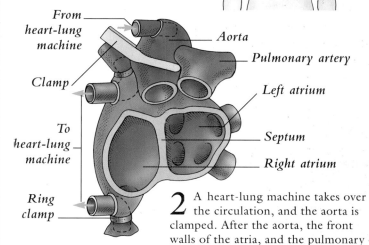

From heart-lung machine
Aorta
Pulmonary artery
Clamp
Left atrium
To heart-lung machine
Septum
Right atrium
Ring clamp

2 A heart-lung machine takes over the circulation, and the aorta is clamped. After the aorta, the front walls of the atria, and the pulmonary artery have been cut, the diseased heart is removed.

3 The back walls of the atria remain in place, and the donor heart is stitched to their free edges as well as to the septal wall (not seen in this illustration).

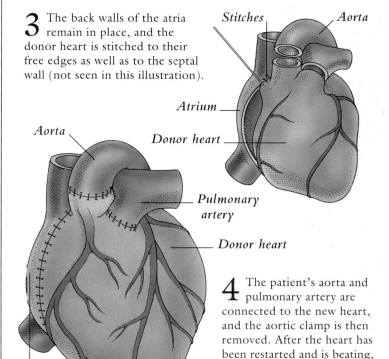

Stitches
Aorta
Atrium
Donor heart
Aorta
Pulmonary artery
Donor heart
Atrium

4 The patient's aorta and pulmonary artery are connected to the new heart, and the aortic clamp is then removed. After the heart has been restarted and is beating, the patient is disconnected from the heart-lung machine.

CIRCULATORY DISORDERS

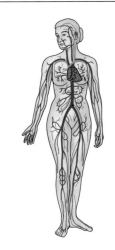

BLOOD FLOW MAY BE IMPEDED OR BLOCKED as it travels through the arteries and veins. Normally, blood flow varies in response to factors such as exercise, digestion, and changes in external temperature, all of which can influence the diameter of the blood vessels. These responses can be impaired by structural defects such as weakened or thickened arterial walls or malfunctioning valves. Blood flow may also be hindered by obstructions such as blood clots or fatty plaques. Treatment is required to forestall strokes or heart attacks.

THROMBOSIS

When a blood vessel is cut, clotting occurs to prevent blood loss. A blood clot can form in an intact vessel, an artery that contains fatty deposits, or an artery or vein that is inflamed. These clots, known as thrombi, cause pain and also damage the tissues served by the vessel; they can be life-threatening in some cases.

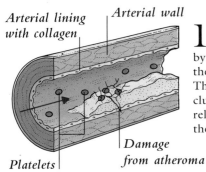

Arterial lining with collagen   *Arterial wall*

Platelets   *Damage from atheroma*

1 When a blood vessel wall is damaged, for example by an atheroma, platelets in the blood adhere to the site. This causes the platelets to clump together and start to release chemicals that initiate the process of blood clotting.

2 The chemicals help to convert fibrinogen, a soluble blood protein, into strands of insoluble fibrin. These strands trap platelets and blood cells to form a clot.

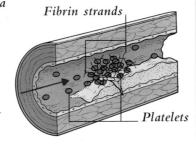

Fibrin strands

Platelets

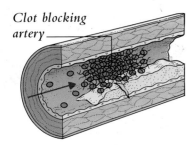

Clot blocking artery

3 Blood flow slows down as the fibrin enmeshes more platelets. The clot increases in size and may eventually block the artery. Particles may break away to become emboli.

EMBOLISM

An embolism occurs when particles traveling in the bloodstream lodge in a vessel away from their point of origin. These fragments may be either parts of a thrombus, or a whole thrombus that has detached from its original site. Emboli may also be composed of atheromatous debris, cholesterol crystals, air, or fat from the marrow of fractured bones. If drugs do not inhibit or dissolve clots, surgery may be needed.

Pulmonary embolism
An embolus may travel from the veins of the leg or pelvis, through the heart, to a pulmonary artery. It may lodge there, thus creating an obstruction that deprives lung tissue of vital oxygen. Known as a pulmonary embolism, the condition is life-threatening.

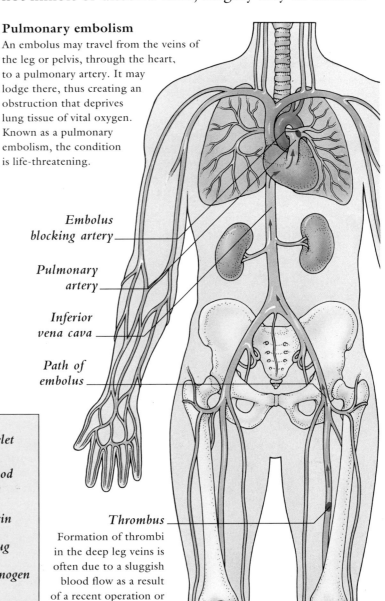

Embolus blocking artery

Pulmonary artery

Inferior vena cava

Path of embolus

Thrombus
Formation of thrombi in the deep leg veins is often due to a sluggish blood flow as a result of a recent operation or prolonged immobility.

ANTICOAGULANTS

Anticoagulants such as warfarin and heparin in part inhibit the conversion of fibrinogen into the fibrin strands that trap platelets. They are given to people whose blood has an increased tendency to clot, such as those with atrial fibrillation or who have just had surgery. These drugs do not dissolve clots, but they stop further growth and prevent new clots from forming.

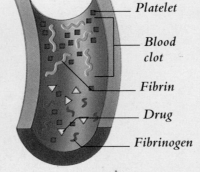

Platelet

Blood clot

Fibrin

Drug

Fibrinogen

ANEURYSM

An aneurysm is an abnormal swelling in a weakened arterial wall. The defect may be due to disease or an injury, or it may be congenital. Although aneurysms may occur anywhere in the body, they most often affect the aorta. In older people, they develop more frequently in the abdominal aorta, at a point just below the kidneys. They may be treated surgically.

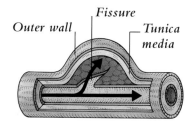

Outer wall — *Fissure* — *Tunica media*

Dissecting aneurysm

A fissure, or split, in the inner lining of an arterial wall allows blood to push through and press against the tunica media (middle wall) and the outer arterial wall. The artery swells and its walls thin and may burst.

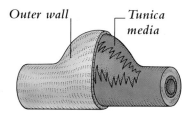

Outer wall — *Tunica media*

Common aneurysm

A common, or saccular, aneurysm forms when the muscle fibers of the tunica media are weak. When blood flows through the affected area, the arterial wall bulges and may rupture.

OPERATION

VARICOSE VEINS

Defective valves in lower-leg veins can cause blood to drain backward and collect in the superficial veins that are nearest the surface of the skin. These veins may become swollen, twisted, and painful, leading to skin ulcers or swelling of the feet. Surgery may be needed.

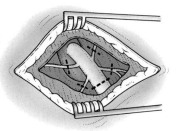

1 A small incision is made in the groin. This exposes the great saphenous vein and its four branches. All are tied and then cut to stop blood flow.

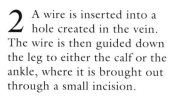

2 A wire is inserted into a hole created in the vein. The wire is then guided down the leg to either the calf or the ankle, where it is brought out through a small incision.

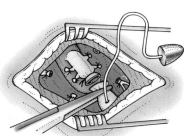

3 The wire has a specially designed, thick top end. This is tied securely to the vein, and the groin incision is closed.

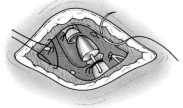

4 The vein is "stripped," or removed, by pulling on the wire exiting from the lower incision. The incision is closed and the leg bandaged.

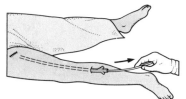

HYPERTENSION

Blood pressure creates the force with which blood flows through the vessels. Hypertension is persistent, abnormally elevated arterial blood pressure. It alone causes no symptoms, but increases the risk of stroke, heart attack, and kidney failure. Blood pressure may be recorded in millimeters of mercury (mmHg) by using a sphygmomanometer.

Blood pressure readings

Normal blood pressure in a healthy young adult is about 110/75mmHg. The first number is the systolic pressure, taken just after the ventricles contract when the pressure is greatest. The second is the diastolic pressure, taken when the ventricles relax and the pressure is lowest.

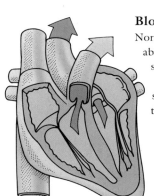

SYSTOLE: VENTRICLES CONTRACT AND FORCE BLOOD OUT

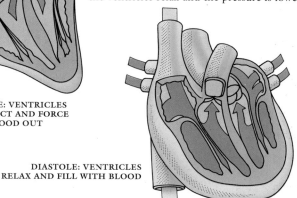

DIASTOLE: VENTRICLES RELAX AND FILL WITH BLOOD

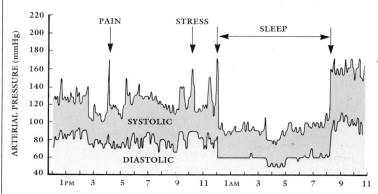

Variations in a 24-hour period

The chart above shows how blood pressure varies greatly in response to various stimuli, such as pain or stress. Variations of this kind are normal.

TREATMENT

Low-sodium, low-fat diets and lifestyle changes to reduce smoking and stress are recommended, and drugs may be prescribed. By inhibiting reabsorption of water and salts, diuretics increase urine excretion. Less water in the blood reduces the heart's workload and lowers blood pressure.

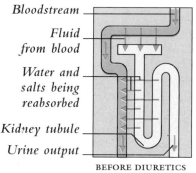

Bloodstream
Fluid from blood
Water and salts being reabsorbed
Kidney tubule
Urine output

BEFORE DIURETICS

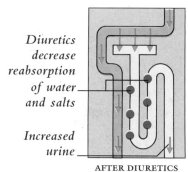

Diuretics decrease reabsorption of water and salts
Increased urine

AFTER DIURETICS

CHAPTER SEVEN

C H A P T E R   7

The IMMUNE SYSTEM

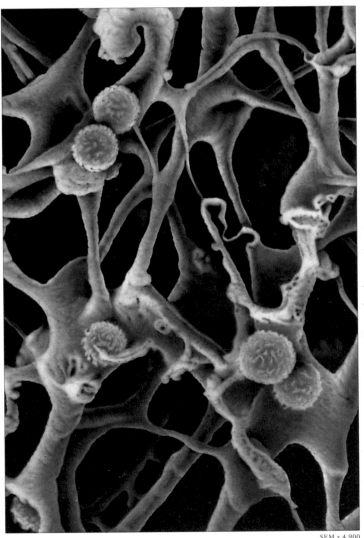

Lymphatic tissue from
inside a lymph node

SEM x 4,900

INTRODUCTION

The body has its very own security force – the immune system – that patrols the body and guards it against invasion from outside and subversion from within. A newborn baby is protected by the antibodies supplied by its mother, partly in breast milk. Soon after birth, the immune defenses take up their lifelong task of recognizing invading disease organisms, such as viruses and bacteria. Certain cells of the immune system, the B and T lymphocytes, have memories, which enable them to respond quickly to any infection they have previously encountered. Immunization is a way of stimulating the body to form more of these defense memories; it plays a crucial role in controlling infectious diseases, such as poliomyelitis and diphtheria. The internal surveillance system is also alert to abnormal cell division within body tissues, and a healthy immune system will eliminate potential cancers at an early stage. Impairment of the immune system by aging, or by diseases such as AIDS, inhibits this response, increasing the risk of malignancy. Some medications, such as the corticosteroid drugs that are given to some patients suffering from chronic diseases, can harm the immune system. Malfunctioning of the immune system may also cause it to turn on itself, resulting in one of the autoimmune disorders, such as rheumatoid arthritis, in which the body selectively attacks its own tissues.

A bacterium

Virus particles escape after attacking a cell

THE IMMUNE SYSTEM

DEFENSES AGAINST INFECTION

A HEALTHY BODY CAN DEFEND ITSELF against most invading organisms that may cause infection or disease. There are two types of defense, innate and adaptive. Innate defenses include mechanical barriers, such as the skin, and chemical defenses such as an antibacterial enzyme in tears. The adaptive system is based on specialized white blood cells called lymphocytes, which respond to invasion of the body by microorganisms. B cells produce proteins called antibodies, which circulate in the blood and attack specific disease-causing organisms; T cells attack the organisms directly. These cells can retain a memory of earlier infections and respond quickly to further attacks.

LYMPH AND LYMPH VESSELS

Lymph is a clear, watery liquid that begins as fluid flowing between cells. This so-called interstitial fluid is not called lymph until it drains into the network of lymph capillaries located in the tissue spaces. From lymph capillaries, lymph flows into larger vessels called lymphatics (seen at right), which are studded with filters called nodes. Lymph is not pumped, but is moved when lymph vessels are compressed by surrounding muscles as they contract during movement.

Supratrochlear node

Lacrimal glands
These glands produce tears that contain a protective enzyme.

Adenoid

Tonsils
These two glands and the adenoids produce antibodies against ingested or inhaled organisms.

Salivary glands

Thymus
Stem cells are produced in bone marrow. They then migrate to the thymus, replicate, and develop into T cells.

Axillary nodes

Cisterna chyli
Lymphatics from the lower body converge to form this vessel.

Lateral aortic nodes

Common iliac nodes

Thoracic duct

Subclavian veins
Lymph drains from the upper right part of the body into the right subclavian vein, while lymph from the rest of the body collects in the thoracic duct, draining from there into the left subclavian vein.

Spleen
Some types of lymphocyte mature in and are then stored in the spleen, the largest of the lymph organs.

Stomach
Acid and enzymes secreted here destroy ingested organisms.

Peyer's patch
Clusters of lymph tissue, called Peyer's patches, are found in the lower part of the small intestine.

LYMPH CAPILLARY

Interstitial fluid enters

Open valve governs direction of flow

Valve closed

Overlapping epithelial cells

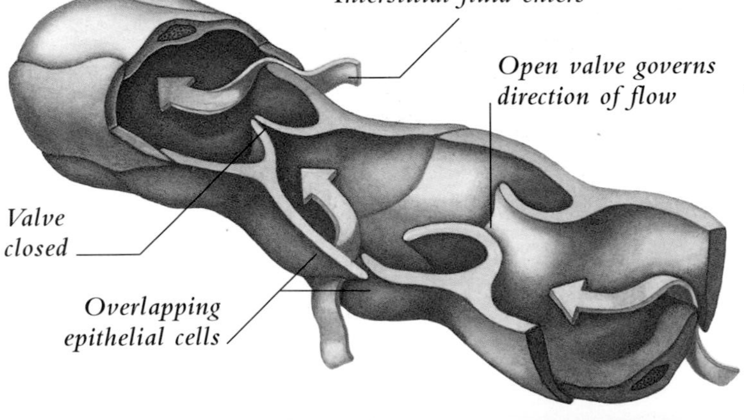

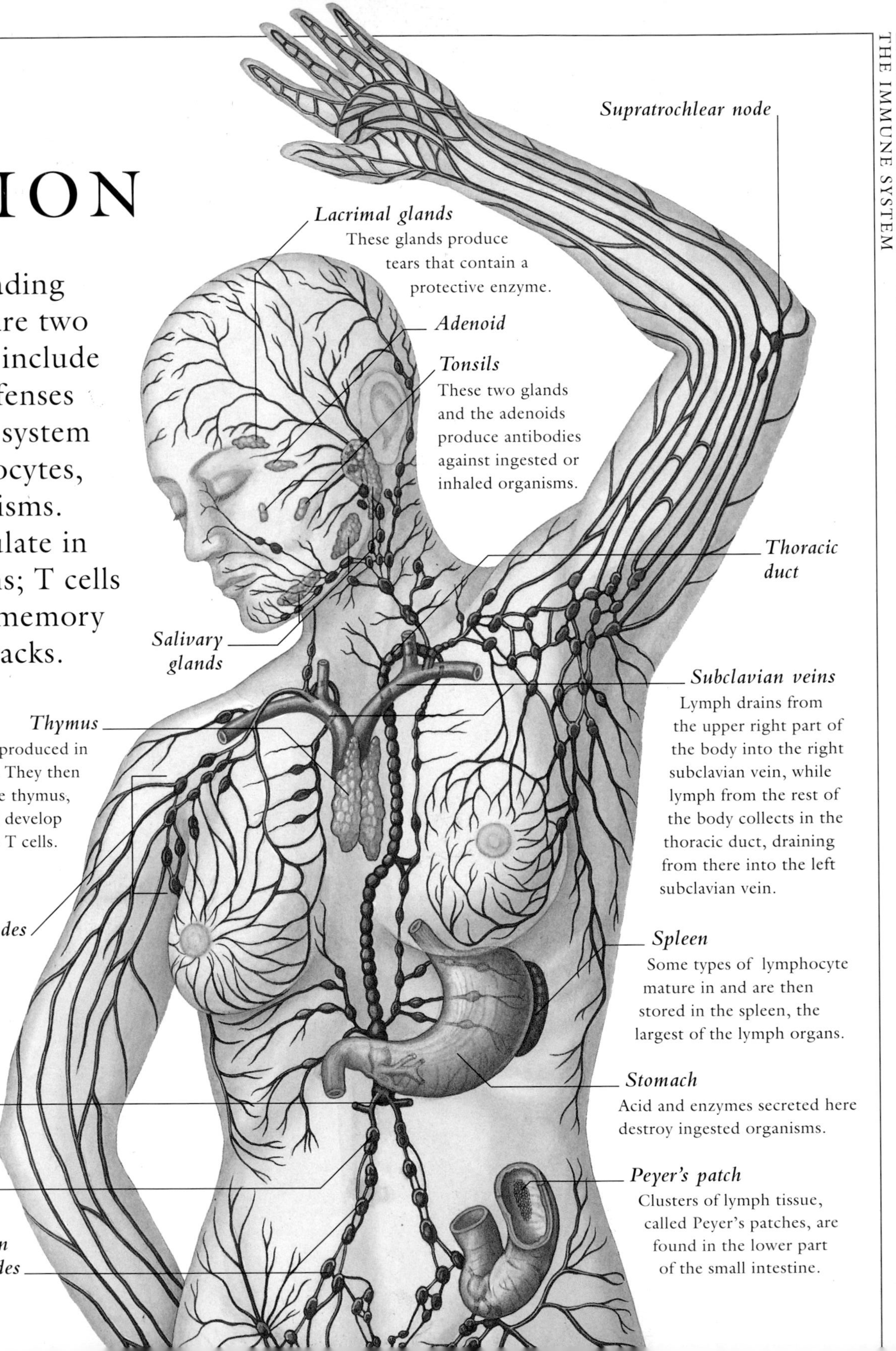

Bone marrow
The lymphocytes begin life as stem cells in the bone marrow. Also generated here are monocytes, the largest of the white blood cells. These migrate from the blood into connective tissues where they develop into scavenger cells called macrophages that ingest bacteria and dead cells.

Popliteal lymph nodes
These nodes drain excess lymph from the legs and feet.

Lymph capillaries
The lymph circulation system is not a closed circuit; instead, capillaries start as blind-ended sacs within tissue spaces before joining larger lymphatics.

Lymphatics
From lymph capillaries, lymph flows into the lymphatics; as the diameter of the vessels increases, the walls become thicker. In the tissue just below the skin, these vessels roughly parallel the path of veins. In the organs, they follow the arteries and may form networks around them.

Skin
The barrier formed by skin is the body's first defense against invading organisms.

External iliac nodes

Deep inguinal nodes

Afferent lymph vessel
A number of vessels transport lymph to a node.

Germinal centers
When an infection occurs, the germinal centers release lymphocytes; as these mature, some move toward the surface of the node, becoming plasma cells that produce antibodies.

Trabecula
Columns composed of fibers divide the node into segments.

Reticular fibers
A meshwork of fibers helps support the cells in the node.

Macrophage

B cell

T cell

LYMPH NODE STRUCTURE

Lymph nodes, also known as lymph glands, are masses of lymph tissue covered by a fibrous capsule. They range up to 0.8in (2cm) in size and contain sinuses (spaces) where many scavenging white blood cells, known as macrophages, ingest bacteria and other foreign matter and debris. The lymph from most tissues or organs crosses one or more lymph nodes to be filtered before draining into the venous bloodstream. Swollen lymph nodes often indicate disease.

Efferent lymph vessel
Filtered lymph is carried away from each lymph node by only one vessel.

Valve

Vein

Artery

Arteriole

Venule

Sinuses
Channels within the sinuses slow down the flow of lymph so that macrophages can ingest bacteria and debris.

Capsule
Lymph nodes are composed of lymph tissue enclosed in a fibrous capsule made of collagen and elastin (types of protein).

Lymphocytes

INFLAMMATORY *and* IMMUNE RESPONSES

IF INVADING INFECTIOUS ORGANISMS BREACH THE SKIN or are not killed by surface chemicals such as enzymes contained in tears or saliva, the inflammatory and immune responses of the body spring into action. Pain, swelling, and fever may be signs of the battle against infection as several types of white blood cell try to prevent infection from spreading.

INFLAMMATORY RESPONSE

Some disease organisms may trigger an inflammatory response in affected tissues. This type of defense is nonspecific: it is not specially tailored to destroy a specific organism, but attacks all invading organisms in the same way. It increases blood flow and brings special cells called neutrophils to the area (seen here are bronchi) to ingest and destroy the organisms.

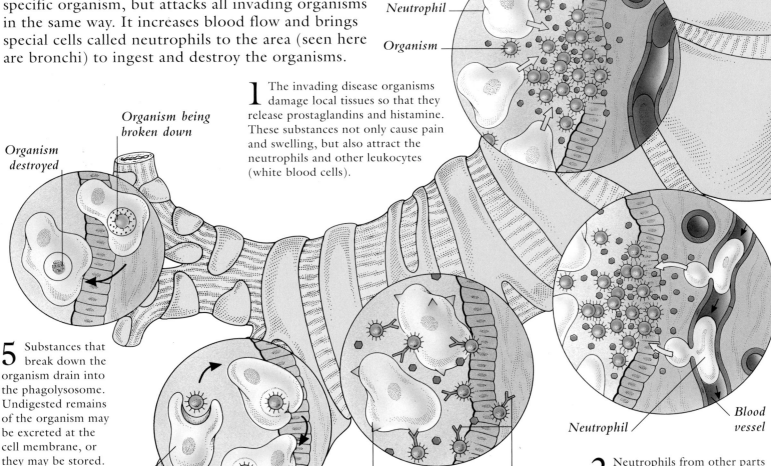

Bronchial lining

Prostaglandin

Neutrophil

Organism

Organism being broken down

Organism destroyed

Neutrophil ingesting organism

Phagolysosome

Receptor　　*Antibodies*

Neutrophil　　*Blood vessel*

1 The invading disease organisms damage local tissues so that they release prostaglandins and histamine. These substances not only cause pain and swelling, but also attract the neutrophils and other leukocytes (white blood cells).

5 Substances that break down the organism drain into the phagolysosome. Undigested remains of the organism may be excreted at the cell membrane, or they may be stored.

4 The neutrophils form pseudopods, or "false feet," that engulf the organism. After it has been ingested, the organism is isolated within a phagolysosome, a small, spherical organelle (a type of cell subunit). This process is known as phagocytosis.

3 Antibodies are specifically created proteins that attach to the invading organisms. The newly arrived neutrophils have receptors so that they recognize the antibodies. Both the antibody and the disease organism then attach to the neutrophils.

2 Neutrophils from other parts of the body are attracted by toxins (poisons) produced by the organisms, as well as by histamine and prostaglandins. Neutrophils squeeze through tiny spaces in the blood vessel walls in order to reach the site of the injured tissues.

SPECIFIC IMMUNE RESPONSES

The rapid, nonspecific inflammatory response may prevent the spread of infection. If infection persists or spreads, however, two types of specific defense, either an antibody or a cellular defense, may be activated. These defenses are called immune responses; they depend on the action of white blood cells, the B and T lymphocytes, and provide protection against future infections.

ANTIBODY DEFENSES

B lymphocytes recognize the foreign proteins, or antigens, of disease organisms since they differ from natural body proteins. Antigens trigger B cells to multiply. Some develop into plasma cells, which secrete antibodies – proteins that attack and destroy only the antigens.

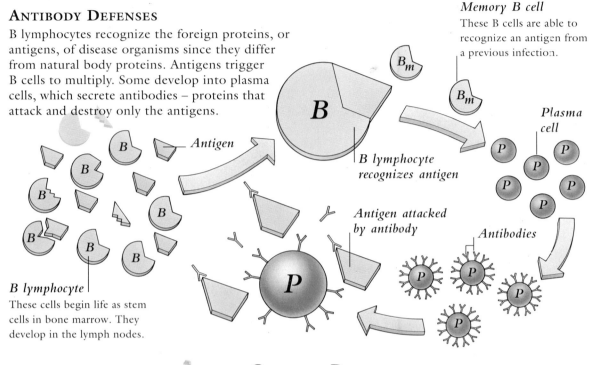

Memory B cell
These B cells are able to recognize an antigen from a previous infection.

B lymphocyte recognizes antigen

Plasma cell

Antigen attacked by antibody

Antibodies

B lymphocyte
These cells begin life as stem cells in bone marrow. They develop in the lymph nodes.

COMPLEMENT SYSTEM

Circulating in the blood are at least 25 inactive proteins. These "complement" proteins are activated by antibodies or certain lymphokines. They help destroy bacteria, neutralize their toxins, and clear away antigen/antibody complexes.

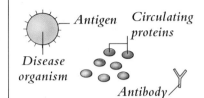

Antigen   Circulating proteins

Disease organism

Antibody

1 In response to the antigens manufactured by a disease organism, such as a virus or a bacterium, the immune system produces specific antibodies.

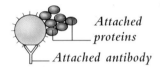

Attached proteins

Attached antibody

2 Antibodies become attached to the antigens, resulting in antigen/antibody complexes that trigger a cascade of complement proteins. These proteins also attach to the disease organism.

CELLULAR DEFENSES

T lymphocytes develop inside the thymus gland. "Killer" T cells react to the remains of destroyed specific antigens, attacking them, as well as any infected cells, with powerful proteins called lymphokines. "Helper" T cells activate B and T cells, while "suppressor" T cells inhibit the response of other cells to the invading antigens.

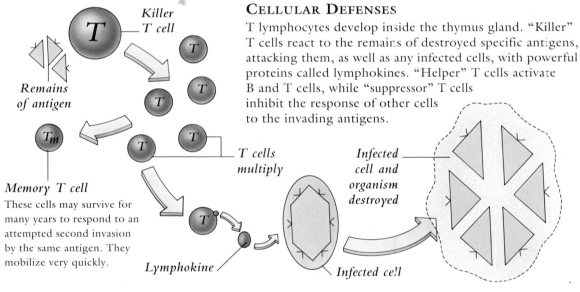

Killer T cell

Remains of antigen

Memory T cell
These cells may survive for many years to respond to an attempted second invasion by the same antigen. They mobilize very quickly.

T cells multiply

Lymphokine

Infected cell

Infected cell and organism destroyed

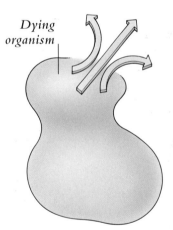

Dying organism

3 The proteins penetrate the cell membrane of the disease organism, which bursts and dies as intracellular fluid rushes in.

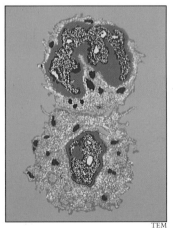

TEM

A T cell in action
The activated cytotoxic ("killer") T lymphocyte at the top of the image has become attached to an infected target cell after having recognized antigens on its surface. As well as attacking specific antigens, T cells also attack cancer cells, slowing tumor growth.

LOCAL AND SYSTEMIC INFECTION

Battles against infection are often simply local skirmishes, such as a swelling in one of the lymph nodes or a slightly infected wound. If local defenses are breached, a global response to spreading infection occurs, signs of which may be fever or a high white blood cell count.

An abscess: isolated war zone
Pus is a collection of dead cells, destroyed bacteria, and dead neutrophils. An abscess is formed when a membrane surrounds pus.

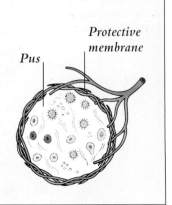

Protective membrane

Pus

INFECTIOUS ORGANISMS *and* IMMUNIZATION

THE HUMAN BODY IS CONSTANTLY INFECTED by organisms – bacteria, viruses, fungi, and protozoa. Some are beneficial, such as the intestinal bacteria that produce some of our vitamins. Others are harmful, causing illnesses that range from mild colds to life-threatening meningitis. The immune system produces antibodies that fight infections, some of which also help prevent reinfection.

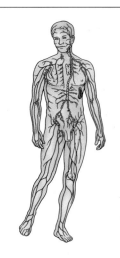

BACTERIA

Bacteria that are present in soil, water, and air can cause serious illnesses such as tetanus, typhoid, and pneumonia. Fortunately, antibiotics are effective against most bacteria; some work by destroying the bacterial cell wall. In addition to antibiotics, some vaccines can prevent certain bacterial infections, such as tetanus and diphtheria.

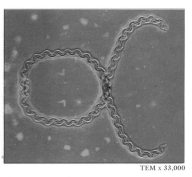

TEM x 33,000

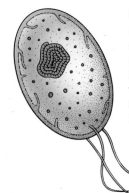

Structure of bacteria
Bacteria are single-celled organisms whose shapes vary greatly. Shown on the far left is a spiral-shaped spirochete, while on the near left an oval-shaped bacillus is seen. Coccal bacteria (not seen) are round.

Flagella help organism to move

HOW BACTERIA DAMAGE TISSUE

Some bacteria adhere to and invade tissue cells, such as the dysentery-causing *Shigella*. Others produce poisonous substances known as toxins. Some toxins are very dangerous: 7lb of botulinum could kill all the people in the world.

1 Toxins may alter certain chemical reactions in cells, so that normal cell function is disrupted or the cell dies. One example is the diphtheria toxin, which damages heart muscle by inhibiting protein synthesis.

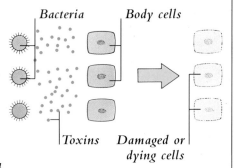

Bacteria   *Body cells*

Toxins   *Damaged or dying cells*

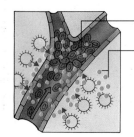

Clot in blood vessel

Toxins from bacteria

2 Some toxins cause blood to clot in small blood vessels. Areas of tissue supplied by these vessels may be deprived of blood and damaged.

3 Toxins can damage the cell walls of tiny blood vessels, and thereby cause leakage. The fluid loss results in decreased blood pressure, and the heart is unable to supply adequate amounts of blood to the vital organs.

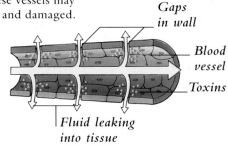

Gaps in wall

Blood vessel

Toxins

Fluid leaking into tissue

RESISTANCE TO ANTIBIOTICS

Since the introduction of penicillin in 1940, many bacteria have developed intricate ways of resisting antibiotics. The most effective mechanism is the rapid transfer of plasmids – small packages of the cell's genetic material, DNA – between bacterial populations that contain resistant genes. Bacteria that receive these plasmids also inherit the resistant genes, and therefore develop the same resistance that the donor bacterium possessed.

The role of plasmids
Plasmids signal bacteria to produce enzymes that can inactivate drugs. They may also stimulate a bacterium to alter its receptor sites (where antibiotics normally bind).

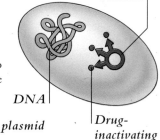

Plasmid

DNA

Drug-inactivating enzyme

Original plasmid

Duplicate plasmid

DNA

DNA

DONOR   *Pilus*

RECIPIENT

Conjugation
Plasmid transfer takes place during a process known as conjugation. The plasmid duplicates itself in a donor bacterium. This copy passes through a tube, the pilus, to the recipient cell.

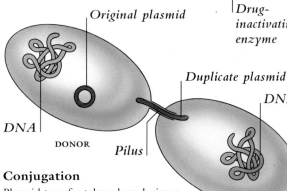

Drug-inactivating enzymes

DNA

DNA

DONOR

RECIPIENT

Drug-resistant strains
Plasmid transfer may produce large populations of bacteria that form enzymes against a range of antibiotics.

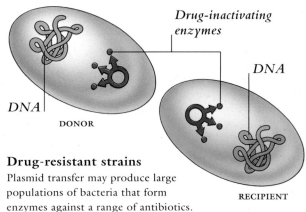

VIRUSES

Infections from these tiny germs cause a wide range of diseases, including colds, hepatitis, and AIDS. A virus cannot be killed by any antibiotic; the body must produce specific antibodies to combat each virus. However, newer antiviral drugs will control some viruses.

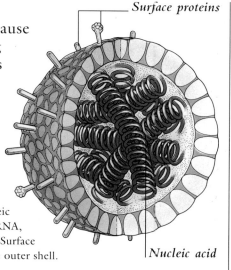

Surface proteins

Nucleic acid

Typical structure

Each virus has a core of nucleic acid, composed of DNA or RNA, and one or two protein shells. Surface proteins, or antigens, stud the outer shell.

HOW VIRAL DISEASES OCCUR

Because viruses have few genes, they are unable to process nutrients nor can they reproduce independently. In order to live and replicate, all viruses must invade host cells, which die or function abnormally. Some viruses provoke the immune system to destroy the body's own normal cells.

HUMAN CELL

Invading virus

Viral nucleic acid

1 Before a virus invades a host cell, its surface proteins must attach to specific receptor sites on the surface. After attaching itself, part or all of the virus penetrates the host cell and sheds its protein coat to release its nucleic acid.

Cytoplasm

Nucleus

Cell membrane

Replicated virus particles

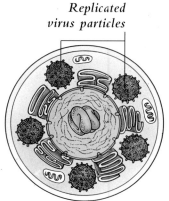

2 The nucleic acid makes copies of itself, using the host cell's raw materials, and sometimes its enzymes, to do so. Replicated nucleic acid generates new virus particles.

DYING CELL

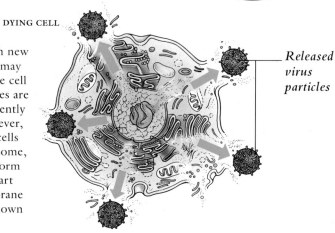

3 The cell swells with new virus particles and may burst and die. When the cell bursts, the virus particles are released; these subsequently infect other cells. However, not all viruses destroy cells as they leave. Instead, some, such as herpesviruses, form buds and take away a part of the host cell's membrane with them. They are known as enveloped viruses.

Released virus particles

ANTIGENIC SHIFT IN INFLUENZA

The three types of influenza viruses are A, B, and C. Each of these changes its structure periodically. The antibodies in a vaccine that defended against a previous viral structure may be ineffective against a new one so that reinfection occurs. This alteration in structure, called the antigenic shift, occurs in the surface proteins (antigens) where antibodies attach.

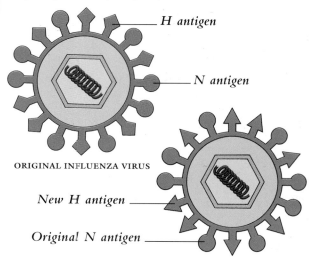

H antigen

N antigen

ORIGINAL INFLUENZA VIRUS

New H antigen

Original N antigen

NEW STRAIN OF VIRUS

TYPES OF VIRUS

Depending on the main constituent of the genetic material, viruses can be broadly classified as RNA viruses or DNA viruses. Further classification is governed by the size, shape, and symmetry of the virus. The most common families are listed below.

FAMILY		TYPE AND DISEASES
Adenoviruses		DNA viruses. Cause infections of the tonsils, respiratory tract, and eyes (such as conjunctivitis).
Papovaviruses		DNA viruses. Initiate benign, or noncancerous, tumors such as warts on the hands and feet.
Herpesviruses		DNA viruses. Cause cold sores, genital herpes, chickenpox, shingles, and glandular fever.
Coronaviruses		RNA viruses. Named for their resemblance to the sun, they cause the common cold.
Picornaviruses		RNA viruses. Cause myocarditis, polio, viral hepatitis, and one form of meningitis.
Retroviruses		RNA viruses that can convert DNA into RNA. They cause AIDS and a type of leukemia.
Reoviruses		RNA viruses. Cause respiratory infections; a group (rotaviruses) causes gastroenteritis.
Orthomyxo-viruses		RNA viruses. Cause influenza, symptoms of which include fever, cough, and a sore throat.
Paramyxo-viruses		RNA viruses. Cause mumps, measles, rubella and respiratory infections such as croup.

PROTOZOA

Protozoa are single-celled, primitive animals, some of which are parasites that can cause disease in humans. Malaria and toxoplasmosis are caused by protozoal parasites, which are particularly common, affecting over a third of the world's population. Parasites employ various mechanisms for evading the immune system. The *Leishmania* parasite, responsible for causing a disease called kala-azar, lives and multiplies within phagocytes – defense cells forming part of the immune system.

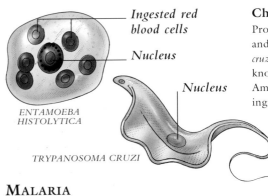

Ingested red blood cells

Nucleus

ENTAMOEBA HISTOLYTICA

Nucleus

Flagellum

TRYPANOSOMA CRUZI

Characteristics of protozoa

Protozoa have a large, well-defined nucleus and no cell wall. Many, such as *Trypanosoma cruzi*, have one or more tail-like appendages known as flagella, which aid movement. Amoebas, such as *Entamoeba histolytica*, can ingest red blood cells and food particles.

MALARIA

Four types of *Plasmodia* protozoan can cause malaria. Spread by the bite of the female *Anopheles* mosquito, most of these have a similar life cycle (shown below) and produce chills and fever. One type, *Plasmodium falciparum*, affects vital organs such as the kidneys and brain, and can be fatal within hours if it is not treated immediately. Because resistance to antimalarial drugs is increasing, scientists are trying to develop vaccines.

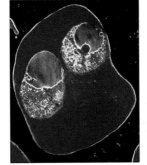

TEM x 5,000

Two *P. falciparum* in a red blood cell

A female Anopheles *mosquito bites, injecting saliva that contains sporozoites, the infective form of the malaria parasite*

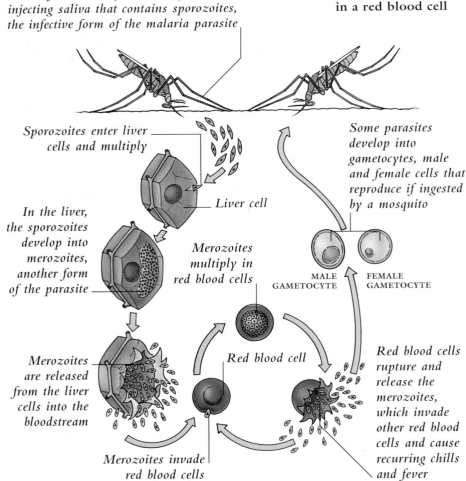

Sporozoites enter liver cells and multiply

In the liver, the sporozoites develop into merozoites, another form of the parasite

Liver cell

Merozoites multiply in red blood cells

Merozoites are released from the liver cells into the bloodstream

Red blood cell

Merozoites invade red blood cells

Some parasites develop into gametocytes, male and female cells that reproduce if ingested by a mosquito

MALE GAMETOCYTE   FEMALE GAMETOCYTE

Red blood cells rupture and release the merozoites, which invade other red blood cells and cause recurring chills and fever

FUNGI

Fungi are simple organisms that scavenge dead or rotting tissue. Some fungi infect humans, causing either harmless superficial diseases of the skin, hair, nails, or mucous membranes, or possibly fatal infections of certain vital organs, such as the lungs.

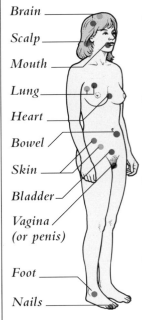

Brain

Scalp

Mouth

Lung

Heart

Bowel

Skin

Bladder

Vagina (or penis)

Foot

Nails

- **Cryptococcosis**
  This infection causes meningitis and pneumonia, and can affect the skin and bones.

- **Aspergillosis**
  This fungus may spread through either ventilation systems or humidifiers to infect the lungs.

- **Dermatophytosis**
  This skin infection, also called tinea, most commonly affects the scalp, feet, or nails.

- **Candidiasis**
  Candida infect the mouth and genitals, and may occur in the heart, bowel, bladder, and brain.

Histoplasmosis

Histoplasmosis is associated with soil that is contaminated by bird droppings. Fungal spores are inhaled by humans and can cause pneumonia. The fungus can spread to and infect organs throughout the body.

HISTOPLASMA CAPSULATUM

TREATMENT

Fungal infections respond in a variety of ways to antifungal drugs. Superficial infections like thrush (oral candidiasis) respond to local applications of antifungal drugs. Deep infections in people with lowered immunity are difficult to cure, and often require prolonged therapy with more toxic drugs.

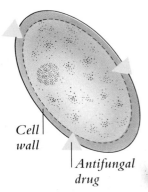

Before drug
Fungal cells, which are harder to treat than bacteria, closely resemble human cells. Drugs must affect fungal cells but not cause excessive damage to body cells in the process.

Cell wall

Antifungal drug

Cell contents leak out

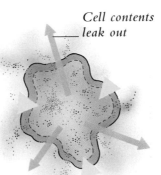

After drug
Most antifungal drugs form channels in the membrane of the fungal cell through which vital cell contents can leak out, causing cell death.

DIAGNOSIS

Diagnosis of infectious diseases is usually confirmed in the laboratory by techniques such as microscopy and culture. Many bacteria are colorless, and consequently require special staining procedures. One commonly used stain is Gram's stain.

Culture

Certain bacteria or fungi are identified when specimens are cultured by growing them on plates until colonies are visible. Viruses are cultured on live cells or eggs.

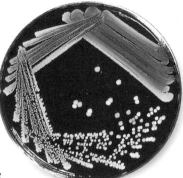

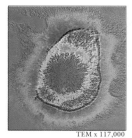

Saturated disk

Antibiotic sensitivity

To find the best treatment, bacterial colonies may be transferred onto plates with disks that are saturated with different antibiotics. No bacteria grow around the most effective antibiotic.

No growth

TESTS FOR VIRUSES

Viruses are too small to see with a light microscope. Instead, infections caused by viruses are diagnosed indirectly by their effects on cells. Some viruses alter the surface of cultured cells, causing them to agglutinate, or stick together.

Single layer of cells

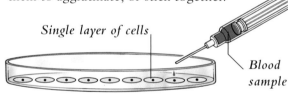

Blood sample

1 A single layer of tissue cells is cultured on a prepared plate. A specimen, such as blood from an infected person, is added to the plate.

Agglutinated cells

2 The presence of a virus can be identified in several ways. In this case, the virus has caused the single layer of tissue cells to clump together.

A virus that agglutinates

Like many viruses, herpes simplex is identified by agglutination, and not by means of microscopy. The high-magnification image reveals the protein coat of the virus (seen here in orange) surrounded by an irregularly shaped membrane.

TEM x 117,000

IMMUNIZATION

Some infectious diseases are common and can recur in the same person. Others occur only once in a lifetime because the immune system can remember the organism and resist subsequent infections. To prevent an epidemic of a serious infectious disease, such as polio, immunization artificially creates a "memory" before the disease can be acquired.

ACTIVE IMMUNIZATION

Vaccine

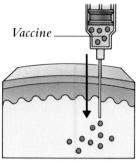

Antibody   *Vaccine*

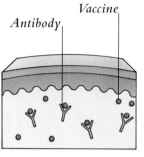

Disease organism

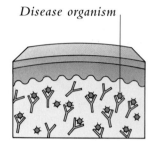

1 A vaccine with dead or harmless living forms of an organism is injected into a person.

2 The vaccine stimulates the immune system to memorize the organism and produce antibodies.

3 In any subsequent infection with this organism, the antibodies stop the infection.

PASSIVE IMMUNIZATION

Antibody

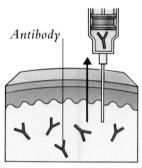

Serum

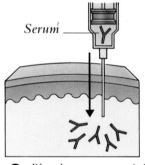

Disease organism

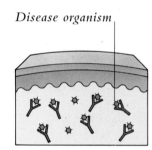

1 Blood with antibodies is taken from humans or animals who have had the infection recently.

2 Blood serum containing antibodies is separated from the blood, processed, and injected.

3 Antibodies either attack a current infection or provide short-term protection.

GENETICALLY ENGINEERED VIRUSES

Genetic engineering is a term that describes a technique that alters the genetic material (DNA) of an organism by inserting the genes from another organism. Viral genes are inserted into the DNA of other organisms. When these organisms multiply, the large amounts of replicated material are used as vaccines or hormones.

Gene inserted into bacterial DNA

Surface antigen gene from DNA

Hepatitis B vaccine

The gene of the surface antigen (protein) of the hepatitis B virus is inserted into a bacterium's DNA. The cell produces viral antigens, which are injected to stimulate an immune response.

Replicated bacterium

Surface antigen

IMMUNE SYSTEM DISORDERS

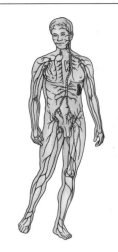

THE IMMUNE SYSTEM PROVIDES DEFENSES for the body against infections and cancers, damage by poisonous and corrosive chemicals, and to some extent, injuries. There are two types of immune system disorder. In allergies and autoimmune diseases, the immune system overreacts; in immunodeficiency diseases, the defense systems are too weak to cope with threats to health.

THE ALLERGIC RESPONSE

Allergy is an inappropriate response by the immune system to a substance that, for most people, is usually harmless. These substances, known as allergens, may be inhaled or swallowed, or they may come into direct contact with the eyes or skin. They may then provoke allergic responses, such as hay fever, asthma, or rashes. Some people are allergic to eggs, milk, and grains.

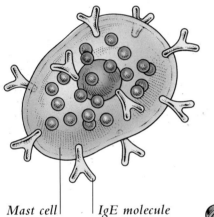

1 Allergens can provoke the immune system to produce an antibody called immunoglobulin E (IgE). The IgE molecules coat the surface of mast cells, which are located in the skin and in the lining of the stomach, lungs, and upper airways.

Mast cell   *IgE molecule*

2 In an allergic person, subsequent exposure to allergens, which may be ingested or inhaled, causes them to bind to the IgE molecules. This is known as cross-linking.

Allergen binds to IgE molecule

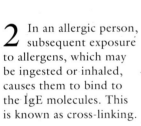

Cross-linked molecule

Granules release histamine and prostaglandins

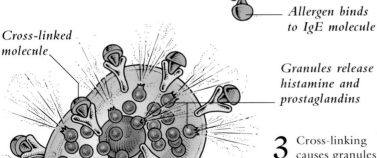

3 Cross-linking causes granules located inside the mast cell to release the inflammatory substances histamine and prostaglandins, which trigger types of allergic response.

AUTOIMMUNE DISORDERS

Sometimes the immune system forms antibodies not against invaders such as bacteria, but against some of the body's own tissues. The mistaken attack may be directed against a particular organ, such as the thyroid gland, or it may cause a more generalized illness (see table below). These diseases become more common in middle age and affect women more often than men.

Vitiligo
Cells called melanocytes produce a skin-darkening pigment called melanin. Vitiligo, an autoimmune disorder, is the result of an absence of these cells. The multiple irregular areas of depigmentation are treated with corticosteroid drugs.

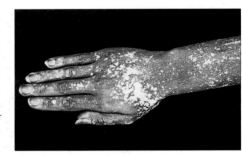

DISORDER	DESCRIPTION
ADDISON'S DISEASE	Damaged adrenal gland cortexes cause low blood pressure and weakness, lowering the body's ability to respond to stress.
INSULIN-DEPENDENT DIABETES MELLITUS	Clusters of pancreatic cells known as islets of Langerhans are destroyed, and are unable to produce sufficient insulin for glucose control.
PERNICIOUS ANEMIA	Pernicious anemia results when the stomach lining cells are attacked and destroyed by antibodies formed elsewhere in the body.
GRAVES' DISEASE	The thyroid becomes overactive and may become enlarged, forming a goiter. There is loss of weight, restlessness, and tremor.
MULTIPLE SCLEROSIS	Damage to nerve fiber coverings causes muscle weakness, disordered sensations, and problems with speech and vision.
MYASTHENIA GRAVIS	Damage to the nerve receptors of the muscles causes muscle weakness, which is especially noticeable in the muscles of the face.
SYSTEMIC LUPUS ERYTHEMATOSUS	Multiorgan damage results in progressive loss of function in the kidneys, lungs, and joints. A typical rash appears over the cheekbones.

AIDS

Acquired immunodeficiency syndrome, or AIDS, is caused by the human immunodeficiency virus (HIV). The virus destroys one type of white blood cell, the CD_4 lymphocyte. As the number of cells declines, the immune system becomes less effective, and death may occur approximately 10 years after infection. HIV is spread by sexual intercourse and contaminated blood.

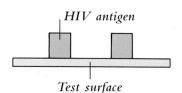

SEM x 16,000

An infected lymphocyte
Small, spherical HIV particles are shown in green on the surface of a CD_4 "helper" lymphocyte. HIV compromises the body's immune system by destroying the CD_4 lymphocytes within which it lives.

ELISA TEST

A blood test for HIV infection looks for the antibodies to the virus, which are easy to detect. The technique employed is the enzyme-linked immunosorbent assay, or ELISA. If antibodies are found, a confirmatory test called the Western Blot is done. If both test results are positive, the person is HIV positive.

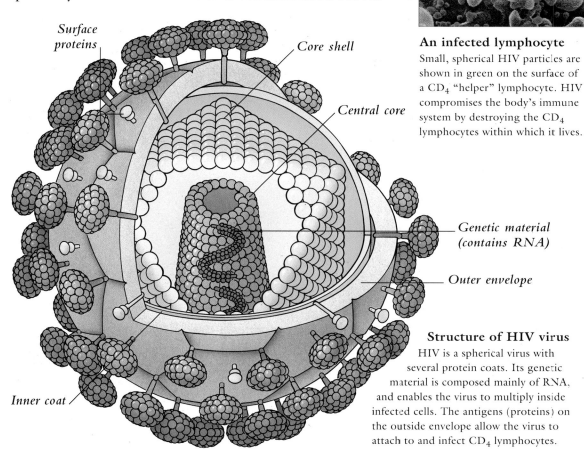

Surface proteins
Core shell
Central core
Genetic material (contains RNA)
Outer envelope
Inner coat

Structure of HIV virus
HIV is a spherical virus with several protein coats. Its genetic material is composed mainly of RNA, and enables the virus to multiply inside infected cells. The antigens (proteins) on the outside envelope allow the virus to attach to and infect CD_4 lymphocytes.

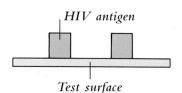
HIV antigen
Test surface

1 Antigens, or surface proteins, from the AIDS virus are first spread on a prepared test surface or on the inside of a test tube.

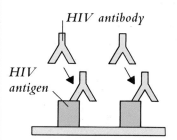
HIV antibody
HIV antigen

2 The test surface is exposed to blood serum. If any HIV antibodies are present, they will bind to the HIV antigens.

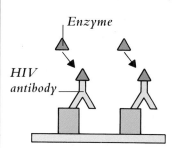
Enzyme
HIV antibody

3 The test surface is washed. A special chemical linked to the enzyme peroxidase, which is known to bind to HIV antibodies, is added to the test surface. The surface is then washed again.

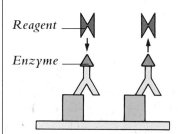
Reagent
Enzyme

4 A reagent – a substance that is used to analyze biological substances – is added to the test surface. Any HIV antibodies will change the color of the reagent, producing a positive test result.

EFFECTS OF AIDS

Many infected people have no symptoms for many years, and are known as "asymptomatic carriers." In later stages, they lose weight and develop night sweats, fevers, and diarrhea. In full-blown AIDS, infected people become susceptible to a wide variety of infections and to types of cancer.

Nervous system
If HIV infection spreads to the brain and nervous system, it causes mental disturbances. Blindness may result from infection with cytomegalovirus.

Lungs
Someone whose immune system has been damaged by HIV often develops lung infections. One type commonly associated with the disease is *Pneumocystis carinii* pneumonis, or PCP.

Skin
Kaposi's sarcoma is the cancer that is often associated with AIDS. Brown or blue patches and nodules appear on the skin. These gradually spread to all parts of the body surface and also affect the internal organs.

Digestive system
Persistent diarrhea is one of the most common features of AIDS. It is caused by infection of the gastrointestinal tract with parasites, for example *Giardia* and *Cryptosporidia,* and with fungi, especially *Candida*.

CHAPTER EIGHT

C H A P T E R   8

The RESPIRATORY SYSTEM

Cilia, tiny hairs projecting from the mucous lining of the respiratory tract

SEM x 7,680

INTRODUCTION

Our lungs are second only to the heart in their work rate: each lung expands and contracts between 12 and 20 times a minute to supply the body with the oxygen it needs and, just as important, to expel carbon dioxide. The air is sometimes contaminated by chemicals and usually contains dust, pollen, bacteria, and viruses. So it is not surprising that, in industrialized countries, respiratory disorders are the most common reason for seeing a doctor. As has occurred in other areas of medicine, the investigation of patients with lung disorders has been transformed by the invention of fiberoptic endoscopes. These enable doctors to inspect the interior of the breathing organs in great detail. Doctors are also able to measure the effectiveness of the lungs very precisely by carrying out tests in a respiratory laboratory. In spite of these decisive advances, the list of important respiratory illnesses is long. Bronchitis due to smoking is the most frequent cause of serious respiratory illness, and lung cancer (which is also usually caused by smoking) remains another leading cause of adult mortality. Pneumonia is often a cause of death in the elderly, while tuberculosis may be about to pose a renewed threat to people of all ages. The number of children suffering from asthma has doubled in the past two decades, although the reason for this is not known.

A narrowed airway due to asthma

SEM x 570

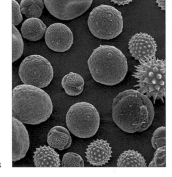

Pollen grains

THE RESPIRATORY SYSTEM

The BODY'S GASEOUS EXCHANGE SYSTEM

TO FUNCTION, BODY CELLS REQUIRE OXYGEN. The respiratory system, which consists of air passages, pulmonary vessels, and the lungs, as well as breathing muscles, supplies fresh oxygen to the blood so that it can be distributed to the rest of the body tissues. It also removes carbon dioxide, a waste product of body processes. Air moves into and out of the lungs as a result of pressure changes brought about by contraction and relaxation of the diaphragm and other breathing muscles. Normal breathing is mainly an involuntary process, controlled by respiratory centers in the brain stem.

AIR PASSAGES

As air is inhaled and passes through the nasal passages, it is filtered, heated, and humidified. The filtering process continues as air flows down through the throat, larynx, trachea, and bronchi to the lungs. Each lung contains a tree of branching tubes that end in tiny air sacs, or alveoli, where gases diffuse into and out of the bloodstream in tiny vessels.

Nasal cavity
A sticky mucous membrane lines the nasal cavity and traps dust particles; its surface hairs, called cilia, move them toward the nose to be either sneezed or blown out. A membrane lining the larynx and the trachea moves particles to the oropharynx to be swallowed.

Nose hairs
Hairs at the entrance to the nose trap large inhaled particles.

THE LARYNX

The larynx plays an essential role in human speech. During sound production, the vocal cords close together and vibrate as air expelled from the lungs passes between them. The false vocal cords have no role in sound production, but help close off the larynx when food is swallowed.

Paranasal sinuses
Air spaces within the skull make it lighter. Sounds echo in these spaces, creating vocal resonance.

Brain stem

Nasopharynx

Oropharynx

Laryngopharynx

Pharynx
The upper part of the pharynx (throat) lets only air pass through. Lower parts permit air, foods, and fluids to pass.

Epiglottis
This flap of cartilage stops food from entering the trachea.

Hyoid bone

Epiglottis

Thyroid cartilage

False vocal cords

Vocal ligament

Vocal cords

Cricoid cartilage

Thyroid gland

Tracheal cartilage

Trachea

FRONT VIEW

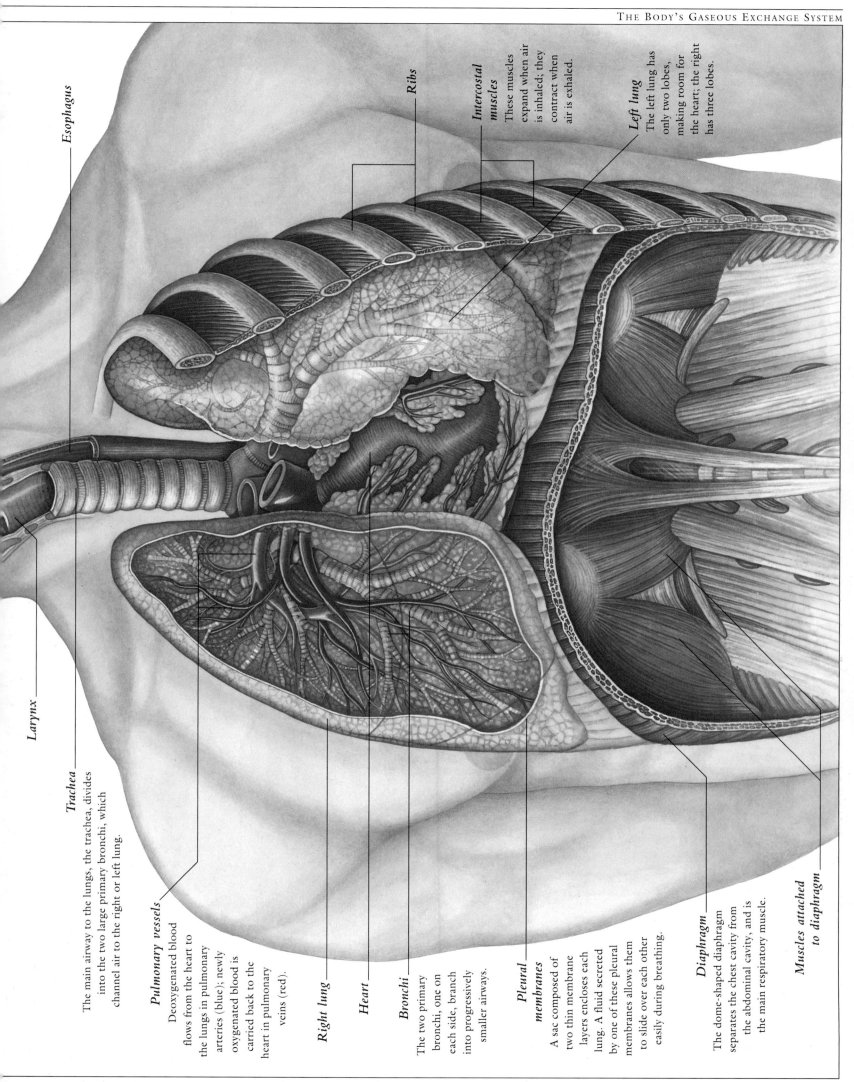

Esophagus

Trachea
The main airway to the lungs, the trachea, divides into the two large primary bronchi, which channel air to the right or left lung.

Larynx

Pulmonary vessels
Deoxygenated blood flows from the heart to the lungs in pulmonary arteries (blue); newly oxygenated blood is carried back to the heart in pulmonary veins (red).

Right lung

Heart

Bronchi
The two primary bronchi, one on each side, branch into progressively smaller airways.

Pleural membranes
A sac composed of two thin membrane layers encloses each lung. A fluid secreted by one of these pleural membranes allows them to slide over each other easily during breathing.

Diaphragm
The dome-shaped diaphragm separates the chest cavity from the abdominal cavity, and is the main respiratory muscle.

Muscles attached to diaphragm

Ribs

Intercostal muscles
These muscles expand when air is inhaled; they contract when air is exhaled.

Left lung
The left lung has only two lobes, making room for the heart; the right has three lobes.

The LUNGS

THE TWO SPONGELIKE LUNGS FILL MOST OF THE CHEST CAVITY and are protected by the flexible rib cage. Together the lungs form one of the largest organs in the body. Their essential function is to provide oxygen from inhaled air to the capillaries and to exhale the carbon dioxide delivered from them. The tissue surface area involved in this exchange is about 40 times greater than the body's outer surface.

Trachea

Hilum

Right primary bronchus

LUNG STRUCTURE

Each cone-shaped lung has a slightly concave base that rests on the diaphragm. Air enters the lungs through a complex of air passages that begins at the trachea just below the larynx. The trachea branches to form two primary bronchi, which enter each lung at the hilum and continue to subdivide into increasingly finer branches until they distribute air to the alveoli.

Lobes of the lung
The right lung is separated by its surface fissures into three lobes, while the smaller left lung is divided into only two lobes. Each lobe is subdivided into segments.

Ribs

THE BRONCHIAL TREE

The intricate network of air passages that supply the lungs looks like an inverted tree, with the trachea resembling the trunk. The illustration below shows a resin cast of this bronchial tree, with each color indicating an individual segment of the lung. Since each segment is aerated by a tertiary (segmental) bronchus, surgical removal of any single segment is possible.

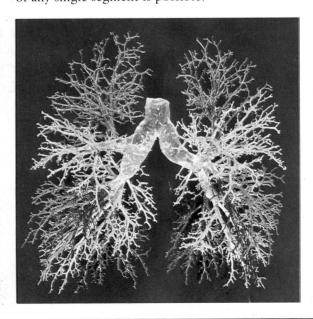

Tertiary bronchus
These branches of the five lobar (secondary) bronchi are also called segmental bronchi because each one aerates an individual segment within each lobe. They may subdivide further into 50 to 80 terminal bronchioles.

Secondary bronchus
The five secondary, or lobar, bronchi are branches of the primary bronchi. They supply the lobes of the lung.

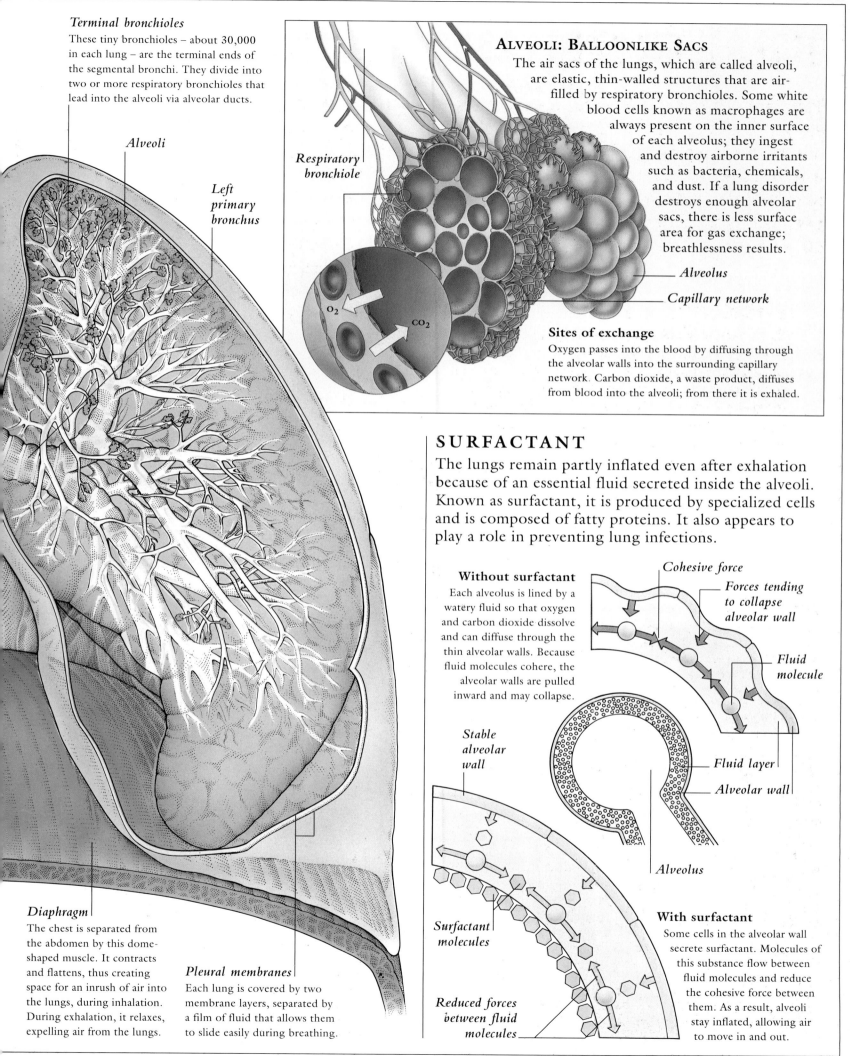

Terminal bronchioles
These tiny bronchioles – about 30,000 in each lung – are the terminal ends of the segmental bronchi. They divide into two or more respiratory bronchioles that lead into the alveoli via alveolar ducts.

Alveoli

Left primary bronchus

Respiratory bronchiole

ALVEOLI: BALLOONLIKE SACS

The air sacs of the lungs, which are called alveoli, are elastic, thin-walled structures that are air-filled by respiratory bronchioles. Some white blood cells known as macrophages are always present on the inner surface of each alveolus; they ingest and destroy airborne irritants such as bacteria, chemicals, and dust. If a lung disorder destroys enough alveolar sacs, there is less surface area for gas exchange; breathlessness results.

O_2

CO_2

Alveolus

Capillary network

Sites of exchange
Oxygen passes into the blood by diffusing through the alveolar walls into the surrounding capillary network. Carbon dioxide, a waste product, diffuses from blood into the alveoli; from there it is exhaled.

SURFACTANT

The lungs remain partly inflated even after exhalation because of an essential fluid secreted inside the alveoli. Known as surfactant, it is produced by specialized cells and is composed of fatty proteins. It also appears to play a role in preventing lung infections.

Without surfactant
Each alveolus is lined by a watery fluid so that oxygen and carbon dioxide dissolve and can diffuse through the thin alveolar walls. Because fluid molecules cohere, the alveolar walls are pulled inward and may collapse.

Cohesive force

Forces tending to collapse alveolar wall

Fluid molecule

Stable alveolar wall

Fluid layer

Alveolar wall

Alveolus

Surfactant molecules

With surfactant
Some cells in the alveolar wall secrete surfactant. Molecules of this substance flow between fluid molecules and reduce the cohesive force between them. As a result, alveoli stay inflated, allowing air to move in and out.

Reduced forces between fluid molecules

Diaphragm
The chest is separated from the abdomen by this dome-shaped muscle. It contracts and flattens, thus creating space for an inrush of air into the lungs, during inhalation. During exhalation, it relaxes, expelling air from the lungs.

Pleural membranes
Each lung is covered by two membrane layers, separated by a film of fluid that allows them to slide easily during breathing.

RESPIRATION *and* BREATHING

THE BODY CANNOT STORE OXYGEN, so we need to breathe day and night to move air into and out of the lungs. The rate and depth of breathing can be consciously modified, but the underlying need to breathe is controlled by involuntary centers in the brain stem. Responses to changes in oxygen and carbon dioxide levels control our breathing, even without our attention.

TWO TYPES OF RESPIRATION

External respiration refers to the exchange of oxygen and carbon dioxide within the lungs. Internal respiration occurs in body tissues when oxygen – carried in blood from the lungs to order to fuel cellular processes – is exchanged for carbon dioxide. Water and carbon dioxide are produced when cells break down nutrients such as glucose. Carbon dioxide travels in blood to the lungs and is exhaled.

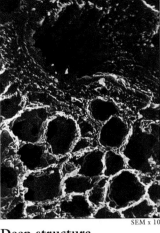

Deep structure
Deep within a lung, a respiratory bronchiole (shown at the top) brings air to the smaller alveoli where gas exchange occurs.

SEM x 10

Oxygen in

Carbon dioxide out

Trachea

Aorta

Pulmonary arteries

Pulmonary veins

Left side of heart

Right side of heart

Alveoli

Lung

Bronchi

Body tissue cell

Blood

Artery

Glucose

Body tissue cells

Capillaries

Capillary wall

Vein

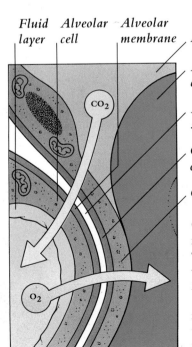

Fluid layer

Alveolar cell

Alveolar membrane

Plasma

Red blood cell

Fluid-filled space

Outer membrane of capillary

Capillary cell

CO₂

O₂

Gas exchange in the lungs
The respiratory membrane, across which gas exchange occurs, has several layers. Despite this, it is extremely thin. Carbon dioxide from the blood diffuses into the alveoli, and oxygen passes from the alveoli into the capillaries, where it is taken up by red blood cells.

KEY

●● Oxygen (O₂)

●●● Carbon dioxide (CO₂)

●• Water (H₂O)

BREATHING

The movement of air into and out of the lungs is generated by differences in pressure inside and outside the body. The diaphragm is the primary muscle involved, and is assisted by the internal and external muscles located between the ribs and in the abdomen. A person inhales about 1pt (500ml) of air 12 to 17 times a minute. The rate and volume increase automatically if the body demands a larger oxygen supply.

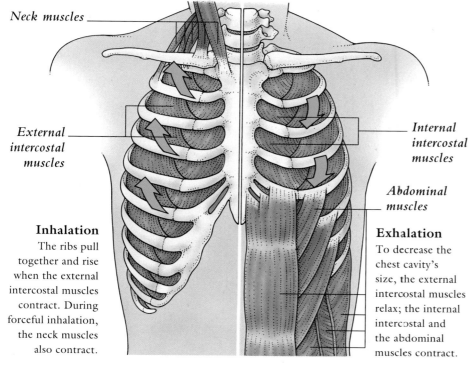

Neck muscles

External intercostal muscles

Internal intercostal muscles

Abdominal muscles

Inhalation
The ribs pull together and rise when the external intercostal muscles contract. During forceful inhalation, the neck muscles also contract.

Exhalation
To decrease the chest cavity's size, the external intercostal muscles relax; the internal intercostal and the abdominal muscles contract.

PRESSURE CHANGES

Atmospheric pressure is about 760mmHg. During an intake of breath, the contracting diaphragm increases the size of the chest cavity, and pressure within the lungs and pleural space drops. Air moves from areas of high to lower pressure and rushes into the lungs. As the diaphragm relaxes, pressure rises in the smaller chest cavity. To equalize pressure air is exhaled.

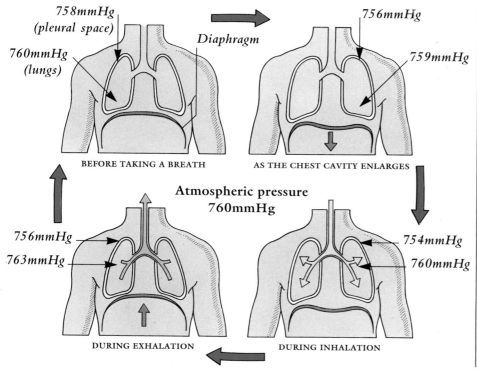

758mmHg (pleural space)

760mmHg (lungs)

Diaphragm

756mmHg

759mmHg

BEFORE TAKING A BREATH

AS THE CHEST CAVITY ENLARGES

Atmospheric pressure 760mmHg

756mmHg

763mmHg

754mmHg

760mmHg

DURING EXHALATION

DURING INHALATION

THE VOCAL CORDS

The vocal cords are paired bands of fibrous tissue at the base of the larynx. Sounds are generated when exhaled air passes through cords that have been brought together and tightened. The greater the tension in the vocal cords, the higher the pitch.

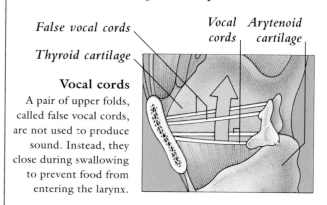

False vocal cords

Thyroid cartilage

Vocal cords

Arytenoid cartilage

Vocal cords
A pair of upper folds, called false vocal cords, are not used to produce sound. Instead, they close during swallowing to prevent food from entering the larynx.

COUGHING

Inhaled particles stimulate nerve cell receptors in the larynx, trachea, and bronchi. Nerve signals are transmitted to the brain stem, which then relays a response to trigger the coughing reflex. This expels irritants, and sometimes mucus, from the body.

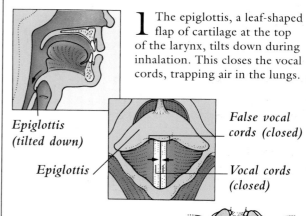

1 The epiglottis, a leaf-shaped flap of cartilage at the top of the larynx, tilts down during inhalation. This closes the vocal cords, trapping air in the lungs.

Epiglottis (tilted down)

Epiglottis

False vocal cords (closed)

Vocal cords (closed)

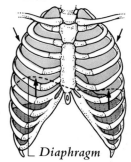

2 The diaphragm rises and muscles in the abdomen contract so that the lungs are increasingly compressed. Because any decrease in the volume of a gas increases its pressure, air in the smaller space of the chest cavity is under greater pressure.

Diaphragm

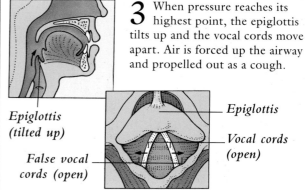

3 When pressure reaches its highest point, the epiglottis tilts up and the vocal cords move apart. Air is forced up the airway and propelled out as a cough.

Epiglottis (tilted up)

False vocal cords (open)

Epiglottis

Vocal cords (open)

RESPIRATORY INFECTIONS

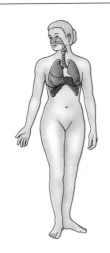

BILLIONS OF MICROORGANISMS – BACTERIA, VIRUSES, AND FUNGI – populate the air. During inhalation, these minute organisms can easily enter the air passages that lead to the lungs, making respiratory infections quite common. Infections of the upper airways, also known as upper respiratory tract infections, may cause a mild illness, such as a common cold or pharyngitis, or a more complex disease such as sinusitis. Other respiratory infections can affect the lower air passages, causing bronchitis, or the lung tissue itself, which is termed pneumonia.

UPPER AIRWAY INFECTIONS

These illnesses include infections of the nasal sinuses, pharynx, and larynx, and are caused when droplets contaminated by viruses and sometimes bacteria are inhaled. Infections often result in the inflammation and swelling of mucous membranes that line these structures. As people grow older, they become immune to most of the common viruses and have fewer infections.

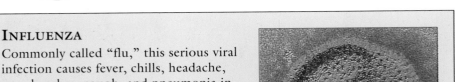

Sinusitis
A bacterial infection may follow a viral infection, causing pus and mucus to accumulate in the nasal sinuses. Fever, headache, stuffy nose, and no sense of smell are common symptoms.

Tonsillitis
Most common in young children, inflamed tonsils may cause a fever, headache, sore throat, discomfort when swallowing, and earache. The lymph nodes in the neck often swell.

Pharyngitis
Inflammation of the pharynx (throat) causes a sore throat, fever, difficulty swallowing, and sometimes an earache and swollen lymph nodes in the neck.

Laryngitis
Usually caused by a virus, this infection can produce hoarseness, loss of voice, a dry cough, and sore throat.

INFLUENZA
Commonly called "flu," this serious viral infection causes fever, chills, headache, muscle aches, cough, and pneumonia in some. It spreads rapidly, often occurring in localized outbreaks, or every few years in epidemics. There are three main types of virus: A, B, and C. Because the viruses can change their structures, a previous immunity to one type may no longer be effective. Influenza is life-threatening to the very young as well as the elderly, and some epidemics kill people of all ages.

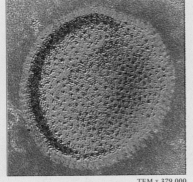

TEM x 379,000

Influenza virus

THE COMMON COLD
Colds are easily transmitted from person to person by virus-containing droplets that are released onto hands or clothing or into the atmosphere when an infected person coughs or sneezes. Approximately 200 different viruses cause colds. Antibiotics have no effect and only symptoms can be treated; the body's immune system must overcome infections.

1 After being carried by infected droplets, virus particles enter the body and invade the cells that line the throat and nose. These virus particles then replicate to produce new viruses, which continue to multiply rapidly.

Body cell

Virus particles

Infected nasal lining

2 The blood supply brings lymphocytes (white blood cells) to the infected mucosa. The blood vessels within the nasal mucosa swell and cause the secretion of excess fluid, resulting in a "runny nose."

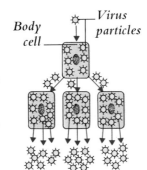

Lymphocyte

Blood vessel

3 Some types of lymphocyte make virus-specific proteins (antibodies) that immobilize the virus particles, while other types secrete chemical substances that can destroy infected cells.

Antibodies

Chemicals

Phagocyte

4 Phagocytes, a type of white blood cell, can engulf and destroy dead viruses, immobilized virus particles, and damaged cells. Symptoms of the cold soon subside.

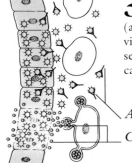

ACUTE BRONCHITIS

This form of bronchitis – which means inflammation of the bronchi – develops suddenly. It can occur as a complication of an upper respiratory tract infection, such as a common cold, or can accompany measles or influenza. This disease is usually caused by a virus, and produces symptoms that include a sputum-producing cough, a fever, and sometimes a slight wheeze.

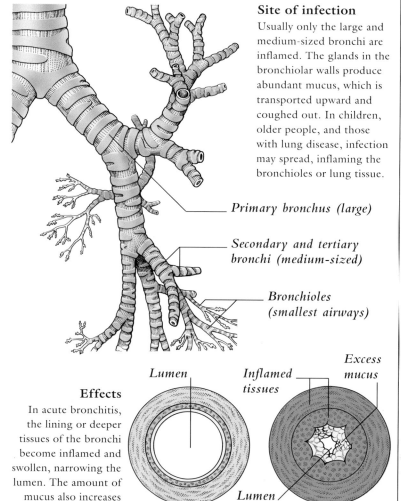

Site of infection

Usually only the large and medium-sized bronchi are inflamed. The glands in the bronchiolar walls produce abundant mucus, which is transported upward and coughed out. In children, older people, and those with lung disease, infection may spread, inflaming the bronchioles or lung tissue.

Primary bronchus (large)

Secondary and tertiary bronchi (medium-sized)

Bronchioles (smallest airways)

Effects

In acute bronchitis, the lining or deeper tissues of the bronchi become inflamed and swollen, narrowing the lumen. The amount of mucus also increases and causes congestion.

Lumen

Inflamed tissues

Excess mucus

Lumen

NORMAL

BRONCHITIS

PNEUMONIA

In pneumonia, the smallest bronchioles and alveolar tissue become inflamed. There are two main types. Lobar pneumonia affects one lobe of the lung, while bronchopneumonia affects patches of tissue in one or both lungs. Usually resulting from a viral or bacterial infection, pneumonia may also be caused by fungi, yeasts, or protozoa. Early symptoms include chills, fever, sweating, joint and muscle pain, and headache. Chest pain, coughing, and breathlessness develop.

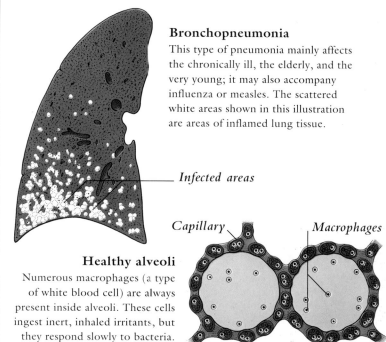

Bronchopneumonia

This type of pneumonia mainly affects the chronically ill, the elderly, and the very young; it may also accompany influenza or measles. The scattered white areas shown in this illustration are areas of inflamed lung tissue.

Infected areas

Capillary

Macrophages

Healthy alveoli

Numerous macrophages (a type of white blood cell) are always present inside alveoli. These cells ingest inert, inhaled irritants, but they respond slowly to bacteria.

Fluid

Neutrophils

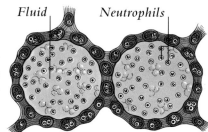

Infected alveoli

The infective process triggers changes in the capillary walls, allowing neutrophils (a type of white blood cell) in to fight the invading organisms. Fluid also flows in and accumulates.

PLEURAL EFFUSION

Pleural inflammation may result from infections, particularly pneumonia or tuberculosis. It may cause excess fluid to build up in the space between the two membrane layers of the pleura. If extensive, this effusion can cause breathlessness. Fluid may need to be removed via a hollow needle or drain inserted through the chest wall.

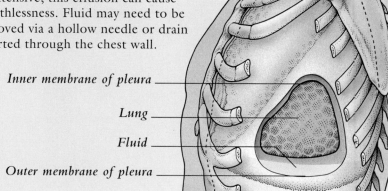

Inner membrane of pleura

Lung

Fluid

Outer membrane of pleura

LEGIONNAIRE'S DISEASE

This rare bacterial condition was first described in 1976, after an outbreak of severe pneumonia that affected war veterans attending an American Legion convention. The disease affects men more often than women. Symptoms include a high fever, chills, muscle aches, a severe headache, abdominal pain, confusion, and diarrhea. Patients usually require hospitalization and intravenous antibiotics such as erythromycin.

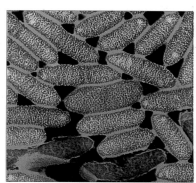

The cause

The bacterium *Legionella pneumophila* is found in small numbers in almost all water supplies. It thrives, however, in water-cooled air conditioning systems, and in plumbing systems where water stagnates.

SEM x 11,230

LUNG DISORDERS

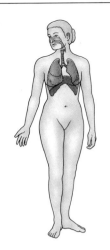

DISORDERS THAT CAUSE BREATHING PROBLEMS may be present from birth or may develop over many years. Others may occur suddenly without warning or after an injury. Inhaled substances, such as gases, fumes, organic chemicals, or mineral dust, can contribute to some disorders, while others have no known cause. Lung disorders may be grouped into those marked by inflammation that is caused by a variety of chemicals, infections, allergies, or other autoimmune disorders; those due to cancers and other growths; and those that are inherited.

PULMONARY HYPERTENSION

Elevated blood pressure in the pulmonary arteries leading to the lungs may be the result of a lung disorder, such as emphysema, or a circulatory disorder affecting the veins in the arms and legs. Left-sided heart failure causing a backup of blood in the lungs also raises pulmonary artery pressure.

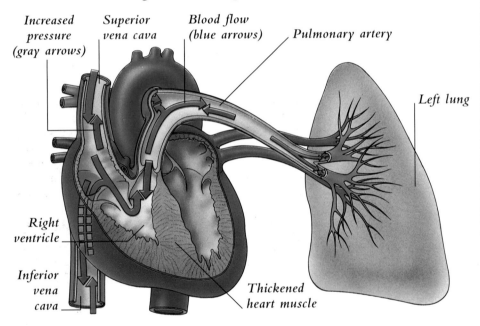

Increased pressure (gray arrows) · *Superior vena cava* · *Blood flow (blue arrows)* · *Pulmonary artery* · *Left lung* · *Right ventricle* · *Inferior vena cava* · *Thickened heart muscle*

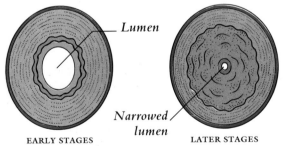

Lumen · *Narrowed lumen*

EARLY STAGES · LATER STAGES

Early and late stages
As the condition develops, the walls of the pulmonary artery thicken with muscle and fibrous tissue. This narrows the lumen, impedes blood flow, and raises pressure in the arteries. Blood volume pumped from the heart becomes progressively reduced.

SARCOIDOSIS
Thought to be due to an extreme immune response, sarcoidosis features multiple areas of inflammation interspersed with fibrous and grainlike tissue. The circular nodules, called granulomas (shown right), are often found in the lungs, lymph nodes, and eyes. Symptoms include breathlessness, fatigue, joint pain, and sometimes a skin rash.

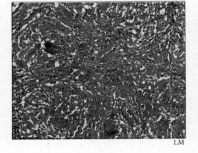

LM

PNEUMOTHORAX

A pneumothorax occurs when one of the pleural membranes ruptures, which allows air to enter the pleural space and cause the lung to collapse. Sometimes a spontaneous pneumothorax occurs, while others are the result of an injury; breathlessness and chest pain are common symptoms. If air is not reabsorbed, it may compress the lungs and heart and must be drained by a needle or tube inserted into the pleural space.

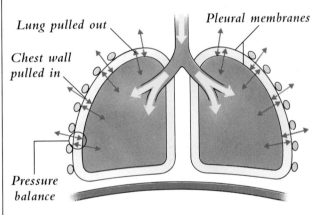

Lung pulled out · *Pleural membranes* · *Chest wall pulled in* · *Pressure balance*

Normal lungs
During normal breathing, the lungs inflate and are pulled out while the chest wall is pulled in. Within the pleural space a balance is maintained between these opposing pressures.

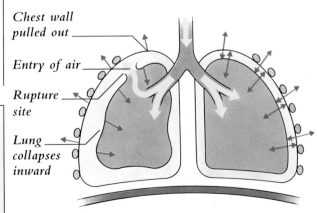

Chest wall pulled out · *Entry of air* · *Rupture site* · *Lung collapses inward*

Pneumothorax
If air enters the pleural space, it changes the pressure balance. This pressure change causes the lungs to collapse inward suddenly.

FIBROSING ALVEOLITIS

Fibrosing alveolitis is sometimes called idiopathic pulmonary fibrosis (IPF), an autoimmune disorder of unknown cause. It also occurs with various other immune disorders, such as rheumatoid arthritis. The disease causes fibrosis (scarring) and thickening of the walls of the lung's air sacs, resulting in severe breathlessness. Corticosteroid drugs may be given.

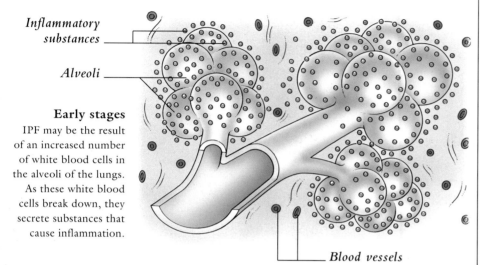

Inflammatory substances

Alveoli

Early stages
IPF may be the result of an increased number of white blood cells in the alveoli of the lungs. As these white blood cells break down, they secrete substances that cause inflammation.

Blood vessels

Growth of fibrous tissue
The inflammatory substances stimulate fibroblasts to produce an overgrowth of fibrous tissue. Thick cuboidal cells replace the thin cells that usually line the bronchi, restricting the passage of oxygen.

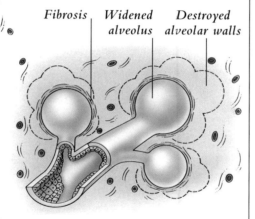

Fibrosis *Widened alveolus* *Destroyed alveolar walls*

Cuboidal cells *Fibrous tissue* *Fibroblast*

Late stages
Formation of scar tissue (fibrosis) occurs, thereby destroying the alveolar walls. The remaining alveoli both widen and thicken, reducing the surface area for gas exchange. Scar tissue also restricts lung expansion.

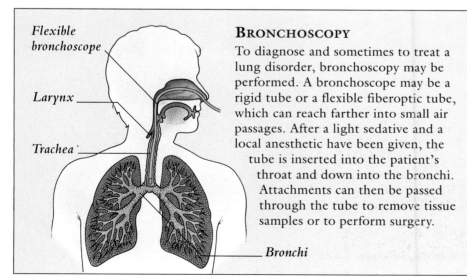

Flexible bronchoscope

BRONCHOSCOPY
To diagnose and sometimes to treat a lung disorder, bronchoscopy may be performed. A bronchoscope may be a rigid tube or a flexible fiberoptic tube, which can reach farther into small air passages. After a light sedative and a local anesthetic have been given, the tube is inserted into the patient's throat and down into the bronchi. Attachments can then be passed through the tube to remove tissue samples or to perform surgery.

Larynx

Trachea

Bronchi

DUST DISEASES

Asbestosis, silicosis, and pneumoconiosis are some of the diseases that are caused by inhalation of dust particles. These inhaled particles inflame lung tissue, thus causing irreversible scarring. Those most at risk are people whose work exposes them to these dusts for several years. Some molds that develop in hay, grain, or straw may cause farmer's lung, an allergic reaction that results in inflammation of the alveoli.

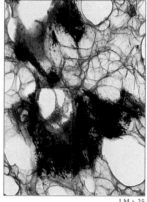

LM x 25

Coal-miner's pneumoconiosis
If coal dust is inhaled over a period of 10 to 15 years, it can lead to pneumoconiosis, or "black lung disease." The dust particles deposited in lung tissue (left) produce inflammatory nodules. Scar tissue formation destroys alveoli and bronchioles.

SILICOSIS

Silicosis is the world's most common occupational disease. It is a form of fibrosis in the lungs caused by silica dust, usually in the form of quartz. Quarry workers, stone masons, coal miners, and others are at risk. Symptoms such as breathlessness may not develop for many years. The disease may lead to lung cancer, especially if an affected person smokes.

1 Inhaled silica particles are deposited in the lungs and ingested by scavenging white blood cells called macrophages.

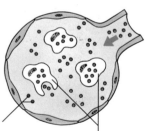

Silica particles

Macrophages

2 Macrophages burst and die, releasing the silica and chemicals. The latter attract fibroblasts, which produce fibrous tissue. Silica is consumed by more macrophages and the process is repeated.

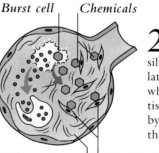

Burst cell *Chemicals*

Fibrous tissue *Fibroblasts*

Dense nodule of scar tissue

3 More fibrous tissue develops, leading to dense nodules of scar tissue. Buildup of this tissue severely restricts effective lung function.

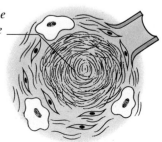

CHRONIC LUNG DISEASES

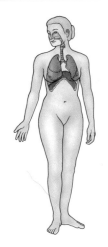

COMMON IN SMOKERS AND IN URBAN OR INDUSTRIALIZED AREAS, chronic lung disorders that obstruct the airways and reduce airflow have increased in every part of the world. The disorders increasingly affect women, partly due to the greater numbers who smoke or who are in industrial workplaces. Known risk factors include passive exposure to cigarette smoke and repeated respiratory infections during childhood, as well as a family member with a similar disease. In the past 20 years, the number of childhood asthma cases has doubled.

CHRONIC BRONCHITIS

Although recurring acute bronchitis caused by a virus or a bacterium may cause chronic inflammation of the bronchi, the most common cause is smoking and chemical irritants. At first the resulting cough is troublesome mostly during the damp, cold months, but eventually it persists all year. Symptoms such as hoarseness and breathlessness also occur.

HOW BRONCHITIS DEVELOPS

If bronchi are irritated by smoking or prolonged exposure to pollutants, they begin to produce too much mucus. This causes a progressively worsening cough in order to clear the airways.

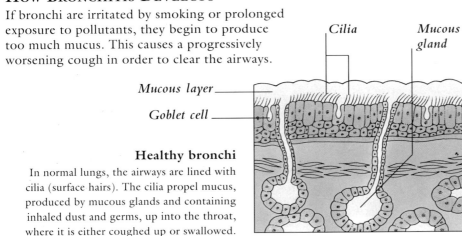

Cilia

Mucous gland

Mucous layer

Goblet cell

Healthy bronchi

In normal lungs, the airways are lined with cilia (surface hairs). The cilia propel mucus, produced by mucous glands and containing inhaled dust and germs, up into the throat, where it is either coughed up or swallowed.

1 Inhaled irritants cause goblet cells to increase in number and mucous glands to enlarge so that more mucus is produced. Damaged cilia cannot propel mucus along.

2 Mucus retained in the airway becomes a breeding ground for bacteria so that inflammation is likely to recur. Cilia are slowly destroyed; more mucus collects.

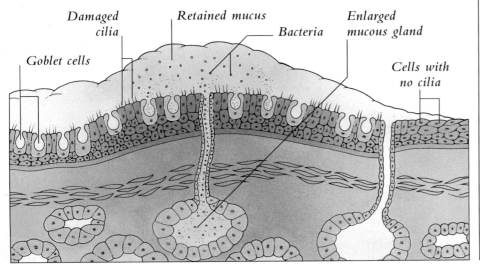

Damaged cilia

Retained mucus

Bacteria

Enlarged mucous gland

Goblet cells

Cells with no cilia

EMPHYSEMA

The lungs are filled with millions of tiny air sacs called alveoli. In emphysema, they become overstretched and rupture. Most people who are severely affected are heavy, long-term smokers, but a rare inherited enzyme deficiency is a known risk factor. At present the disorder is incurable, but stopping smoking slows its progression.

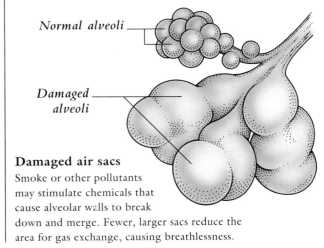

Normal alveoli

Damaged alveoli

Damaged air sacs

Smoke or other pollutants may stimulate chemicals that cause alveolar walls to break down and merge. Fewer, larger sacs reduce the area for gas exchange, causing breathlessness.

DEATHS FROM SMOKING

A comparison of death rates in nonsmokers and smokers from both chronic bronchitis and emphysema is shown in the graph below.

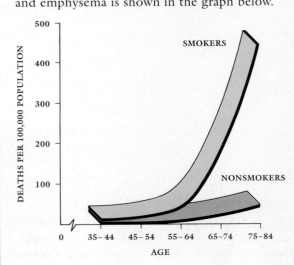

SMOKERS

NONSMOKERS

DEATHS PER 100,000 POPULATION

500

400

300

200

100

0   35–44   45–54   55–64   65–74   75–84

AGE

ASTHMA

Asthma attacks involve both wheezing and breathlessness, varying in intensity, and caused by constricted airways. Allergic asthma often develops in childhood and may be accompanied by eczema. Asthma is confirmed by lung function tests, and by skin and blood tests to identify substances triggering these attacks. In some forms of the disease, there is no specific trigger and no known cause.

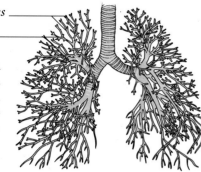

Bronchioles

Tertiary bronchi

Airways affected in asthma
The smaller bronchi and the bronchioles (the smallest airways) become constricted, inflamed, and congested with mucus. As a result, breathing becomes difficult.

Blood vessels

Mucus

Relaxed smooth muscle

Normal airway
Normally, the smooth muscle in the bronchiolar walls is relaxed, creating a wide lumen, or space, in the center of the air passage. This allows air to flow easily and steadily in order to meet the oxygen requirements of the body.

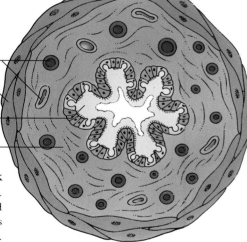

Inflammatory substances widen the blood vessels

Smooth muscle contracts

Increased mucus

Inflammation and swelling

During an asthma attack
Contracted muscle walls narrow airways. Further narrowing is caused by increased mucus and inflammation due to chemicals released during an allergic response.

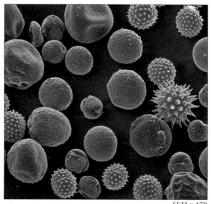

THE ROLE OF ALLERGENS
Allergens are substances that trigger an allergic response. Common allergens that may trigger or intensify attacks of asthma include molds, pollens, animal dander, dust, and some foods and drugs. Anxiety, vigorous exercise in cold weather, and respiratory infections are other factors.

A specific allergen
In some people, exposure to certain pollens triggers asthma attacks. A variety of pollen grains is illustrated on the left.

SEM x 470

TREATMENT OF ASTHMA

Obstruction of airways may be relieved by inhaled steroids to suppress inflammation, and by bronchodilator drugs, which relax bronchiolar walls. Beta adrenergic receptor stimulants and anticholinergic drugs are also used for the relief of asthma attacks. Both the frequency and severity of allergic asthma attacks can be reduced by avoiding the specific allergens that trigger attacks.

MAST-CELL STABILIZERS
Mast cells play a critical role in the allergic response. Allergens (antigens) attach to these cells, stimulating them to produce histamine. Mast-cell stabilizers can inhibit the production of histamine by bronchiolar mast cells, which helps reduce inflammation of the airways.

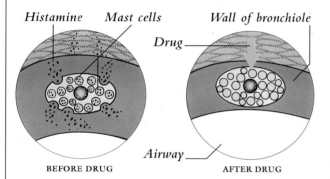

Histamine   Mast cells   Wall of bronchiole

Drug

Airway

BEFORE DRUG   AFTER DRUG

BRONCHODILATORS AND STEROIDS
Bronchodilator drugs work by affecting the nerve signals that control the contraction and relaxation of bronchiolar muscles. They do not reduce inflammation of the mucous lining. Corticosteroid drugs, usually inhaled, widen bronchioles by reducing inflammation.

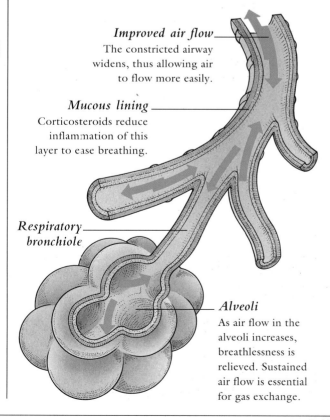

Improved air flow
The constricted airway widens, thus allowing air to flow more easily.

Mucous lining
Corticosteroids reduce inflammation of this layer to ease breathing.

Respiratory bronchiole

Alveoli
As air flow in the alveoli increases, breathlessness is relieved. Sustained air flow is essential for gas exchange.

LUNG CANCER

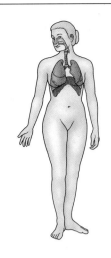

THE MOST COMMON CAUSE OF LUNG CANCER – about 87 percent of all cases in the US – is tobacco smoking. In the past, lung cancer was far more common in men than women because more men than women smoked in the first half of this century; now its incidence in women is rapidly rising and it has overtaken breast cancer as the most common cancer causing death in women. Other causes of lung cancer include exposure to coal dust, asbestos, and silica. Lung cancer is more common in industrial areas of the world than in rural areas.

CAUSES OF LUNG CANCER

Many inhaled irritants trigger the growth of abnormal cells in the lungs. Cigarette smoke, however, contains thousands of known carcinogenic (cancer-causing) substances, and is the main cause of lung cancer. Diagnostic tests may include a chest X-ray, biopsy, and bronchoscopy (examining the bronchi through a viewing tube).

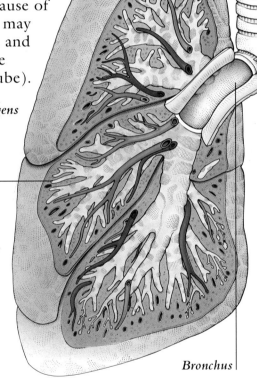

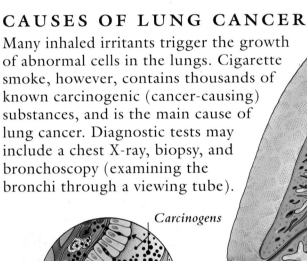

Carcinogens

Capillary

Respiratory airway

Bronchus

Spreading carcinogens

Each lung contains millions of air spaces. Carcinogenic substances from tobacco smoke, notably tar, are able to pass from these directly into the bloodstream.

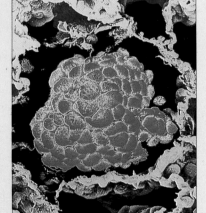

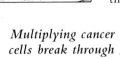

SEM x 230

GROWTH OF LUNG CANCER

In about 95 percent of lung cancer cases, the tumor starts to grow in the bronchi, where it may enlarge or bleed and obstruct breathing. Some cells of a bronchial tumor may break away and infiltrate other parts of the lung, or spread to other organs either directly or via the bloodstream. The cancerous tissue that develops at the new site is known as a metastasis.

Tumor in an alveolus

A tiny tumor fills a single alveolus. A few of the cancer cells (shown in red) have broken away and begun to spread.

HOW SMOKING DAMAGES THE LUNGS

Tobacco smoke is a complex mixture of over 3,000 different substances, and burning cigarette tar is strongly carcinogenic. Some risk factors known to predispose toward the development of lung cancer include the number of cigarettes smoked per day, their tar content, the number of years a person has smoked, and the depth of inhalation.

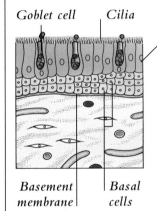

Goblet cell   *Cilia*   *Columnar cell*

Basement membrane   *Basal cells*

1 Columnar cells topped by cilia (tiny hairs) line healthy bronchi. Under this layer are basal cells, which constantly divide to replace damaged columnar cells. Mucus produced by goblet cells lubricates the bronchi.

Squamous cells

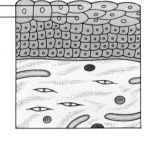

2 Over a number of years, columnar cells damaged by smoke flatten and turn into squamous cells, which gradually lose their cilia.

Basal cells become cancerous

3 In an attempt to replace the damaged squamous cells, the basal cells change and multiply at an increased rate (dysplasia); some of these become cancer cells.

Multiplying cancer cells break through

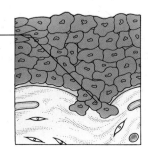

4 The cancer cells start to replace healthy cells. If these cells break through the basement membrane, they can metastasize, establishing new sites of cancer.

SYMPTOMS

A persistent cough is usually the earliest symptom of lung cancer. Because most people who develop lung cancer are smokers, this is often dismissed as simply a "smoker's cough." Other symptoms of lung cancer include coughing up blood, wheezing, persistent hoarseness, headache, weight loss, and chest pain.

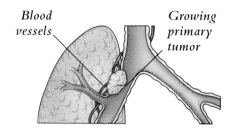

Blood vessels *Growing primary tumor*

Symptoms from tumor growth
A tumor that grows may obstruct a bronchus, causing shortness of breath and chest pain. Sometimes a tumor invades the esophagus, thus making swallowing difficult.

Symptoms of spreading cancer
Lung cancer metastasizes (or spreads) to other parts of the body and causes a variety of symptoms. Metastases in bones may cause pain and fractures; in the brain they may cause paralysis and confusion; in the liver they may cause nausea and weight loss.

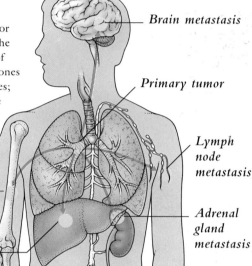

Brain metastasis
Primary tumor
Lymph node metastasis
Adrenal gland metastasis
Bone metastasis
Liver metastasis

DRUG TREATMENT
Some anticancer drugs will relieve symptoms of certain lung cancers, or cause those symptoms to temporarily disappear. Since these drugs damage normal cells as well, they are given at intervals of 3 to 4 weeks to allow healthy tissue to recover between treatments. Vomiting, diarrhea, and hair loss may be side effects.

Cytotoxic antibiotics
For cells to multiply, DNA must replicate to create chromosomes in the new cell. Cytotoxic antibiotics prevent replication, halting growth of normal and cancer cells.

Duplicated chromosome
Spindle fibers
Threadlike chromosomes

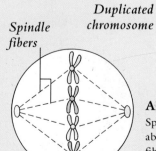

Alkylating drugs
Spindle fibers form in cells that are about to divide. By breaking up these fibers, alkylating drugs interfere with rapidly reproducing cancer cells.

OPERATION

LOBECTOMY
If diagnostic tests confirm the presence of lung cancer, a lobectomy, or the removal of a lobe of the lung, may be performed. This operation is only appropriate in certain circumstances. The tumor must be small and confined to a localized area; breakaway cancer cells must not have spread to other parts of the body; and the patient must also be reasonably well. For the few for whom the operation is suitable, it offers relief of symptoms as well as the chance of a cure.

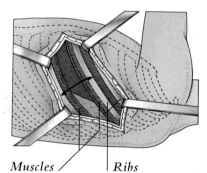

1 After administration of a general anesthetic, the surgeon makes an incision around the side of the chest. The muscles between the ribs are cut and the ribs are then separated in order to expose the affected lung, which is covered by its pleura.

Muscles *Ribs*

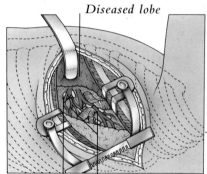

Diseased lobe

2 The diseased lobe is moved to one side so that the vessels and lymph nodes can be easily seen. Lymph node samples are obtained. These specimens are examined immediately under a microscope to find out whether the cancer has spread to the nodes.

Lymph nodes *Blood vessels*

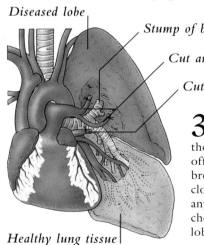

Diseased lobe *Stump of bronchus* *Cut arteries* *Cut vein*

3 The arteries, veins, and main bronchus that supply the affected lobe are first tied off and then cut. The remaining bronchial stump is permanently closed off with sutures to stop any air from leaking into the chest cavity, and the diseased lobe is cut away and removed.

Healthy lung tissue

4 Two chest drains are inserted into the chest cavity before it is closed. The drains, which remove excess blood and fluid from the area around the lungs, are left in place for 3 to 5 days and then removed.

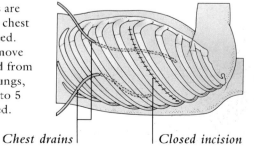

Chest drains *Closed incision*

CHAPTER NINE

C H A P T E R 9

The DIGESTIVE SYSTEM

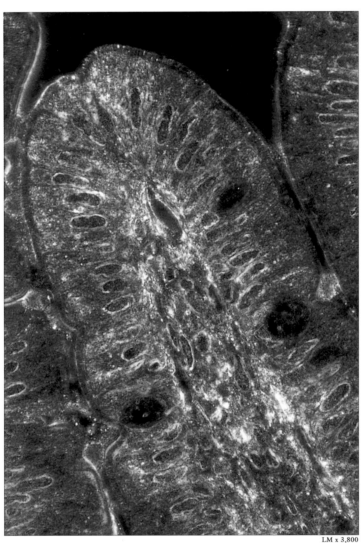

A cross-section
of a villus in the
small intestine

LM x 3,800

INTRODUCTION

More than any other system in the human body, the digestive organs make us aware when they are working well, or when they are upset and need attention. Hunger and the need to empty the bowels are just two of the messages we cannot ignore. Digestive disorders such as indigestion, peptic ulcers, inflammatory bowel disease, and irritable colon syndrome can often be partly due to psychological causes. So it is hardly surprising that these problems occur often, or that treatment may entail psychological

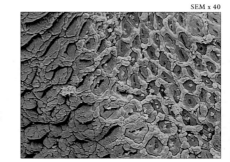

SEM x 40

The junction of the small and large intestines

support as well as physical intervention. During the last two decades, our understanding of these disorders has increased. The importance of a fiber-rich diet has been recognized, so constipation has become a less troublesome problem. Another important recent advance is the identification of the species of bacterium that causes many peptic ulcers, and this has made it possible to use drugs to cure them. Lastly, the range of diagnostic tests available to gastroenterologists has been totally transformed. Today, endoscopy – the examination of the internal organs by means of viewing tubes – is the primary method of inspecting areas like the esophagus, stomach, intestines, and bile ducts. This technique has made it possible to recognize cancers at an early stage, and to monitor their treatment.

Endoscopic examination of the bile ducts

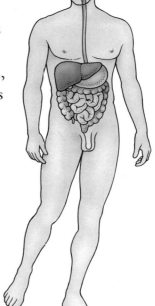

THE DIGESTIVE SYSTEM

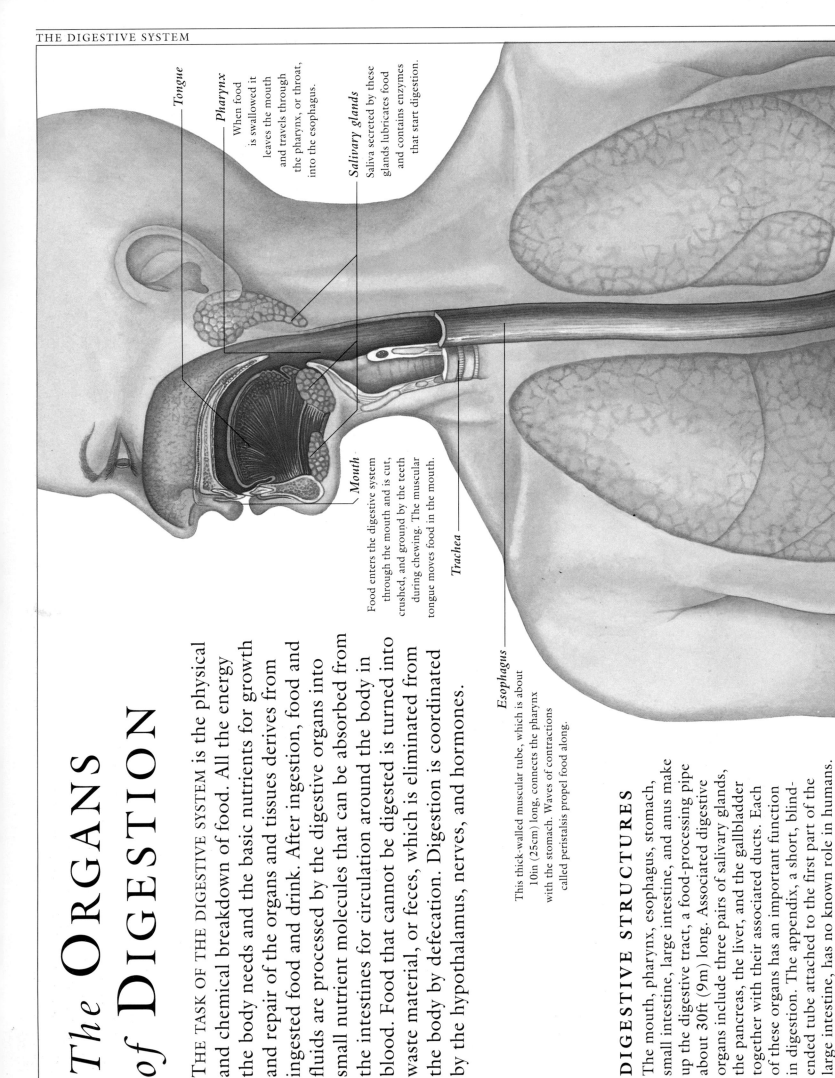

Tongue

Pharynx
When food is swallowed it leaves the mouth and travels through the pharynx, or throat, into the esophagus.

Salivary glands
Saliva secreted by these glands lubricates food and contains enzymes that start digestion.

Mouth
Food enters the digestive system through the mouth and is cut, crushed, and ground by the teeth during chewing. The muscular tongue moves food in the mouth.

Trachea

Esophagus
This thick-walled muscular tube, which is about 10in (25cm) long, connects the pharynx with the stomach. Waves of contractions called peristalsis propel food along.

The ORGANS of DIGESTION

THE TASK OF THE DIGESTIVE SYSTEM is the physical and chemical breakdown of food. All the energy the body needs and the basic nutrients for growth and repair of the organs and tissues derives from ingested food and drink. After ingestion, food and fluids are processed by the digestive organs into small nutrient molecules that can be absorbed from the intestines for circulation around the body in blood. Food that cannot be digested is turned into waste material, or feces, which is eliminated from the body by defecation. Digestion is coordinated by the hypothalamus, nerves, and hormones.

DIGESTIVE STRUCTURES

The mouth, pharynx, esophagus, stomach, small intestine, large intestine, and anus make up the digestive tract, a food-processing pipe about 30ft (9m) long. Associated digestive organs include three pairs of salivary glands, the pancreas, the liver, and the gallbladder together with their associated ducts. Each of these organs has an important function in digestion. The appendix, a short, blind-ended tube attached to the first part of the large intestine, has no known role in humans.

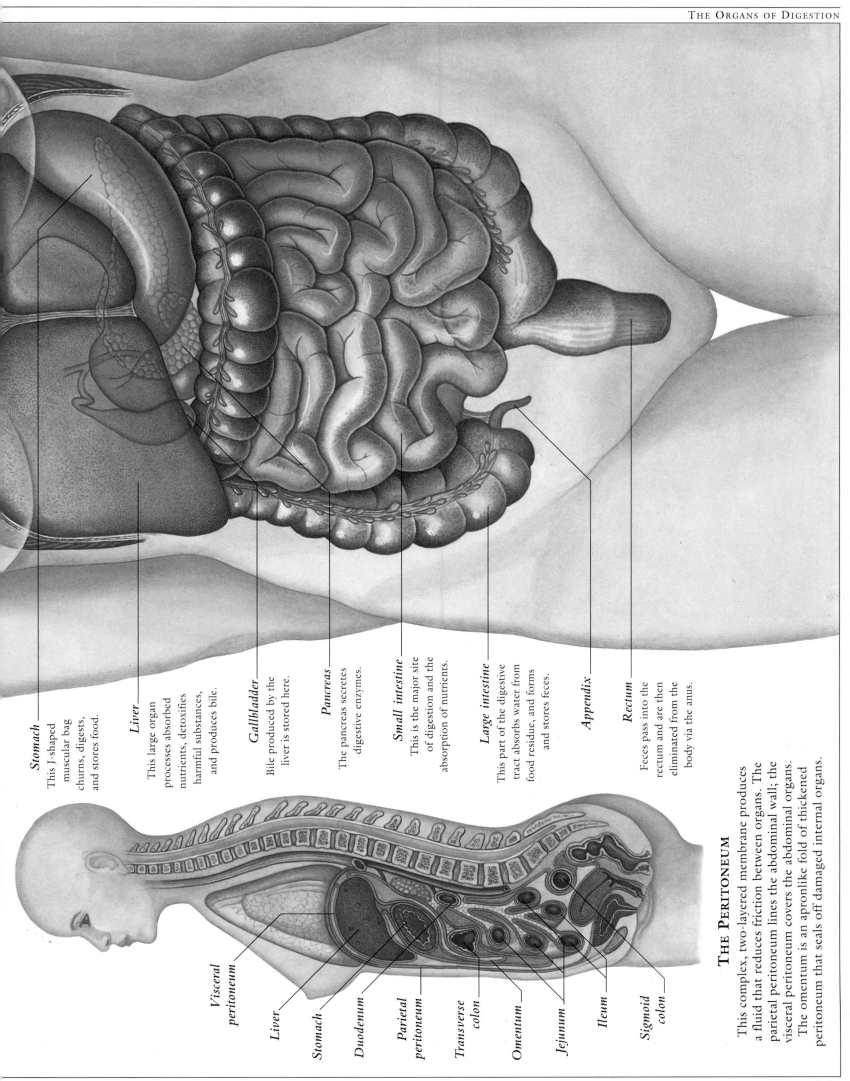

Stomach
This J-shaped muscular bag churns, digests, and stores food.

Liver
This large organ processes absorbed nutrients, detoxifies harmful substances, and produces bile.

Gallbladder
Bile produced by the liver is stored here.

Pancreas
The pancreas secretes digestive enzymes.

Small intestine
This is the major site of digestion and the absorption of nutrients.

Large intestine
This part of the digestive tract absorbs water from food residue, and forms and stores feces.

Appendix

Rectum
Feces pass into the rectum and are then eliminated from the body via the anus.

THE PERITONEUM

This complex, two-layered membrane produces a fluid that reduces friction between organs. The parietal peritoneum lines the abdominal wall; the visceral peritoneum covers the abdominal organs. The omentum is an apronlike fold of thickened peritoneum that seals off damaged internal organs.

Visceral peritoneum

Liver

Stomach

Duodenum

Parietal peritoneum

Transverse colon

Omentum

Jejunum

Ileum

Sigmoid colon

The DIGESTIVE PROCESS

THE DIGESTIVE TRACT IS A MUSCULAR TUBE extending from the mouth through the stomach and intestines to the anus. Food moves along the digestive tract while it is changed into substances that can be absorbed into the bloodstream for distribution. The pancreas, the salivary glands, and the biliary system all connect to the digestive tract, producing various substances that are essential to healthy digestion.

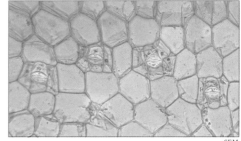

Dietary fiber (cellulose) SEM

1 IN THE MOUTH AND ESOPHAGUS

Food is chewed by the teeth and mixed with saliva. The enzyme amylase, present in saliva, begins the breakdown of starch into sugar. Each soft lump of food, called a bolus, is swallowed and propelled by contractions down the esophagus into the stomach.

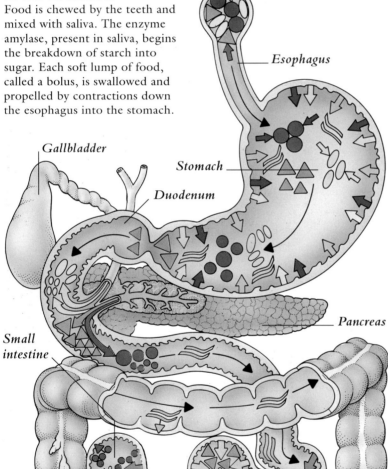

Esophagus

Gallbladder

Stomach

Duodenum

Pancreas

Small intestine

Large intestine

THE BREAKDOWN OF FOOD

Certain nutrients, such as salts and minerals, can be absorbed directly into the circulation. Proteins, fats, and carbohydrates, however, must be broken down into smaller molecules before they can be absorbed. Food is broken down both by mechanical action and by the chemical action of digestive enzymes. Fats are split into glycerol and fatty acids; carbohydrates into monosaccharide sugars; and proteins into shorter chains and subsequently into individual amino acids.

2 IN THE STOMACH

Pepsin is an enzyme produced when gastric acid catalyzes inactive pepsinogen. It breaks down proteins into smaller units called polypeptides and peptides.
Lipase breaks down a small proportion of fats into glycerol and fatty acids.
Hydrochloric acid is produced by the lining. Its acidity, needed for the action of pepsin, can also kill certain bacteria.

3 IN THE DUODENUM

Lipase, a pancreatic enzyme, breaks down fats into glycerol and fatty acids.
Amylase, another enzyme produced by the pancreas, breaks down starch into maltose, which is a disaccharide sugar.
Trypsin and **chymotrypsin** are powerful pancreatic enzymes that split proteins into polypeptides and peptides.

4 IN THE SMALL INTESTINE

Maltase, sucrase, and **lactase** are enzymes secreted by certain glands in the intestinal wall. They convert disaccharide sugars into monosaccharide sugars.
Peptidase, another enzyme secreted by glands in the intestinal wall, splits large peptides into smaller peptides and then into individual amino acids.

5 IN THE LARGE INTESTINE

Undigested food enters the large intestine, where water and salt are absorbed by the intestinal lining. The residue, together with waste pigments, dead cells, and bacteria, is pressed into feces and then stored for excretion.

KEY

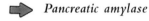

- **Salivary amylase**
- **Pancreatic amylase**
- **Maltase, sucrase, and lactase**
- **Pepsin**
- **Trypsin and chymotrypsin**
- **Peptidase**
- **Lipase**
- **Bile salts**
- **Hydrochloric acid**
- **Starch**
- **Disaccharides (maltose, sucrose, and lactose)**
- **Monosaccharides (glucose, fructose, and galactose)**
- **Proteins**
- **Peptides**
- **Amino acids**
- **Fats**
- **Fatty acids**
- **Glycerol**
- **Water**

COMPONENTS OF FOOD

Food contains carbohydrates, fats, and proteins as well as vitamins, minerals, water, and fiber. Starchy and sugary foods are rich in carbohydrates, which, along with fats, are the body's main source of energy. Fats and protein are used for cell growth and repair.

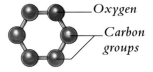

Oxygen
Carbon groups

Monosaccharides
These single sugar units have a hexagonal structure. They form the building blocks of the more complex carbohydrates.

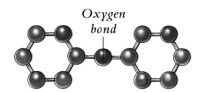

Oxygen bond

Disaccharides
These molecules are formed when two single saccharide units chemically bond together. The most common ones are sucrose, maltose, and lactose.

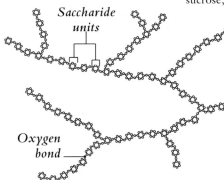

Saccharide units

Oxygen bond

Polysaccharides
Polysaccharides consist of long chains of saccharide units. Starch and glycogen, both storage carbohydrates, are examples. Cellulose is another polysaccharide and is also the main component of dietary fiber.

Fats
Most dietary fats consist of three fatty acids linked by oxygen bonds to a glycerol molecule. Depending on the type and number of oxygen bonds, fatty acids are either saturated (solid at room temperature) or unsaturated.

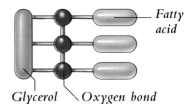

Fatty acid

Glycerol   Oxygen bond

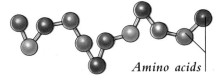

Amino acids

Proteins
Proteins are complex molecules with long chains of amino acids. These acids link in various ways to form many different proteins.

THE ROLE OF FIBER

Fiber, the indigestible part of plant foods, adds bulk to feces and speeds their passage through the bowel. It delays sugar absorption, so the sugar level in blood is better controlled. Fiber also binds with cholesterol and bile acids, which may be implicated in causing colon cancer, and may reduce the cholesterol level in blood.

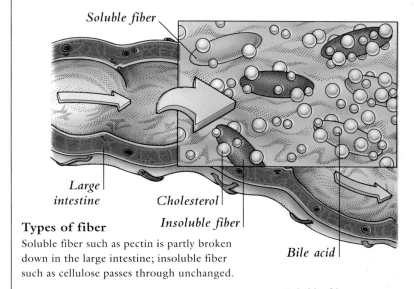

Soluble fiber
Large intestine
Cholesterol
Insoluble fiber
Bile acid

Types of fiber
Soluble fiber such as pectin is partly broken down in the large intestine; insoluble fiber such as cellulose passes through unchanged.

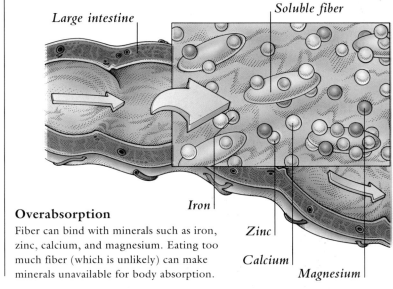

Large intestine
Soluble fiber
Iron
Zinc
Calcium
Magnesium

Overabsorption
Fiber can bind with minerals such as iron, zinc, calcium, and magnesium. Eating too much fiber (which is unlikely) can make minerals unavailable for body absorption.

HOW FOOD PROVIDES ENERGY

The breakdown products of food provide the fuel needed for the building and repair of body cells. Energy is released inside body cells by a complex chain of chemical reactions, which include the Krebs cycle. The energy released is then stored as chemical energy in the form of phosphate bonds. Splitting of these phosphate bonds releases the energy that is needed to power cell activities.

2 Energy is released when the body's main energy-carrying chemical, called adenosine triphosphate (ATP), converts to adenosine diphosphate (ADP). The ADP is continuously converted back into ATP, using energy that is released from glucose and fatty acids.

1 Glucose and fatty acids are the main fuels used by the Krebs cycle and linked reactions to produce energy. Amino acids may be used if the primary fuels are lacking.

ATP split, forming ADP and releasing energy

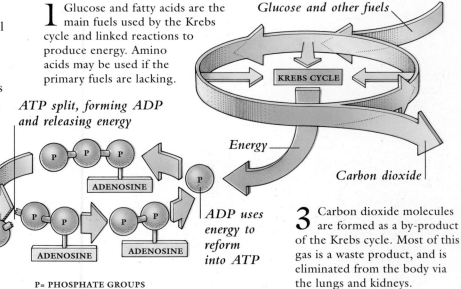

Glucose and other fuels
KREBS CYCLE
Energy
Carbon dioxide
ADP uses energy to reform into ATP
ENERGY
ADENOSINE
P= PHOSPHATE GROUPS

3 Carbon dioxide molecules are formed as a by-product of the Krebs cycle. Most of this gas is a waste product, and is eliminated from the body via the lungs and kidneys.

155

The MOUTH, PHARYNX, *and* ESOPHAGUS

THE PROCESS OF DIGESTION starts as soon as food enters the mouth. The food is chewed, lubricated and slightly digested by saliva, and moved back by the tongue. In about a minute, the food is formed into a soft, moist, round lump, the bolus. Each bolus is swallowed through the pharynx and moved into the esophagus, a muscular tube that passes food down to the stomach in about 1 to 2 seconds.

SWALLOWING

Swallowing begins as a voluntary process when food passes from the mouth into the pharynx. Automatic reflexes take over to control the subsequent stages of swallowing (see right): the muscles of the pharynx contract, moving the food along, then squeeze the food so that it moves into the top of the esophagus.

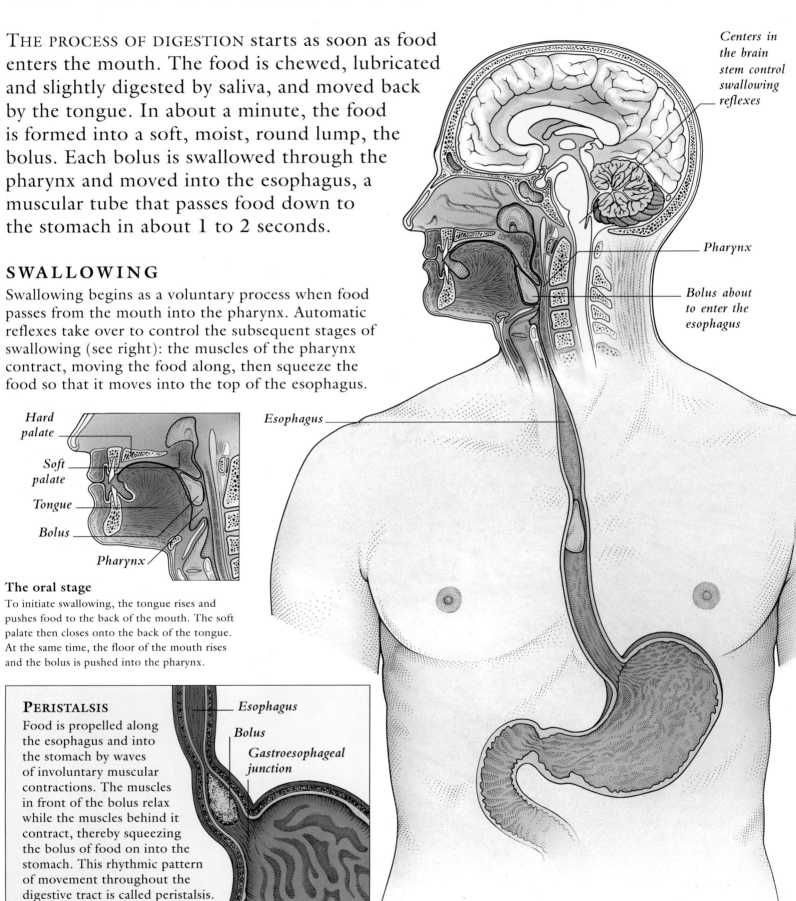

Centers in the brain stem control swallowing reflexes

Pharynx

Bolus about to enter the esophagus

Esophagus

The oral stage

To initiate swallowing, the tongue rises and pushes food to the back of the mouth. The soft palate then closes onto the back of the tongue. At the same time, the floor of the mouth rises and the bolus is pushed into the pharynx.

Hard palate
Soft palate
Tongue
Bolus
Pharynx

PERISTALSIS

Food is propelled along the esophagus and into the stomach by waves of involuntary muscular contractions. The muscles in front of the bolus relax while the muscles behind it contract, thereby squeezing the bolus of food on into the stomach. This rhythmic pattern of movement throughout the digestive tract is called peristalsis.

Esophagus
Bolus
Gastroesophageal junction

THE SALIVARY GLANDS

Saliva is produced by three pairs of salivary glands: the parotid, submandibular, and sublingual. There are also numerous small accessory glands in the mucous membrane lining the mouth and tongue. Saliva carried by ducts from the glands contains amylase, a digestive enzyme; it also makes chewing and swallowing easier.

The parotid glands

The parotid glands are the largest pair of salivary glands. Each gland is located in front of the ear, and has a duct that opens inside the cheek opposite the second upper molar tooth. The small accessory parotid glands lie just above the parotid duct.

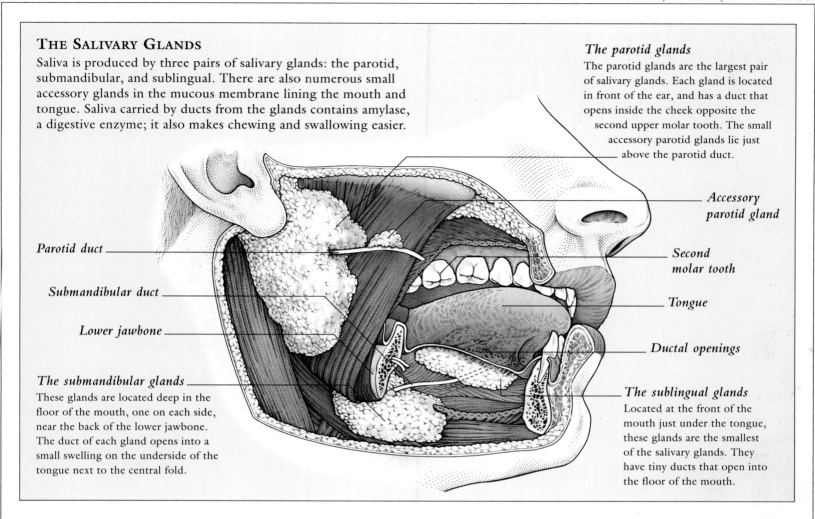

Parotid duct

Submandibular duct

Lower jawbone

The submandibular glands

These glands are located deep in the floor of the mouth, one on each side, near the back of the lower jawbone. The duct of each gland opens into a small swelling on the underside of the tongue next to the central fold.

Accessory parotid gland

Second molar tooth

Tongue

Ductal openings

The sublingual glands

Located at the front of the mouth just under the tongue, these glands are the smallest of the salivary glands. They have tiny ducts that open into the floor of the mouth.

THE ROLE OF THE TEETH

The teeth are made of hard bonelike material and are set in shock-absorbent gums and bone. The incisors are chisel-shaped with sharp edges for cutting, while the pointed canines are designed for tearing food. Premolars, with two ridges, and flatter molars (the largest and strongest teeth) crush and grind food.

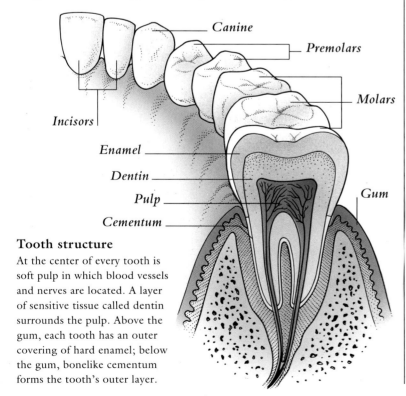

Canine

Premolars

Molars

Incisors

Enamel

Dentin

Pulp

Cementum

Gum

Tooth structure

At the center of every tooth is soft pulp in which blood vessels and nerves are located. A layer of sensitive tissue called dentin surrounds the pulp. Above the gum, each tooth has an outer covering of hard enamel; below the gum, bonelike cementum forms the tooth's outer layer.

BREATHING AND SWALLOWING

The pharynx is a channel for air as well as food. It leads into both the larynx (voice box) for breathing and the esophagus for swallowing. Reflex control of the epiglottis prevents food from entering the larynx during swallowing. If food goes down the wrong way, irritation of the airway lining triggers the coughing reflex to expel inhaled material and prevent choking.

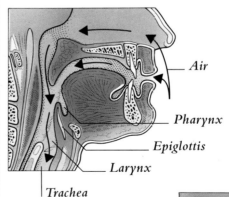

Air

Pharynx

Epiglottis

Larynx

Trachea

Breathing

During breathing, the vocal cords at the entrance to the larynx are relaxed and open, which creates a space between them known as the glottis. Air passes from the pharynx through the glottis into the trachea during inhalation, and passes from the trachea into the pharynx during exhalation

Swallowing

During swallowing, the flap of cartilage called the epiglottis tilts and the larynx rises up. The vocal cords are pressed together, closing the glottis and sealing off the entrance to the larynx. When food has entered the esophagus, the glottis reopens.

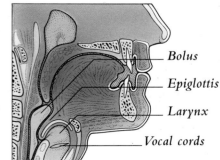

Bolus

Epiglottis

Larynx

Vocal cords

The STOMACH and SMALL INTESTINE

THE STOMACH IS A HOLLOW, ELASTIC SAC where food is churned and mixed thoroughly with juices secreted by the stomach lining. This process begins as soon as food enters the stomach. Processed food is released gradually into the small intestine, a coiled tube approximately 16ft (5m) long. The chemical breakdown of food is completed here by a variety of enzymes. The nonlipid products of digestion are absorbed through the intestinal lining into the bloodstream for processing by the liver.

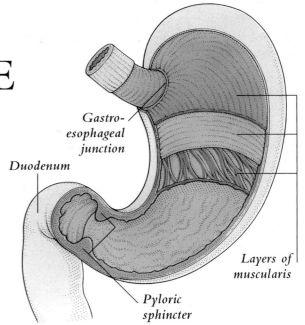

Gastro-esophageal junction

Duodenum

Layers of muscularis

Pyloric sphincter

STRUCTURE OF THE STOMACH

The stomach is J-shaped and forms the widest part of the digestive tube. At the gastroesophageal junction food enters the stomach; when it reaches a muscular ring known as the pyloric sphincter, it passes into the duodenum. The stomach wall is composed of four main layers: the serosa, muscularis, submucosa, and mucosa.

Mucosa
The mucosa, which contains gastric glands, lines the stomach. The surface of the mucosa is heavily folded and covered with numerous gastric pits.

Gastric pits
Three to seven gastric glands open into the bottom of each of these small indentations.

Gastric glands
Gastric glands produce about 6pt (3L) of gastric juice each day. Deep in the glands are specialized cells that secrete acid and enzymes, all of which play an essential part in the digestive process.

Muscular layers of mucosa
Two layers of muscle are located underneath the glands of the mucosa.

Submucosa
This layer of loose tissue connects the mucosa and muscularis.

Longitudinal layer of muscularis

Circular layer of muscularis

Oblique layer of muscularis

Subserous layer
This layer of loose tissue connects the serosa and muscularis.

Serosa
The outer surface of the stomach is coated with this clear membrane.

Lymph nodule

KEY

Acid-secreting cell (secretes hydrochloric acid)

Pepsinogen-secreting cell

Gastrin-secreting cell

Lipase-secreting cell

Mucus-secreting cell

SMALL INTESTINE

The duodenum, jejunum, and ileum together make up the small intestine. Short and curved, the duodenum receives secretions from the liver and pancreas. The jejunum and ileum are both long and coiled, but the jejunum is slightly thicker, redder, and shorter than the ileum. In the small intestine, food is broken down by pancreatic juice, bile, and intestinal secretions so that nutrients can be absorbed and utilized.

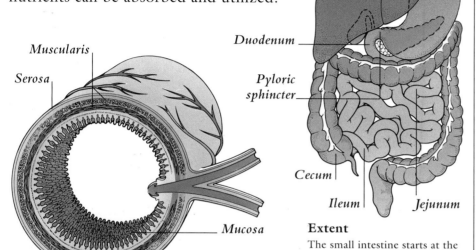

Intestinal cross-section

The intestinal wall has four layers. The outermost, protective coat is known as the serosa. Next is the muscularis, which contains outer longitudinal and inner circular muscle fibers. Adjoining this is the submucosa, a loose layer carrying vessels and nerves. The innermost layer is known as the mucosa.

Extent

The small intestine starts at the pyloric sphincter and ends at the pouchlike cecum, which is the beginning of the large intestine.

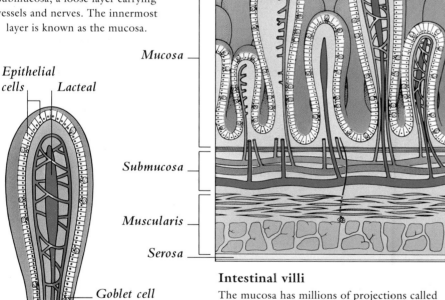

Intestinal villi

The mucosa has millions of projections called villi; each is covered by an epithelium, or cell layer, that absorbs nutrients. The epithelial cells have projections called microvilli. The villi and microvilli increase the surface area of the small intestine for efficient absorption.

Structure of a villus

The central core of each villus contains a lacteal, or lymph vessel, and a network of minute blood vessels. Goblet cells scattered throughout the epithelium secrete mucus.

MOVEMENT OF FOOD

The voluntary act of swallowing moves the bolus of food into the esophagus, which in turn propels food into the stomach. Waves of muscular contractions, called peristalsis, move food through the stomach, and then squirt small amounts into the duodenum. Peristalsis within the small intestine propels food toward the large intestine.

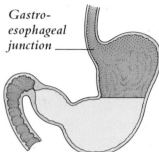

1 The muscular action of the stomach wall mixes food with gastric juice and churns it to form a thick, creamy substance called chyme.

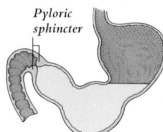

2 Peristaltic waves are most marked in the lower half of the stomach. These move the stomach contents toward the still-closed pyloric sphincter.

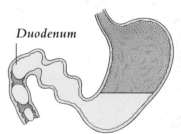

3 The valvelike pyloric sphincter, stimulated by chyme, opens to allow only small quantities of food to pass into the duodenum at a time.

INTESTINAL MOVEMENT

The contents of the small intestine are moved by peristalsis, pendular movements, and segmentation. Coordinated contraction waves are involved in peristalsis. Ringlike, evenly spaced contractions are known as segmentation. Short intestinal sections lengthen and shorten in pendular movements.

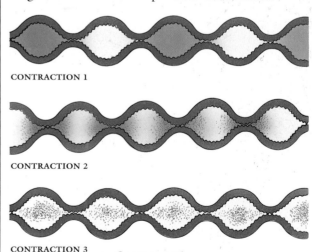

CONTRACTION 1

CONTRACTION 2

CONTRACTION 3

Segmentation

This series of concentric contractions, mixing chyme up to 12 times a minute, is the small intestine's main movement.

The LIVER, PANCREAS, *and* GALLBLADDER

THE LIVER, PANCREAS, AND GALLBLADDER are all closely connected to the digestive tract. The liver is the largest and also one of the most important internal organs. It is similar to a chemical processing factory and has many functions, including the production of the digestive liquid called bile. The gallbladder stores and concentrates this bile, while the pancreas produces a juice that contains powerful digestive enzymes. Each organ secretes these substances into the duodenum, the first part of the small intestine.

Canaliculi

LIVER FUNCTIONS

The liver produces cholesterol and bile from the breakdown of dietary fat and old red blood cells. Using amino acids, it makes proteins and stores iron, glycogen, and vitamins. It also removes substances such as poisons and waste products from the blood, excreting or converting them to safer substances.

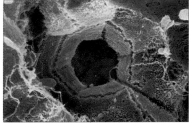

Cells of liver lobule   SEM x 1,050

LIVER STRUCTURE

The wedge-shaped liver is divided by a ligament into two lobes, with the left lobe smaller than the right. It is composed of thousands of distinctive hexagonal lobules made up of billions of cells. Tiny tubes called bile ducts form a network throughout the liver.

Branch of portal vein

Branch of hepatic artery

Red blood cell

Fat-storing cell

White blood cell

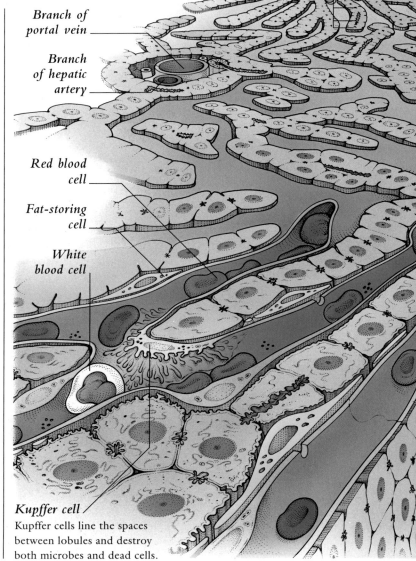

Kupffer cell
Kupffer cells line the spaces between lobules and destroy both microbes and dead cells.

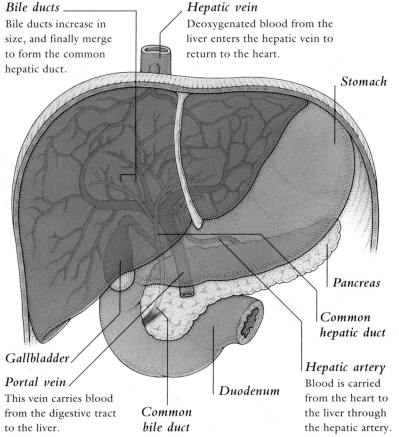

Bile ducts
Bile ducts increase in size, and finally merge to form the common hepatic duct.

Hepatic vein
Deoxygenated blood from the liver enters the hepatic vein to return to the heart.

Stomach

Pancreas

Common hepatic duct

Hepatic artery
Blood is carried from the heart to the liver through the hepatic artery.

Gallbladder

Portal vein
This vein carries blood from the digestive tract to the liver.

Common bile duct

Duodenum

PANCREAS: ROLE IN DIGESTION

The pancreas secretes a juice in response to food in the upper digestive tract. This juice is rich in enzymes that break down proteins, fats, and carbohydrates, and it also contains sodium bicarbonate, which neutralizes stomach acid. The enzymes are secreted into ductules that converge to form the pancreatic duct. This duct transports the enzymes to the duodenum.

Acinar cells of the pancreas
Grapelike clusters of cells in the pancreas, known as acini, contain globules of pancreatic enzymes.

SEM x 900

THE BILIARY SYSTEM

Ducts from the liver and the gallbladder, as well as the gallbladder itself, form the biliary system. It transports bile produced in the liver to the small intestine, where fats are digested. Two ducts, one from the gallbladder and one from the pancreas, meet at a point that is called the ampulla of Vater, the entry into the duodenum.

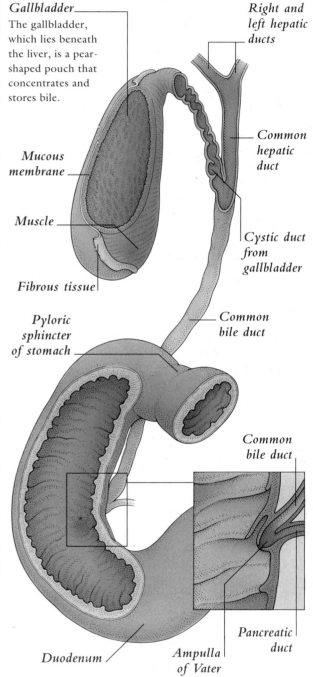

Gallbladder
The gallbladder, which lies beneath the liver, is a pear-shaped pouch that concentrates and stores bile.

Right and left hepatic ducts

Mucous membrane

Common hepatic duct

Muscle

Cystic duct from gallbladder

Fibrous tissue

Common bile duct

Pyloric sphincter of stomach

Common bile duct

Duodenum *Ampulla of Vater* *Pancreatic duct*

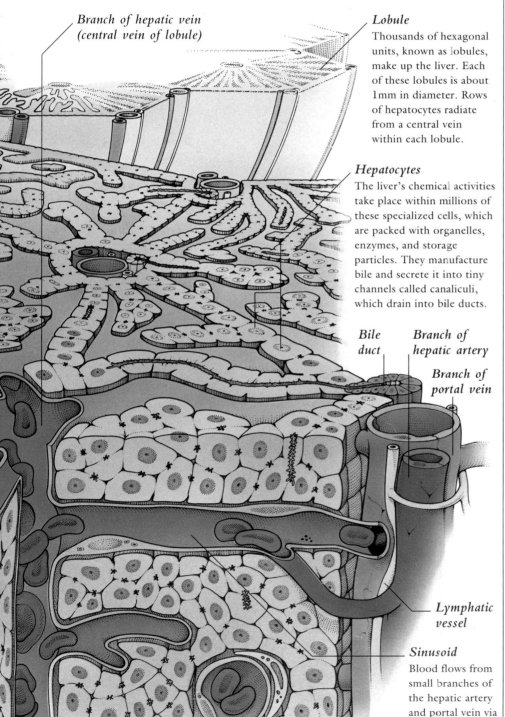

Branch of hepatic vein (central vein of lobule)

Lobule
Thousands of hexagonal units, known as lobules, make up the liver. Each of these lobules is about 1mm in diameter. Rows of hepatocytes radiate from a central vein within each lobule.

Hepatocytes
The liver's chemical activities take place within millions of these specialized cells, which are packed with organelles, enzymes, and storage particles. They manufacture bile and secrete it into tiny channels called canaliculi, which drain into bile ducts.

Bile duct *Branch of hepatic artery*

Branch of portal vein

Lymphatic vessel

Sinusoid
Blood flows from small branches of the hepatic artery and portal vein via sinusoids into the central vein.

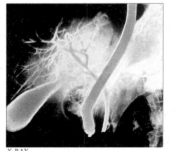

Site of stored bile
The saclike gallbladder, shown in orange, squeezes stored bile into the cystic duct when food is eaten. This duct then joins the common bile duct. A dye was introduced through the tube shown in green to create this image.

X-RAY

The COLON, RECTUM, and ANUS

THE LAST PART OF THE DIGESTIVE TRACT is composed of the colon, rectum, and anus. A short pouch called the cecum connects the small intestine to the colon. The cecum, colon, and rectum form the large intestine, which is 5ft (1.5m) long. By the time digested food reaches the colon, the nutrients essential for bodily functions have been absorbed. The digestive waste products are changed by the colon into feces that can be excreted via the rectum and anus.

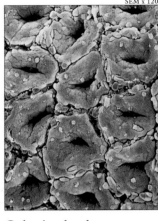

SEM x 120

Colonic glands
This microscopic image shows the openings to several of the tubular glands lining the colon. They absorb water from feces.

DIGESTIVE TRANSIT TIMES

After the first, voluntary stage of swallowing, the passage of food through the different parts of the digestive system is governed entirely by reflex actions. Shown in the illustration below is the approximate time that food takes to pass through each part. The time that food spends in the stomach and the colon depends on its type and on the individual person. People with longer transit times may be more prone to such disorders as cancer of the colon.

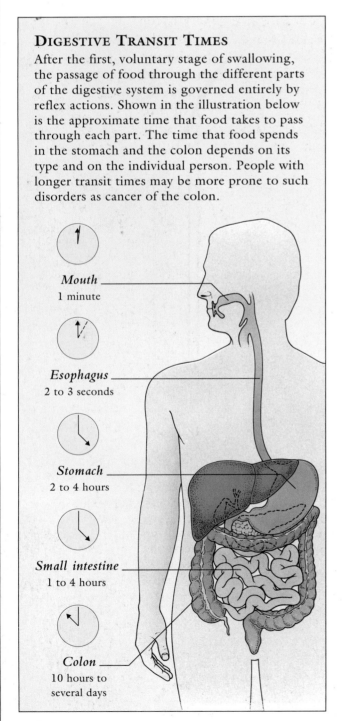

Mouth
1 minute

Esophagus
2 to 3 seconds

Stomach
2 to 4 hours

Small intestine
1 to 4 hours

Colon
10 hours to several days

ACTIVITY IN THE COLON

The main function of the colon is to convert the liquid that is digested food, from which all nutrients have been extracted in the small intestine, into feces. Vitamins K, some B vitamins, hydrogen, carbon dioxide, hydrogen sulfide, and methane are produced by the billions of bacteria that live in the colon. Mucus secreted by the colon lining contains antibodies that protect the colon from disease.

Absorption of water from feces
Sodium, chloride, and water are absorbed through the lining of the colon into the circulation so that feces become drier. Both bicarbonate and potassium are secreted by the colon to replace sodium and chloride.

Transverse colon

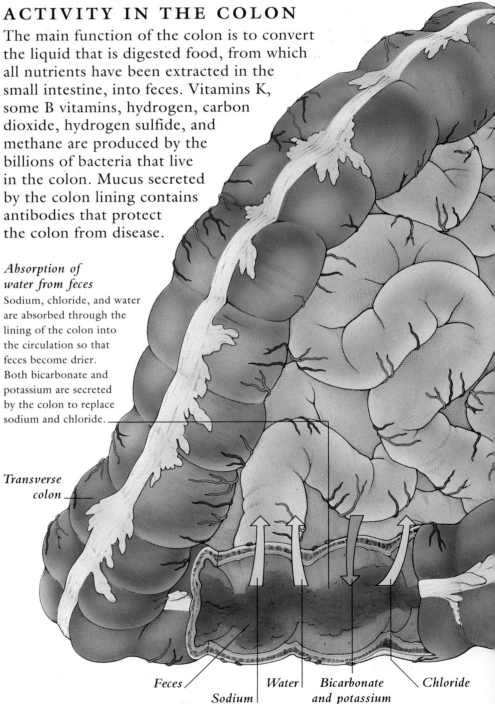

Feces

Sodium

Water

Bicarbonate and potassium

Chloride

COLONIC MOVEMENT

Muscular movement in the wall of the colon mixes and propels feces through the colon to the rectum. The movement of feces within the colon varies in rate, intensity, and nature. Types of movement include segmentation, mass movements, and peristaltic contractions. Feces move at a slower rate through the colon than they do through the small intestine. This permits the colon to reabsorb approximately 2.5pt (1.4L) of water every day.

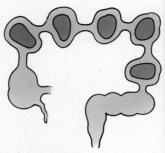

Segmentation
Segmentation describes a series of ringlike contractions that occur at regular intervals; these churn and mix feces, but do not move them.

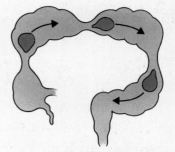

Peristaltic contractions
Waves of peristaltic contractions propel feces toward the rectum. Muscles behind food contract, while the muscles in front relax.

Mass movements
Mass movements are strong peristaltic waves that propel feces relatively long distances two or three times a day.

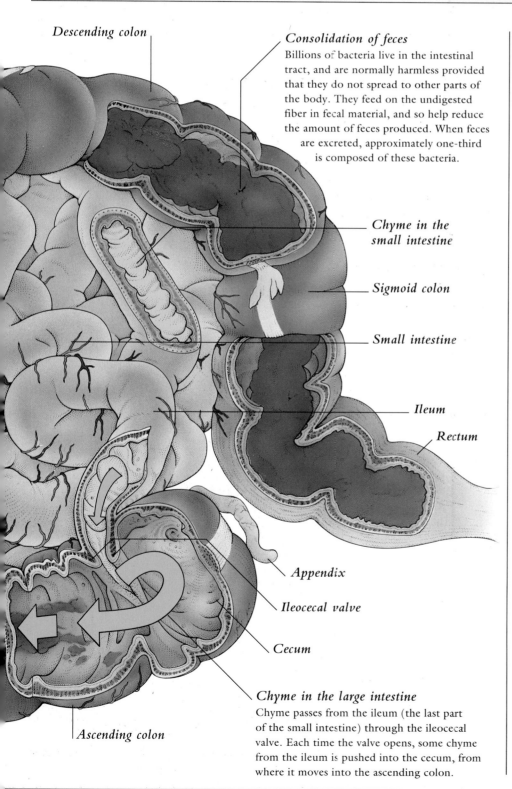

Descending colon

Consolidation of feces
Billions of bacteria live in the intestinal tract, and are normally harmless provided that they do not spread to other parts of the body. They feed on the undigested fiber in fecal material, and so help reduce the amount of feces produced. When feces are excreted, approximately one-third is composed of these bacteria.

Chyme in the small intestine

Sigmoid colon

Small intestine

Ileum

Rectum

Appendix

Ileocecal valve

Cecum

Chyme in the large intestine
Chyme passes from the ileum (the last part of the small intestine) through the ileocecal valve. Each time the valve opens, some chyme from the ileum is pushed into the cecum, from where it moves into the ascending colon.

Ascending colon

THE RECTUM AND ANUS

The rectum is about 5in (12cm) long, and it is normally empty except just prior to defecation. Just below the rectum lies the anal canal, which is about 1½in (4cm) long and lined with vertical ridges called anal columns. In the walls of the anal canal are two circular sheets of muscles called the internal and external sphincters, which act like valves and relax during defecation.

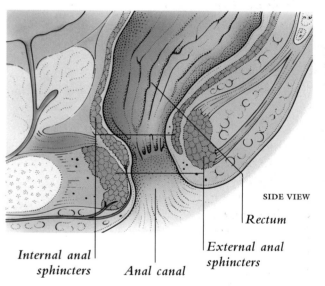

SIDE VIEW

Rectum

External anal sphincters

Internal anal sphincters

Anal canal

DEFECATION

Peristaltic waves in the colon push feces into the rectum, triggering the defecation reflex. Contractions raise the pressure in the rectum until the anal sphincters relax to allow feces to leave the body. Defecation is aided by voluntary contraction of the abdominal muscles, or is consciously overridden, a cause of constipation.

Feces

Anal canal

FRONT VIEW

STOMACH *and* DUODENAL DISORDERS

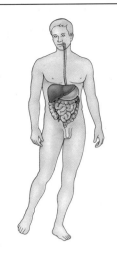

BENIGN ULCERS OF THE STOMACH AND DUODENUM, also called peptic ulcers, are among the most common conditions affecting the upper digestive tract. Other conditions include hiatal hernia, inflammation (such as gastritis and duodenitis), and cancer of the stomach. Doctors often investigate such disorders by endoscopy, in which a flexible fiberoptic viewing tube is passed down through the mouth.

HERNIAS

A hiatal hernia is the protrusion of part of the stomach through a normal opening in the diaphragm into the chest. It occurs most frequently in overweight, middle-aged, or elderly people, especially women. Abdominal organs may protrude through an abnormal opening elsewhere in the diaphragm in a diaphragmatic hernia.

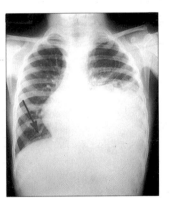

X-RAY

Diaphragmatic hernia

This type may be diagnosed shortly after birth either by ultrasound or X-ray. The opaque area located above the diaphragm (indicated by arrow) shows abdominal organs protruding into the chest cavity. The disorder may be life-threatening and is treated immediately by surgery.

Sliding hiatal hernia

This is the most common type of hiatal hernia. It occurs when the lower esophagus and the upper stomach slide into the chest cavity through an opening, or hiatus, in the diaphragm. As a result, normal pressures at the gastroesophageal junction are disturbed, permitting acid reflux and heartburn.

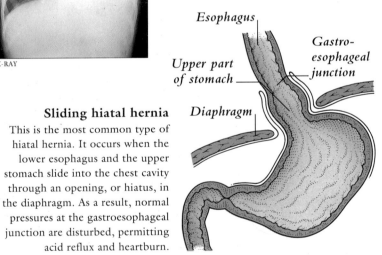

Esophagus

Upper part of stomach

Diaphragm

Gastro-esophageal junction

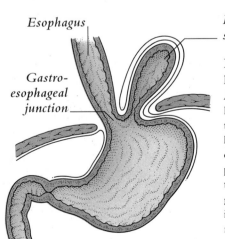

Esophagus

Gastro-esophageal junction

Pouch of stomach

Paraesophageal hiatal hernia

About 10 percent of all hiatal hernias are of this type. Part of the stomach lies adjacent to the lower esophagus after being pushed upward through the diaphragm. Since the gastroesophageal junction is not disturbed, there is no acid reflux.

STOMACH CANCER

The incidence of stomach cancer has declined, but it still accounts for almost 3 percent of US cancer deaths. It appears most commonly in men between the ages of 50 and 70 and the symptoms, if present, are similar to those of a peptic ulcer. *Helicobacter pylori* has been implicated as a cause. Gastrectomy offers a possible cure if the cancer has not yet spread to other organs.

Total gastrectomy

Rarely, the entire stomach is removed, and the esophagus joined to the jejunum. The cut end of the duodenum is closed. This operation prevents normal nutrient absorption, and the patient may become anemic without vitamin B_{12} injections.

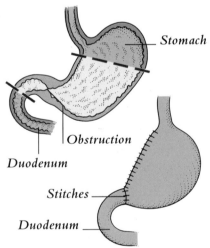

Stomach

Obstruction

Duodenum

Stitches

Duodenum

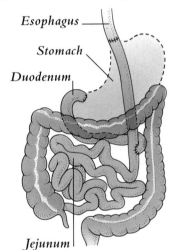

Esophagus

Stomach

Duodenum

Jejunum

Partial gastrectomy

Even if a complete cure is not possible, a partial gastrectomy (surgical removal of part of the stomach) may help if a cancer is obstructing the upper or, as shown here, the lower stomach.

GASTRITIS

The possible causes of gastritis, or inflammation of the stomach lining, include irritation caused by alcohol, nonsteroidal anti-inflammatory drugs (NSAIDs), and/or smoking tobacco. Recent research has focused on the role of infection with the bacterium *Helicobacter pylori* (shown at right) as another cause. Gastritis may appear suddenly or develop slowly over time. Symptoms may include nausea, upper abdominal pain, and indigestion.

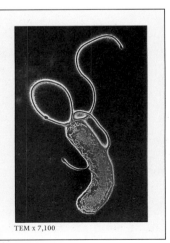

TEM x 7,100

PEPTIC ULCER

About 10 percent of people in developed countries may have experienced peptic ulceration during their lifetime. There appears to be no single cause, but the bacterium *Helicobacter pylori* is associated with 90 percent of duodenal ulcers and 70 percent of gastric ulcers. Other contributory factors include NSAIDs, smoking, stress, alcohol, and family history. The primary symptom is recurring upper abdominal pain relieved by food or antacid.

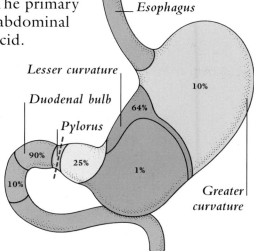

Sites of peptic ulcers

Most peptic ulcers are discovered in the duodenum, especially in the bulb. The lesser curvature is the most common site in the stomach. These ulcers may also occur in the lower esophagus and small intestine. The location frequency for the US population is shown at right.

DEVELOPMENT OF A PEPTIC ULCER

A layer of mucus secreted by the mucosal cells normally protects the lining of the stomach and duodenum from attack by hydrochloric acid, pepsin (a digestive enzyme), and other substances that are potentially harmful.

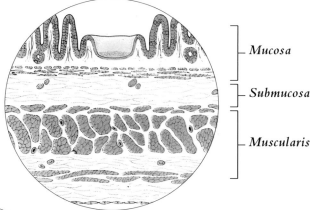

1 Damage occurs when the protective mucus barrier breaks down and stomach juice comes into contact with cells of the lining. In the early stages, the mucosa is only partly destroyed, producing a shallow area of damage called an erosion.

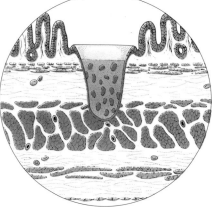

2 A true ulcer penetrates the entire mucosal layer, and usually also the submucosa and muscularis layers. Peptic ulcers tend to be round or oval. They can become chronic, destroying tissue and healing it with scar formation.

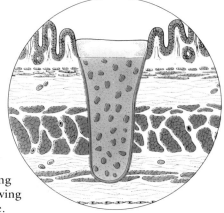

3 A peptic ulcer may burrow through the entire thickness of the wall. This can result in erosion into a large artery, causing a hemorrhage, or perforation of the wall and inflammation of the abdominal lining (peritonitis). Another complication is narrowing of the outlet from the stomach by scar tissue.

DRUG TREATMENT

Various drugs can help treat a peptic ulcer. Antacids neutralize stomach acid while mucosal protectors provide a protective coat. Histamine (H_2) blockers and proton pump inhibitors block the production of acid by cells in the stomach lining. Newer treatments eliminate *H. pylori* by using combinations of bismuth and a variety of antibacterial agents.

Action of antacids

Antacids containing compounds of magnesium, aluminum, and calcium are alkaline substances that neutralize stomach acid. They temporarily relieve ulcer pain and can promote healing.

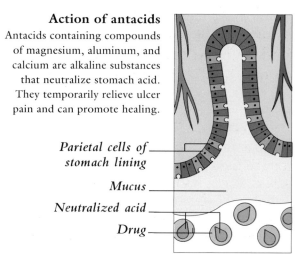

Parietal cells of stomach lining

Mucus

Neutralized acid

Drug

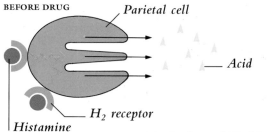

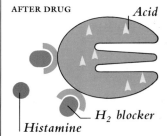

Action of H_2 blockers

Normally, histamine (H_2) stimulates the secretion of acid by occupying specific H_2 receptors on the surface of parietal cells in stomach mucosa. H_2 blockers occupy these receptors, inhibiting the action of histamine and reducing acid secretion.

SURGICAL TREATMENT

In spite of improved drug therapy for peptic ulcers, some still require surgery in the event of persistent ulceration, perforation, bleeding, and obstruction due to scarring. A partial gastrectomy can involve removing the acid-secreting portion of the stomach and joining the remnants to the small intestine.

Vagotomy

Vagotomy reduces stomach acid secretion by preventing nerve signals stimulating acid secretion from reaching the stomach. In truncal vagotomy, the whole vagus nerve is cut. In selective types of vagotomy, only certain nerve branches are cut (as shown here).

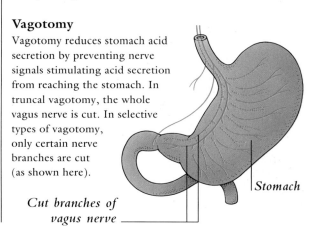

Cut branches of vagus nerve

LIVER DISORDERS

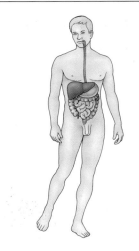

A COMMON CAUSE OF LIVER DISEASE in developed countries is the excessive consumption of alcohol. Other causes include viral infection of the liver and adverse reactions to drugs and chemicals. Damage from any of these can lead to cirrhosis. Cancer can arise in the liver, but more often spreads from another body part, such as the digestive tract or lungs. In a rare condition that is called biliary atresia, the bile ducts are abnormal at birth. An affected infant becomes jaundiced and develops serious liver damage if not treated within a few weeks.

ALCOHOLIC LIVER DISEASE

Persistent alcohol abuse leads to cirrhosis, which is irreversible. Alcoholic liver disease may produce no symptoms in early stages, but blood tests can reveal abnormal liver function. As time goes on, jaundice and severe fluid retention develop. Alcohol is partly broken down by stomach enzymes; because women have fewer enzymes, they are at greater risk.

PROGRESSION OF THE DISEASE

Alcohol damage to the liver initially causes an abnormal accumulation of fat in liver cells (fatty liver). Sometimes damage causes inflammation, or alcoholic hepatitis. If people who have either fatty liver or alcoholic hepatitis continue to abuse alcohol, cirrhosis and liver failure can develop.

How damage occurs
Some alcohol is excreted in the urine or breath unchanged, but most is converted by enzymes in the liver into acetaldehyde. Both alcohol and acetaldehyde are toxic to liver cells.

Alcohol   *Liver cell*

Fat-laden cells

Acetaldehyde   *Water*

Fatty liver
Liver cells become infiltrated with globules of fat, which results in an enlarged and tender liver.

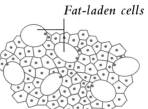

Alcoholic hepatitis
As a result of the production of acetaldehyde, liver cells become acutely inflamed and damaged so that liver function is impaired.

Damaged tissue

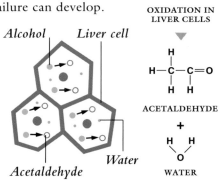

Scar tissue

Cirrhosis
In cirrhosis, bands of scar tissue separate nodules of overgrown cells. At this stage damage is irreversible, but if sufficient functioning liver remains, it can be life saving to stop drinking.

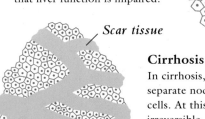

```
    H   H
    |   |
H — C — C — OH
    |   |
    H   H
   ALCOHOL
```
▼
**OXIDATION IN
LIVER CELLS**
▼
```
    H   H
    |   |
H — C — C = O
    |
    H
 ACETALDEHYDE
```
+
```
    H   H
     \ /
      O
    WATER
```

PORTAL HYPERTENSION

Cirrhotic scar tissue can obstruct blood flow in the liver, leading to a rise in pressure in the portal vein. Increased pressure causes reversal of blood flow in some portal system vessels with enlargement of veins in the lower esophagus and upper stomach. The veins may burst, thereby causing a hemorrhage.

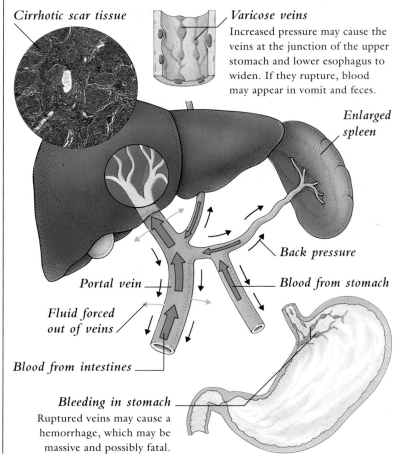

Cirrhotic scar tissue

Varicose veins
Increased pressure may cause the veins at the junction of the upper stomach and lower esophagus to widen. If they rupture, blood may appear in vomit and feces.

Enlarged spleen

Back pressure

Portal vein

Blood from stomach

Fluid forced out of veins

Blood from intestines

Bleeding in stomach
Ruptured veins may cause a hemorrhage, which may be massive and possibly fatal.

SIGNS OF LIVER DISEASE

An important sign of liver disease, jaundice, causes yellow discoloration of the eyes and skin (seen at right). Other symptoms and signs of liver disease include nausea, loss of both appetite and weight, swelling of the abdomen, abnormal blood clotting, dilation of tiny arterioles in the skin, and breast enlargement in men.

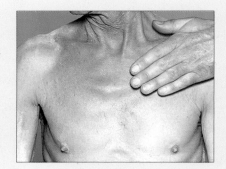

HEPATITIS

Hepatitis, or liver inflammation, is commonly caused by a viral infection, especially with hepatitis A, B, or C viruses. Viral hepatitis is usually an acute, short-lived illness. Chronic hepatitis B or C, however, can lead to cirrhosis; it also increases the risk of liver cancer. Hepatitis A and B can be prevented by immunization. Some other causes of hepatitis include certain herbal medicaments, poisoning by toxic chemicals, including drugs and alcohol, and bacterial infection.

Hepatitis B virus

A protein coat with surface antigens (other proteins) surrounds the DNA core. This infection is transmitted through contaminated blood or drug needles, sexual intercourse, or from an infected mother to her baby at birth.

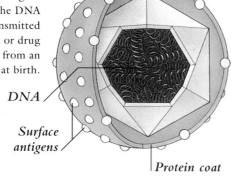

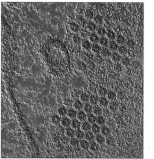

TEM x 139,000

DNA

Surface antigens

Protein coat

Hepatitis A virus

Hepatitis A virus particles (the red circles) are small, 20-sided structures containing RNA. Hepatitis A is usually contracted by ingesting contaminated food or water.

LIVER ABSCESS

An abscess, or collection of pus, in the liver can be caused by infection with bacteria or amoebas. Bacteria may have spread from an infected site in other parts of the body, such as the appendix. Amoebic liver abscess is common in tropical countries and may be preceded by diarrhea. Liver abscesses cause fever, chills, weight loss, liver enlargement, and chest and abdominal pain.

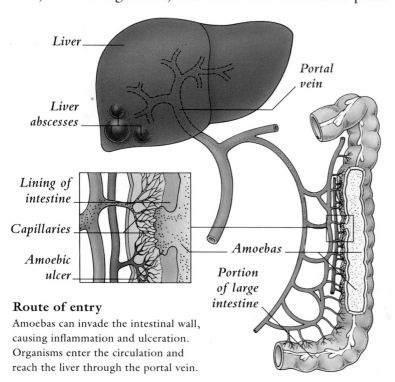

Liver

Liver abscesses

Portal vein

Lining of intestine

Capillaries

Amoebic ulcer

Amoebas

Portion of large intestine

Route of entry

Amoebas can invade the intestinal wall, causing inflammation and ulceration. Organisms enter the circulation and reach the liver through the portal vein.

OPERATION

LIVER TRANSPLANT

The replacement of a diseased liver with one from a donor may benefit people who have a life-threatening liver disease such as acute hepatitis or cirrhosis with liver failure, primary biliary cirrhosis (an autoimmune disorder), or congenital biliary atresia. Patients should have no major infection, and no heart or lung disease. Immunosuppressant drugs, to prevent organ rejection, must be taken for the rest of the patient's life.

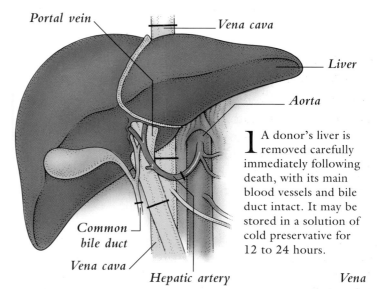

Portal vein

Vena cava

Liver

Aorta

Common bile duct

Vena cava

Hepatic artery

1 A donor's liver is removed carefully immediately following death, with its main blood vessels and bile duct intact. It may be stored in a solution of cold preservative for 12 to 24 hours.

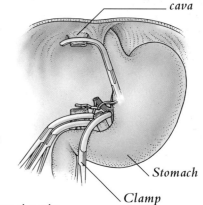

Vena cava

Stomach

Clamp

2 The recipient's abdomen is opened, and the vena cava, the major vein in the abdomen, is clamped above and below the liver. During the operation, venous blood is diverted via a bypass. The hepatic artery, the bile duct, and the portal vein are all cut so that the entire diseased liver can be removed.

3 The donor's liver is connected to the vena cava and other vessels, and the cut ends of the bile duct are joined. A T-shaped tube is temporarily inserted in the reconstructed bile duct, permitting drainage while tissue healing occurs.

Retractors

Stomach

Liver

Gallbladder

T-tube splint

Large intestine

Common bile duct

Portal vein

Vena cava

Aorta

GALLBLADDER *and* PANCREATIC DISORDERS

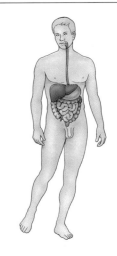

MOST DISORDERS OF THE GALLBLADDER are due to the presence of gallstones. However, people with gallstones do not necessarily experience any trouble. Inflammation of the pancreas may be due to gallstones, alcohol abuse, or viral infections. Other pancreatic disorders are cancer and pseudocysts. Damage to the pancreas can cause diabetes mellitus and impair digestion of food.

CANCER OF THE PANCREAS

Pancreatic cancer occurs most frequently in elderly people. The main symptom of cancer of the pancreatic body is dull, upper abdominal pain that penetrates to the back. Other symptoms include loss of appetite and weight, and jaundice. The cancer may spread from the head of the pancreas directly into the duodenum, and through the bloodstream to the liver and lungs.

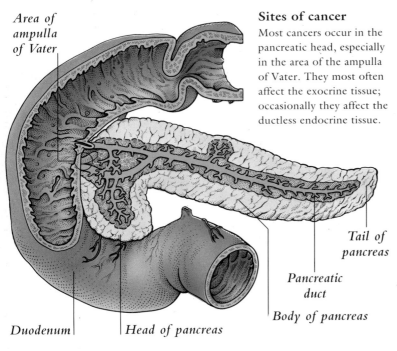

Area of ampulla of Vater

Duodenum *Head of pancreas*

Body of pancreas

Pancreatic duct

Tail of pancreas

Sites of cancer
Most cancers occur in the pancreatic head, especially in the area of the ampulla of Vater. They most often affect the exocrine tissue; occasionally they affect the ductless endocrine tissue.

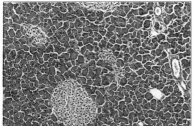
LM

Healthy tissue
The duct-filled exocrine tissue is composed of cell clusters known as acini, which secrete digestive enzymes. The large, pale circle is an islet of Langerhans, endocrine tissue that secretes hormones directly into the bloodstream.

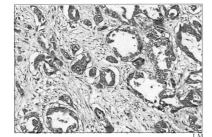
LM

Cancerous tissue
In pancreatic cancer, groups of irregularly shaped malignant cells with enlarged nuclei replace the smaller, healthy cells. The tissue lacks the organized appearance of normal pancreatic tissue.

OPERATION

DRAINING A PSEUDOCYST

A pancreatic pseudocyst is a fluid-filled sac that usually develops between the pancreas and the stomach. These sacs are often a result of inflammation of the pancreas, called pancreatitis. Symptoms include nausea, fever, and swelling in the upper abdomen. Many cysts disappear without treatment, but others require surgery.

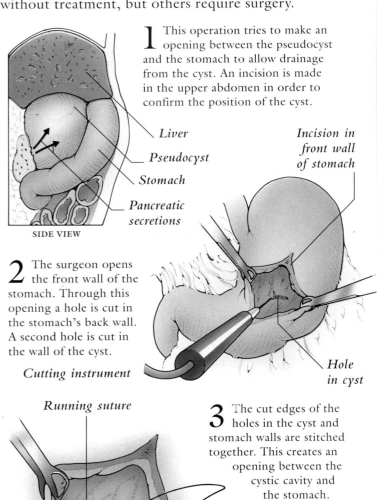

1 This operation tries to make an opening between the pseudocyst and the stomach to allow drainage from the cyst. An incision is made in the upper abdomen in order to confirm the position of the cyst.

Liver
Pseudocyst
Stomach
Pancreatic secretions
SIDE VIEW

Incision in front wall of stomach

2 The surgeon opens the front wall of the stomach. Through this opening a hole is cut in the stomach's back wall. A second hole is cut in the wall of the cyst.

Cutting instrument

Hole in cyst

Running suture

3 The cut edges of the holes in the cyst and stomach walls are stitched together. This creates an opening between the cystic cavity and the stomach.

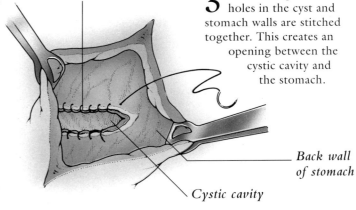

Back wall of stomach

Cystic cavity

GALLSTONES

Most gallstones, which are made of bile pigment and cholesterol, result from an imbalance in the chemical composition of bile. These gallstones are particularly common among overweight, middle-aged women. Gallstones can travel from the gallbladder into the cystic duct, but may then fall back into the cavity of the gallbladder, pass through the common bile duct into the duodenum, or become impacted in ducts.

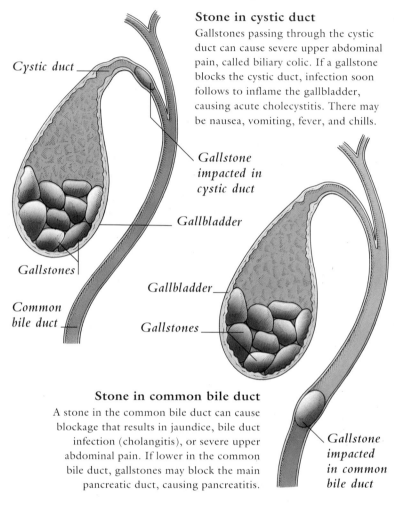

Stone in cystic duct

Gallstones passing through the cystic duct can cause severe upper abdominal pain, called biliary colic. If a gallstone blocks the cystic duct, infection soon follows to inflame the gallbladder, causing acute cholecystitis. There may be nausea, vomiting, fever, and chills.

Cystic duct

Gallstone impacted in cystic duct

Gallbladder

Gallstones

Common bile duct

Gallbladder

Gallstones

Gallstone impacted in common bile duct

Stone in common bile duct

A stone in the common bile duct can cause blockage that results in jaundice, bile duct infection (cholangitis), or severe upper abdominal pain. If lower in the common bile duct, gallstones may block the main pancreatic duct, causing pancreatitis.

LESS COMMON COMPLICATIONS

An inflamed gallbladder may become filled with pus, a condition called empyema, or it may perforate and leak, resulting in bile peritonitis. A mucocele forms when the gallbladder becomes distended with mucus. If a fistula forms between the gallbladder and intestine, a gallstone can cause obstruction of the bowel. Repeated episodes of cholecystitis can scar and shrink the gallbladder.

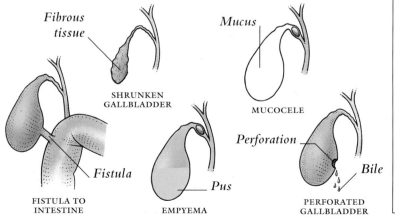

Fibrous tissue

SHRUNKEN GALLBLADDER

Fistula

FISTULA TO INTESTINE

Pus

EMPYEMA

Mucus

MUCOCELE

Perforation

Bile

PERFORATED GALLBLADDER

TREATMENT OF GALLSTONES

Gallstones require treatment only if they are causing symptoms, in which case removal of the gallbladder (cholecystectomy) and of the stones obstructing the common bile duct is necessary. Most cholecystectomies no longer require an operation in which the abdomen is opened, but may be carried out laparoscopically. Small or medium-sized stones may be dissolved by certain drugs.

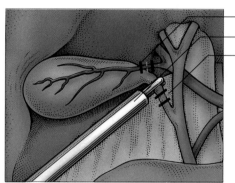

Clips

Grasping forceps

Incision

Laparoscopic cholescystectomy

An endoscope, a fiberoptic viewing instrument, is inserted through a small incision made into the upper abdomen. The gallbladder is severed from its connections and removed.

OPERATION

REMOVING A GALLSTONE

Gallstones can be removed from the lower end of the common bile duct by a flexible endoscope, which is a fiberoptic viewing instrument. The tube is passed into the mouth, down the esophagus, through the stomach, and finally into the duodenum. Special fine instruments are passed down the endoscope through the ampulla of Vater into the common bile duct.

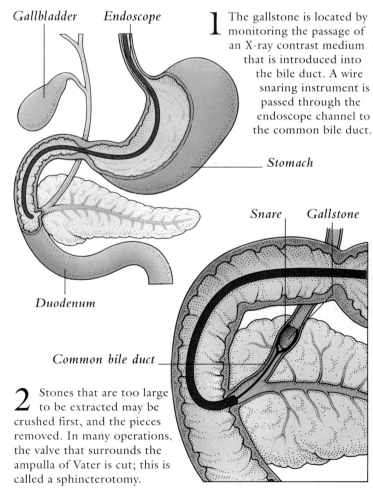

Gallbladder   *Endoscope*

1 The gallstone is located by monitoring the passage of an X-ray contrast medium that is introduced into the bile duct. A wire snaring instrument is passed through the endoscope channel to the common bile duct.

Stomach

Snare   *Gallstone*

Duodenum

Common bile duct

2 Stones that are too large to be extracted may be crushed first, and the pieces removed. In many operations, the valve that surrounds the ampulla of Vater is cut; this is called a sphincterotomy.

INTESTINAL, RECTAL, *and* ANAL DISORDERS

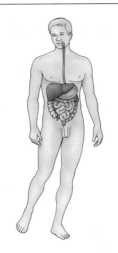

INTESTINAL INFECTIONS are the most common disorders of the digestive tract. In developing regions of the world, many children die from these disorders; in industrialized countries, where cancer and chronic intestinal inflammation are more serious conditions, intestinal infections usually cause only minor problems. Fever, chills, diarrhea, constipation, pain, and rectal bleeding are symptoms.

INFLAMMATORY BOWEL DISEASE

The term inflammatory bowel disease includes ulcerative colitis and Crohn's disease; both result in chronic intestinal inflammation. They may be caused by the immune system attacking the body's own tissues; other family members may also be affected. Bleeding, abdominal pain, fever, and diarrhea are the chief symptoms. A diagnosis is usually made by colonoscopy and microscopy of bowel tissue specimens. The treatment usually includes anti-inflammatory drugs.

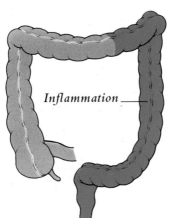

Inflammation

Ulcerative colitis
Ulcerative colitis affects all or part of the rectum and colon. The main symptom is the passage of rectal blood in small or massive quantities. Those affected are at increased risk of colorectal cancer and should have regular checkups.

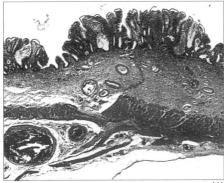

Inflamed membrane
This sample of tissue from a colon affected by ulcerative colitis shows the typical pattern of discontinuity and inflammation of the mucosa.

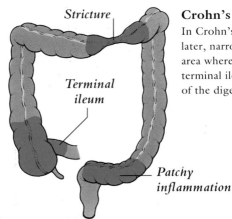

Stricture

Terminal ileum

Patchy inflammation

Crohn's disease
In Crohn's disease, patches of inflammation and, later, narrowing may develop in the intestine. The area where the large and small intestines meet (the terminal ileum) is often affected, although any part of the digestive tract may be involved.

Inflamed wall
Inflammation of the intestinal wall, with ulceration that extends into the submucous layer (arrowed), is a typical effect of Crohn's disease.

DIVERTICULAR DISEASE

Diverticular disease most often affects the lower colon in elderly people. The disease includes diverticulosis (the appearance of pouches, or diverticula, in the intestinal wall) and diverticulitis (inflammation of these pouches). Many affected people have no symptoms, but some have abdominal pain and swelling, diarrhea, constipation, gas, and rectal bleeding. Low-fiber diets and constipation are contributing factors.

Hard, dry feces

Wall of colon

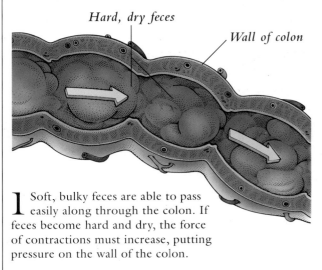

1 Soft, bulky feces are able to pass easily along through the colon. If feces become hard and dry, the force of contractions must increase, putting pressure on the wall of the colon.

Hard, dry feces

Diverticula push through weak parts of muscle walls

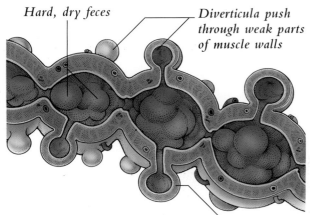

Pouches can become inflamed

2 Eventually, the increased pressure pushes the intestinal lining through points of weakness in the muscle of the intestinal wall. The pouches that form trap feces and bacteria and may become inflamed, causing pain and fever.

IRRITABLE COLON SYNDROME

Irritable colon syndrome may affect up to 40 percent of the population, but relatively few people consult a physician. This chronic condition is the result of a disturbance of muscular movement within the large intestine. Symptoms can be aggravated by anxiety and may include alternating constipation and diarrhea, gas, bloating, and abdominal pain. Treatments include a high-fiber diet, relaxation, and antispasmodic drugs.

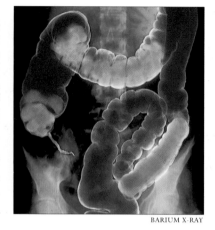

BARIUM X-RAY

Making a diagnosis

Before the diagnosis of irritable colon syndrome is made, other conditions must be excluded. A flexible sigmoidoscopy (using a viewing tube for examining the rectum and sigmoid colon) and barium X-ray help distinguish the disorder from inflammatory bowel disease, diverticular disease, and cancer of the digestive tract. The colon at left appears normal.

CANCER OF THE COLON

Colon cancer is one of the most common cancers in industrialized countries. Risk factors are family history, inflammatory bowel disease, intestinal polyps, and increasing age. Common symptoms are fecal blood, a change in bowel habits, and onset of abdominal pain. People over 50 should have flexible sigmoidoscopy. If polyps or cancer is found, colonoscopy should follow.

Colonoscopy

A colonoscope (flexible viewing tube) is passed through the anus into the colon to help establish the cause of intestinal symptoms, to locate tumors and inflamed areas, and to obtain intestinal tissue samples for examination.

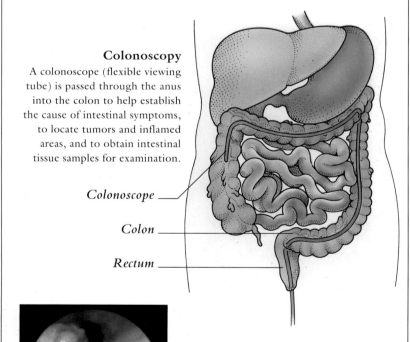

Colonoscope

Colon

Rectum

Polyp in colon

Polyps are abnormal growths projecting from a mucous membrane, such as the lining of the colon shown at left. They may cause rectal bleeding, resulting in anemia, or undergo malignant change. They are removed during colonoscopy.

ENDOSCOPIC VIEW

OPERATION

PARTIAL COLECTOMY

Removing the cancer, a border of normal intestine above and below it, and all regional lymph nodes is the current cancer treatment. The remaining healthy colon is then rejoined. A temporary colostomy, in which a small opening is made in the abdominal wall to permit feces to be excreted, precedes colectomy when the cancer is large enough to cause obstruction.

1 Cancer of the colon often begins as a polyp in the mucous glands of the lining of the intestine. The cancer may invade the intestinal wall and may also spread to nearby lymph nodes, and to more distant organs.

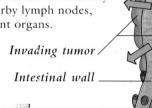

Invading tumor

Intestinal wall

Midline incision

Descending colon

Ascending colon

Tumor

2 An incision is made in the abdominal wall to enter the abdominal cavity. All abdominal organs are inspected, and the tumor's position is confirmed.

Incision   *Clamp*   *Tumor*

Transverse colon tied off

Blood vessels

Mesentery

3 Clamps are placed above and below the diseased area, and the section of colon containing the tumor is cut out. The mesentery (with its lymphatic vessels and glands and blood vessels) attached to the cancer is also removed.

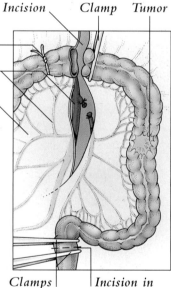

Clamps   *Incision in sigmoid colon*

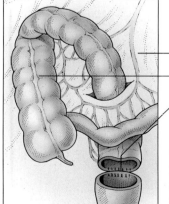

Mesentery

Ascending colon

Cut ends rejoined

4 The healthy cut ends of the colon are sutured together. In some cases, the surgeon uses a special type of surgical stapling instrument to connect the colon directly to the rectum.

INTESTINAL INFECTIONS

The most common intestinal infection is viral gastroenteritis, but infection can also be caused by bacteria and protozoa. Most forms of this disease are transmitted via contaminated water or food. The common symptoms include vomiting, diarrhea, and cramps. Viral gastroenteritis usually clears up within a few days; replacement of lost fluid is essential. Bacterial and parasitic infections are treated with antimicrobials.

Giardia

Giardia lamblia is a protozoan that is shaped like a pear. It attaches to the upper small intestine. Infection is a risk for people traveling to areas of the world where the water supplies are contaminated.

SEM x 2,070

Salmonella

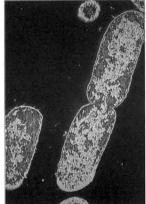

Salmonella typhimurium (seen at right) causes outbreaks of gastroenteritis that may be triggered by consumption of infected eggs or poultry. Salmonella are able to enter the bloodstream, resulting in fever, chills, or abscesses.

SEM x 21,250

APPENDICITIS

Inflammation of the appendix is common, especially in children. It causes acute pain and tenderness in the lower right part of the abdomen. Other symptoms are loss of appetite, nausea, vomiting, and mild fever. Surgical removal of the appendix, called appendectomy, is the usual treatment. An untreated appendix can rupture, causing peritonitis (inflammation of the abdominal lining) and the formation of abscesses.

Inflamed appendix

A short, closed-ended tube, the appendix projects from a cul-de-sac called the cecum. Appendicitis can be triggered by blockage of the appendix or by ulceration of its lining.

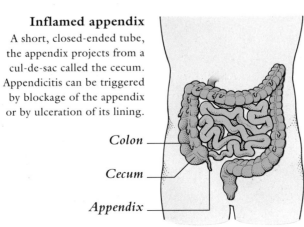

Colon

Cecum

Appendix

INTESTINAL OBSTRUCTION

An obstruction in the intestine can be a surgical emergency. It can cause abdominal pain and distention, the absence of fecal excretion, and sometimes vomiting and dehydration. Obstruction can be confirmed by taking abdominal X-rays. Treatment usually includes the administration of intravenous fluids and the suctioning of fluid from the bowel.

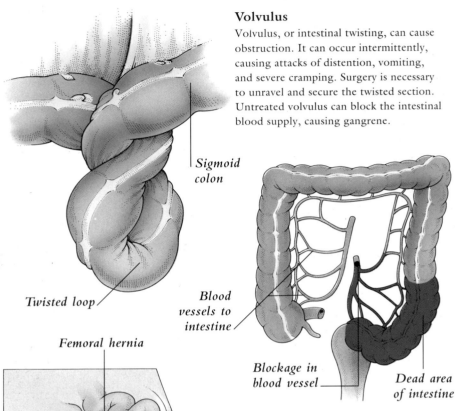

Volvulus

Volvulus, or intestinal twisting, can cause obstruction. It can occur intermittently, causing attacks of distention, vomiting, and severe cramping. Surgery is necessary to unravel and secure the twisted section. Untreated volvulus can block the intestinal blood supply, causing gangrene.

Sigmoid colon

Twisted loop

Blood vessels to intestine

Blockage in blood vessel

Dead area of intestine

Femoral hernia

Intestinal infarction

The blockage of a blood vessel in the intestinal membrane – the mesentery – deprives a segment of intestine of its blood supply. This condition, known as intestinal infarction, results in a bowel obstruction. The affected segment of intestine becomes gangrenous unless surgery is performed.

Hernia

A hernia is an abnormal protrusion through a weakness in the abdominal wall. In the type shown, intestine passes through the femoral canal, and can be trapped, causing severe pain and obstruction. Inguinal hernia also occurs.

BLOCKAGE IN CHILDREN

Intestinal obstruction in young children can be the result of intussusception, a condition in which a segment of intestine telescopes in on itself to form a tube within a tube. Symptoms include abdominal pain and the passage of feces that resemble redcurrant jelly. Intussusception may be life-threatening, but it can be unblocked by surgery or sometimes by a barium enema.

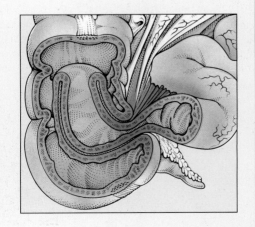

RECTAL AND ANAL DISORDERS

Common symptoms of disorders affecting the rectum and anus are bleeding, pain, constipation, discharge, itching, and the presence of a lump. Rectal and anal infections, including sexually transmitted ones, occur most often among people who have anal intercourse. Anal cancer is less common than rectal cancer and can follow infection by anal wart viruses.

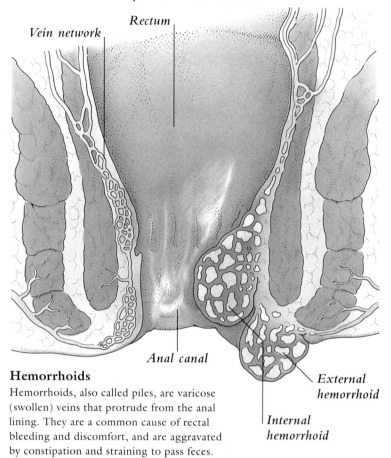

Vein network

Rectum

Anal canal

External hemorrhoid

Internal hemorrhoid

Hemorrhoids

Hemorrhoids, also called piles, are varicose (swollen) veins that protrude from the anal lining. They are a common cause of rectal bleeding and discomfort, and are aggravated by constipation and straining to pass feces.

RECTAL CANCER

Rectal cancer accounts for one-fifth to one-quarter of cancers of the large intestine, and is more common in people between the ages of 50 and 70. Usually a rectal examination, followed by a tissue biopsy, helps confirm the diagnosis. Surgery is required to remove the cancer

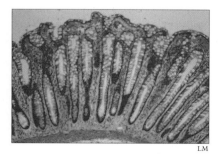

LM

Normal rectal cells

This section of normal rectal tissue shows the organized structure of the mucous membrane lining. Parallel columns of gland cells lie perpendicular to the surface.

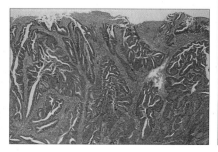

Cancerous rectal cells

The normal arrangement of glandular tissue in the rectal lining has been completely disrupted by cancerous cells so that the specimen has a disorganized appearance.

OPERATION

ANTERIOR RESECTION OF RECTUM

Cancer of the upper rectum may be treated by anterior resection. This operation removes the tumor and some normal bowel on either side. The remaining rectum is then joined to the colon. Cancer of the lower part of the rectum requires removal of the entire rectum along with the anus; a colostomy needs to be carried out.

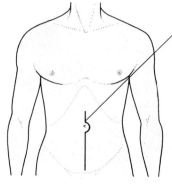

Incision

1 The patient is prepared for surgery with antibiotics, and a catheter is inserted into the bladder. After the patient has been given an anesthetic, a vertical incision is made in the abdominal wall to gain access to the diseased rectum.

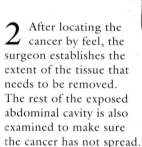

Catheter in bladder　*Rectum*　*Tumor*

2 After locating the cancer by feel, the surgeon establishes the extent of the tissue that needs to be removed. The rest of the exposed abdominal cavity is also examined to make sure the cancer has not spread.

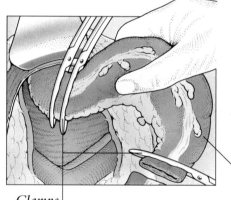

3 Clamps are placed on either side of the cancer. The isolated part of the intestine is then cut away from the surrounding structures and carefully removed.

Diseased rectum

Clamps

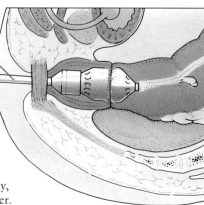

Staple gun

4 A staple gun is used to rejoin the cut ends of healthy intestine. The gun is inserted into the rectum through the anus and delivers a ring of tiny metal staples that hold the edges firmly together. Alternatively, the ends are sutured together.

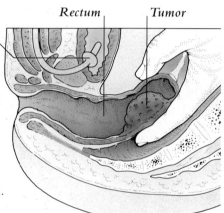

CHAPTER 10

The URINARY SYSTEM

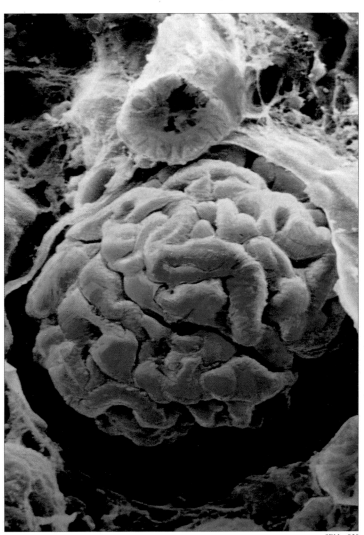

A glomerulus – one of the tiny filtering units in the kidney

SEM x 930

INTRODUCTION

Water is an essential constituent of all living things, and accounts for about 60 percent of an adult's body weight. Different types of tissue contain different amounts of water: fat holds little moisture, while blood, skeletal muscle, and skin have the highest concentrations. Water helps distribute heat around the body, it transports nutrients and hormones both within and between body cells, and it is also the vital medium in which chemical reactions occur. Working in conjunction with the kidneys, water helps dilute toxic substances and absorb waste products. To stay in good health, an adult needs to drink a minimum of 2 quarts (about 2L) of fluid a day. This replaces water that has been lost in exhaled air and sweat, and passed in feces. It also enables the kidneys to produce sufficient urine to keep the body's inner chemistry in balance. Under the influence of hormones, the kidneys regulate the volume, acidity, and salinity of the urine. After being formed in the kidneys, urine is stored in the bladder, which is normally emptied three or four times a day. Certain kidney disorders cause malfunction and this may lead to chronic kidney failure. Symptoms such as an increase in frequency of urination, discomfort or pain felt during urination, or foul odor or discoloration of the urine, suggest a urinary tract disorder, and should be investigated promptly.

A kidney

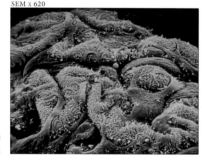

SEM x 620

Cells inside a glomerulus

THE URINARY SYSTEM

ANATOMY *of the* URINARY SYSTEM

THE URINARY SYSTEM REGULATES THE VOLUME AND COMPOSITION of fluids in the body and removes waste products and excess fluid. Waste products are filtered from the blood by the kidneys and excreted in the urine, which descends through the ureters into the urinary bladder. Normally, urine is stored in the bladder until a convenient time, at which point the muscles at the bladder outlet relax, allowing it to be expelled through the urethra.

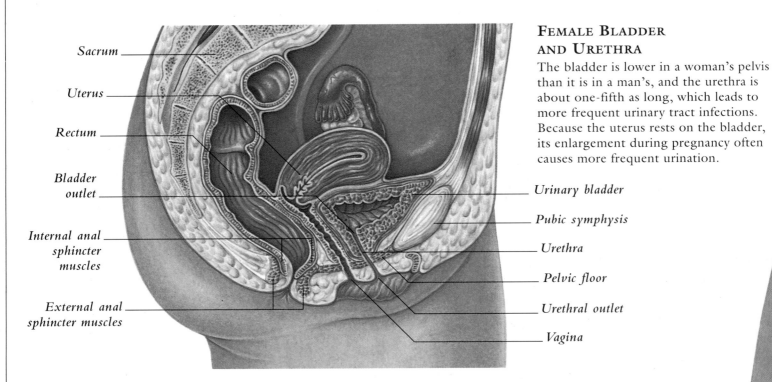

Sacrum

Uterus

Rectum

Bladder outlet

Internal anal sphincter muscles

External anal sphincter muscles

FEMALE BLADDER AND URETHRA

The bladder is lower in a woman's pelvis than it is in a man's, and the urethra is about one-fifth as long, which leads to more frequent urinary tract infections. Because the uterus rests on the bladder, its enlargement during pregnancy often causes more frequent urination.

Urinary bladder

Pubic symphysis

Urethra

Pelvic floor

Urethral outlet

Vagina

MALE BLADDER AND URETHRA

A man's urethra is about 8in (20cm) long, and is made up of three sections named for the surrounding tissue: the spongy urethra, membranous urethra, and prostatic urethra. It is the channel through which urine and semen pass out of the body. The prostate gland encircles the urethra at the bladder's base; as men get older, its enlargement may compress the urethra and cause problems with urination.

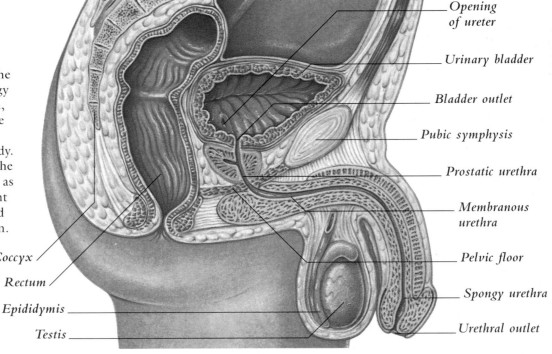

Coccyx

Rectum

Epididymis

Testis

Opening of ureter

Urinary bladder

Bladder outlet

Pubic symphysis

Prostatic urethra

Membranous urethra

Pelvic floor

Spongy urethra

Urethral outlet

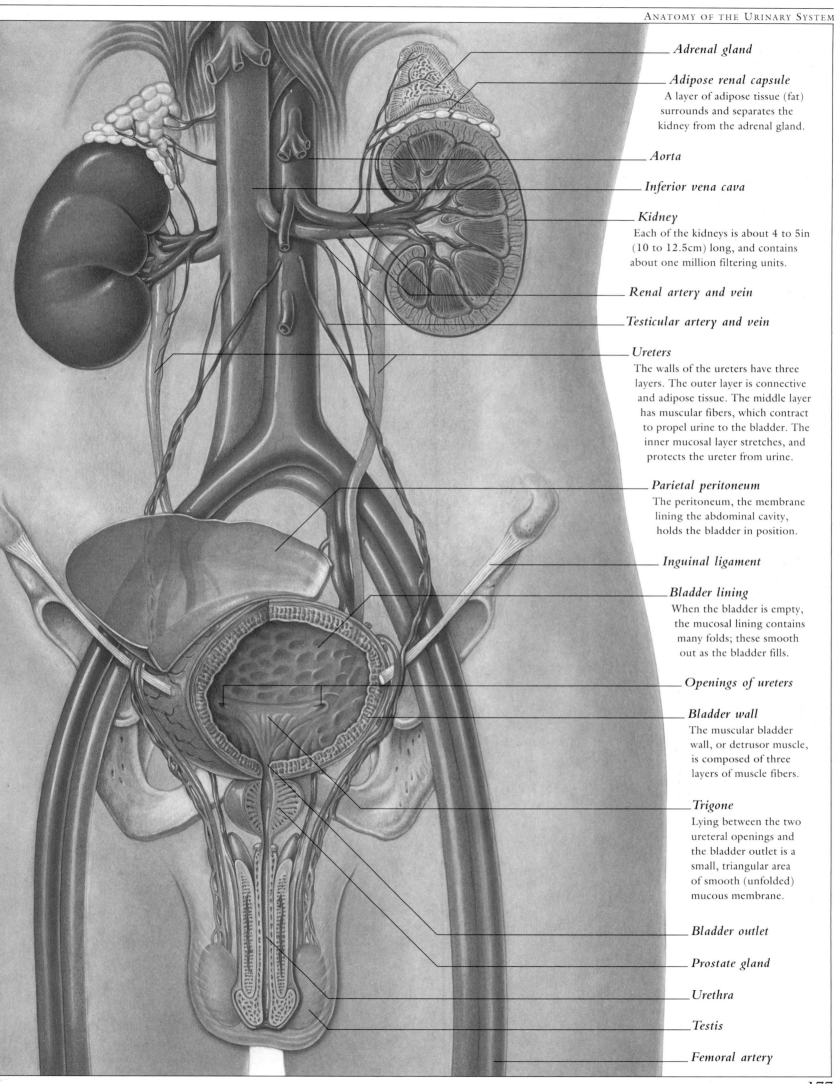

Adrenal gland

Adipose renal capsule
A layer of adipose tissue (fat)
surrounds and separates the
kidney from the adrenal gland.

Aorta

Inferior vena cava

Kidney
Each of the kidneys is about 4 to 5in
(10 to 12.5cm) long, and contains
about one million filtering units.

Renal artery and vein

Testicular artery and vein

Ureters
The walls of the ureters have three
layers. The outer layer is connective
and adipose tissue. The middle layer
has muscular fibers, which contract
to propel urine to the bladder. The
inner mucosal layer stretches, and
protects the ureter from urine.

Parietal peritoneum
The peritoneum, the membrane
lining the abdominal cavity,
holds the bladder in position.

Inguinal ligament

Bladder lining
When the bladder is empty,
the mucosal lining contains
many folds; these smooth
out as the bladder fills.

Openings of ureters

Bladder wall
The muscular bladder
wall, or detrusor muscle,
is composed of three
layers of muscle fibers.

Trigone
Lying between the two
ureteral openings and
the bladder outlet is a
small, triangular area
of smooth (unfolded)
mucous membrane.

Bladder outlet

Prostate gland

Urethra

Testis

Femoral artery

KIDNEY STRUCTURE *and* FUNCTION

THE KIDNEYS ARE PAIRED, REDDISH BROWN ORGANS in back of the abdominal cavity on either side of the spinal column. Their main function is to regulate the amount of water in the body as well as to keep the body fluids at a constant concentration and acid-base level. They achieve this function by filtering blood and excreting waste products and excess water as urine.

ANATOMY OF A KIDNEY

Each kidney has an outer rim, the renal cortex; this rim surrounds an inner region, the renal medulla, which is composed of many conical segments known as renal pyramids. Kidney tissue consists of numerous urine-making units, known as nephrons, and urine-collecting tubules. Urine drains from these small tubules into wider tubes called ducts of Bellini. These open at the tips of the renal pyramids into calyces (cavities).

Distal convoluted tubule

Glomerulus of nephron

Proximal convoluted tubule

Cortex

Duct of Bellini

Medulla

Loop of Henle

Papilla (tip of pyramid)

Renal pyramid

Renal column

Arcuate arteries and veins

Inter-lobular arteries and veins

Renal artery
The kidney is supplied with blood by the renal artery, which branches directly from the aorta (the main artery of the body).

Renal vein
Blood drains from the kidney by way of the renal vein, which feeds into the vena cava (the main vein in the body).

Ureter
Urine travels in this tube from the renal pelvis to the bladder.

Renal capsule
This thin capsule of white fibrous tissue surrounds the kidney.

Adipose tissue
The kidney and its blood vessels are embedded in a cushion of adipose (fatty) connective tissue.

Minor calyx
The ducts of Bellini drain urine into the base of the renal pyramids where short cavities, called minor calyces, are located.

Major calyx
The minor calyces merge to form larger branches called major calyces.

Renal pelvis
The renal pelvis is a funnel-shaped tube that divides into two or three branches called major calyces.

STRUCTURE OF THE NEPHRON

The kidney contains over one million nephrons. Each nephron contains a glomerulus (a rounded tuft of tiny capillary blood vessels) and a long, thin renal tubule. One end of the renal tubule is a cup-shaped membrane, the Bowman's capsule, which envelops the glomerulus. The other end joins a straight urine-collecting tubule. The glomeruli are located mainly in the renal cortex, and the tubules in the medulla.

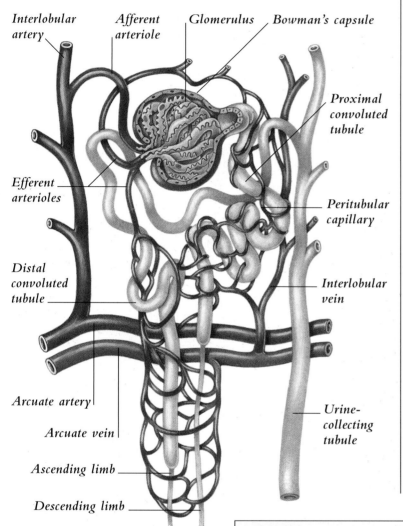

Interlobular artery · Afferent arteriole · Glomerulus · Bowman's capsule · Proximal convoluted tubule · Efferent arterioles · Peritubular capillary · Distal convoluted tubule · Interlobular vein · Arcuate artery · Arcuate vein · Ascending limb · Descending limb · Loop of Henle · Urine-collecting tubule

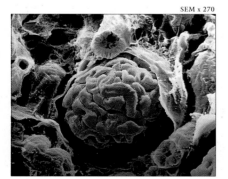

Site of blood filtration
Located beneath the Bowman's capsule is the glomerulus, a mass of capillaries seen here in red. The filtrate produced by the glomerulus is collected by each capsule and distributed to a network of tubules, two of which are shown in blue.

GLOMERULAR FILTRATION

Blood passing through the glomerular capillaries is filtered under pressure into the Bowman's capsule. This filtered fluid is primarily water, with sodium bicarbonate, potassium, glucose, amino acids, and the waste products urea and uric acid. Blood cells and other large particles stay in the capillaries.

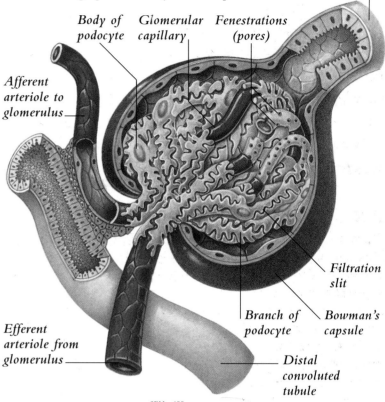

Body of podocyte · Glomerular capillary · Fenestrations (pores) · Proximal convoluted tubule · Afferent arteriole to glomerulus · Filtration slit · Efferent arteriole from glomerulus · Branch of podocyte · Bowman's capsule · Distal convoluted tubule

SEM x 600

Podocytes
Podocytes are specialized branching cells that rest on the glomerular surface. These cells aid filtration by restricting the size of molecules that pass through capillary membranes. Filtration slits are located between the branches of the podocytes.

URINE FORMATION

Water and other substances are reabsorbed from the filtrate as it passes along the coiled renal tubules. Surplus acids and, in only one area, potassium, are secreted. The kidneys can vary the amount of a substance that is reabsorbed or secreted, and thus alter both the volume and composition of urine.

KEY
- Glucose
- Sodium
- Potassium
- Bicarbonate
- Urea
- Water
- Acid
- Blood cells
- Protein

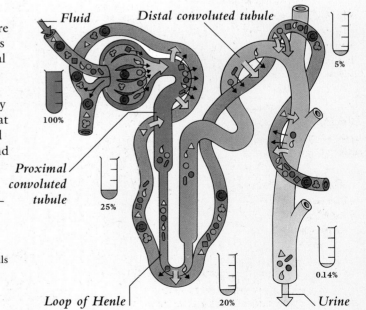

Fluid · Distal convoluted tubule · Proximal convoluted tubule · Loop of Henle · Urine · 100% · 25% · 5% · 20% · 0.14%

URINARY TRACT DISORDERS

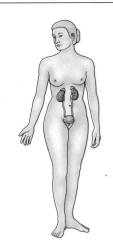

THE URINARY TRACT IS SUSCEPTIBLE TO INFECTION and to chronic, debilitating disorders. Women are especially prone to lower urinary tract infections. Kidney failure, once a major cause of death in early adult life, can now be remedied by dialysis or transplantation. Such common urinary symptoms as incontinence remain troublesome despite substantial advances in their management.

SITES OF DISORDERS

Although each of the urinary organs is affected by its own characteristic diseases, a disorder of any single organ can also affect other parts of the system. For example, stones that pass from the kidney may damage the ureters, and obstruction to urine outflow may damage kidneys through infection or back pressure.

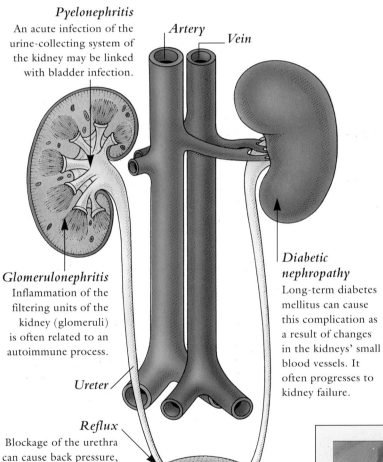

Pyelonephritis
An acute infection of the urine-collecting system of the kidney may be linked with bladder infection.

Artery

Vein

Glomerulonephritis
Inflammation of the filtering units of the kidney (glomeruli) is often related to an autoimmune process.

Diabetic nephropathy
Long-term diabetes mellitus can cause this complication as a result of changes in the kidneys' small blood vessels. It often progresses to kidney failure.

Ureter

Reflux
Blockage of the urethra can cause back pressure, which forces urine up the ureters. Such reflux can damage the kidneys. Reflux can also occur if the openings of the ureters are too relaxed.

Cystitis
An inflammation inside the bladder caused by infection, cystitis affects both sexes but is more common in women.

Opening of ureter

Urethra

INCONTINENCE

Involuntary leakage of urine is common, and occurs more in women than men, primarily because women often have a weakness in the pelvic floor muscles due to childbearing and menopause. Incontinence often accompanies dementia as well as brain or spinal cord damage. Muscle-tightening operations are done or a collagen protein can be injected at the bladder outlet.

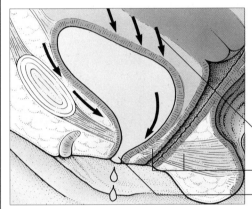

Stress incontinence
Weak pelvic floor muscles may allow a small amount of urine to escape during exertion, such as running, or less strenuous activities, such as coughing.

Pressure

Pelvic floor muscles

Urethra

Bladder contractions

Urge incontinence
An urgent desire to urinate is sometimes triggered by a sudden change of body position. Once urination starts, the bladder contracts involuntarily until empty.

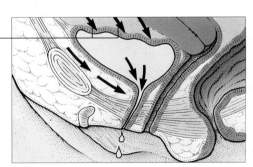

KIDNEY STONES

Concentrated substances in the urine may precipitate to form kidney stones. These may occur in the urine-collecting part of the kidney, the ureters, or the bladder, and cause great pain.

Imaging a kidney stone
After a dye is injected, an X-ray (pyelogram) can reveal stones, such as the one shown here in orange.

X-RAY

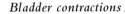

KIDNEY FAILURE

Serious kidney disease may so severely damage the kidneys that they are no longer capable of carrying out their function of removing waste products from the blood and maintaining the fluid balance. The failure or loss of one kidney does not endanger life. However, if both of the kidneys fail, dialysis or transplantation of a healthy organ is required.

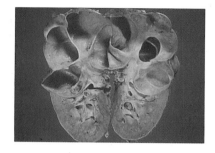

POLYCYSTIC KIDNEY

The presence of multiple cysts (left) in the kidney may occur as a genetic abnormality that may cause death in the first year of life. Adult polycystic disease can also be due to a hereditary cause. Congenital small cysts gradually enlarge, which may cause high blood pressure and loss of kidney function. Half the sufferers eventually require dialysis.

DIALYSIS

In the technique of dialysis, the blood is filtered by passing it through a semipermeable membrane immersed in a special solution known as dialysate. Smaller molecules, such as urea and other waste products, pass through this membrane into the dialysate for disposal; larger molecules such as proteins are retained. The most common procedure is hemodialysis.

Hemodialysis

Blood from an artery passes through a coiled membrane tube and back into a vein. The tube is immersed in a tank filled with dialysate. Waste products filter out into the dialysate.

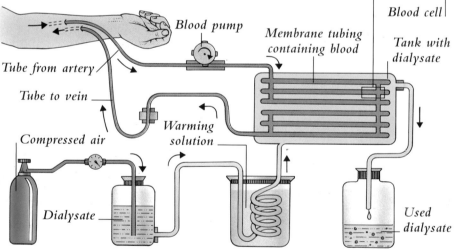

Dialysate

Waste products

Membrane

Blood cell

Blood pump

Membrane tubing containing blood

Tank with dialysate

Tube from artery

Tube to vein

Compressed air

Warming solution

Dialysate

Used dialysate

Peritoneal dialysis

To perform this procedure, 4pt (2L) of dialysate are run into the peritoneal cavity and changed in about 4 hours. Waste products pass from the capillaries lining the cavity through the peritoneal membrane into the dialysate.

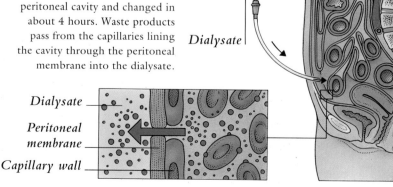

Dialysate

Dialysate

Peritoneal membrane

Capillary wall

OPERATION

KIDNEY TRANSPLANT

The definitive treatment for failure of both kidneys is a kidney transplant. This, the most successful of any organ transplant procedure, is often performed using a kidney donated by a close relative. Alternatively, a computer can find a tissue match, often with a person who has just suffered accidental death.

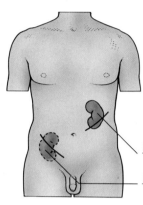

Incision sites
One or both diseased kidneys are removed via an incision made under the ribs. The donated kidney is inserted low in the pelvis through an incision in the groin.

Removed kidney

Donated kidney

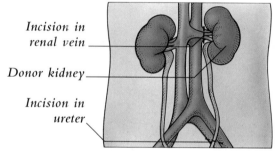

Incision in renal vein

Donor kidney

Incision in ureter

1 The donor's renal artery and vein are cut. The left kidney is usually removed because, being located higher in the body, its longer ureter allows better repositioning of the kidney in the recipient.

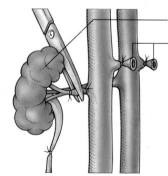

Diseased kidney

Cut blood vessels

2 The diseased kidney(s) may be removed since a diseased kidney can cause hypertension. The ureters and renal blood vessels are tied and cut.

3 The donated kidney is positioned in the pelvis. The cut end of the longer ureter is pushed through a stab incision into the bladder and is stitched in position. The clamps are removed, and the incision in the lower abdomen is closed.

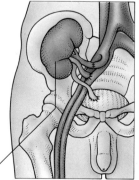

Donated kidney

CHAPTER 11

The REPRODUCTIVE SYSTEM

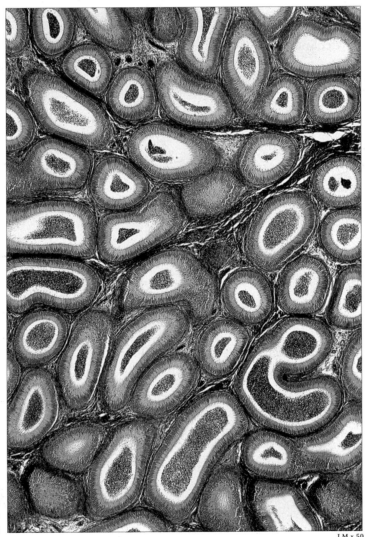

The epididymis (seen in cross-section), the convoluted duct in which sperm mature

LM x 50

INTRODUCTION

In biological terms, one of the primary functions of the human body is to replicate itself, and the sexual and parenting instincts are among the strongest of our basic drives. In this century, sexual behavior in western society has been transformed by the availability of reliable contraception. This has enabled many couples to delay childbearing until relatively late in their lives, when fertility has already begun to decline, and a whole new science of assisted reproduction has evolved to help them become parents. Other research has looked at the hormones controlling the development of sperm and eggs, and the process of conception. This has led

Cross-section of an ovary, showing a maturing egg

to improvements in the treatment of many disorders of the reproductive system. But cancers of the male and female reproductive systems (mostly those of the prostate, breast, ovary, and uterus) still account for a large proportion of cancer deaths. Today, researchers are seeking ways of detecting these disorders at an earlier stage, and better treatments for them. The social upheavals of the twentieth century have had a marked effect on the incidence of sexually transmitted diseases (STDs). These were spread rapidly in World War II, but declined later as antibiotics were developed. Some STDs, such as herpes and acquired immune deficiency syndrome (AIDS), are incurable.

SEM x 2,520

A group of sperm

THE REPRODUCTIVE SYSTEM

The MALE REPRODUCTIVE ORGANS

THE TESTES, A PAIR OF ROUNDED GLANDS that lie in a pouch called the scrotum, produce sperm and male sex hormones. From each testis, sperm pass into a long coiled tube, the epididymis, where they mature and are stored until they are ejaculated or reabsorbed by the body. During sexual arousal, the spongy tissue in the penis fills with blood and the organ becomes erect. Before ejaculation, sperm are propelled along a long duct called the vas deferens. Fluid produced by the two seminal vesicles and the prostate gland is added to the sperm to produce semen, which is ejaculated through the urethra.

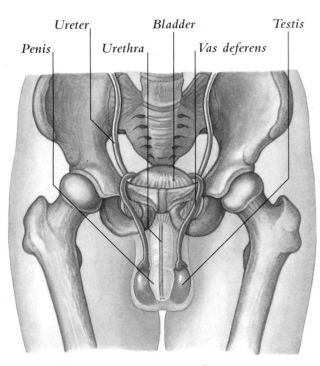

Ureter *Bladder* *Testis*

Penis *Urethra* *Vas deferens*

LOCATION OF THE ORGANS

Unlike the reproductive glands of the female, those of the male, the testes, do not lie within the pelvis. Instead, they are located externally in the scrotum. This arrangement maintains sperm at slightly below body temperature, which is necessary for them to survive. The male pelvis is narrower and deeper, and has thicker, stronger bones, than the female pelvis, which is more adapted for pregnancy and childbirth.

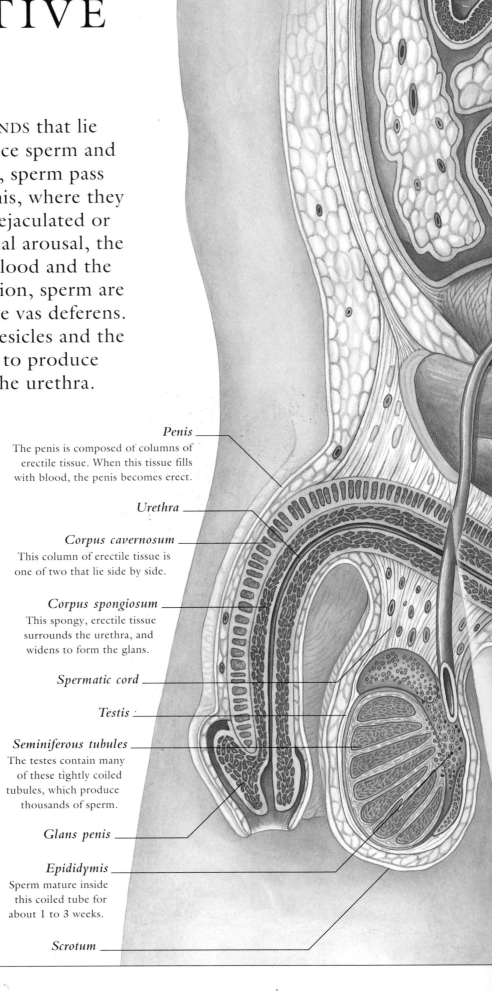

Penis
The penis is composed of columns of erectile tissue. When this tissue fills with blood, the penis becomes erect.

Urethra

Corpus cavernosum
This column of erectile tissue is one of two that lie side by side.

Corpus spongiosum
This spongy, erectile tissue surrounds the urethra, and widens to form the glans.

Spermatic cord

Testis

Seminiferous tubules
The testes contain many of these tightly coiled tubules, which produce thousands of sperm.

Glans penis

Epididymis
Sperm mature inside this coiled tube for about 1 to 3 weeks.

Scrotum

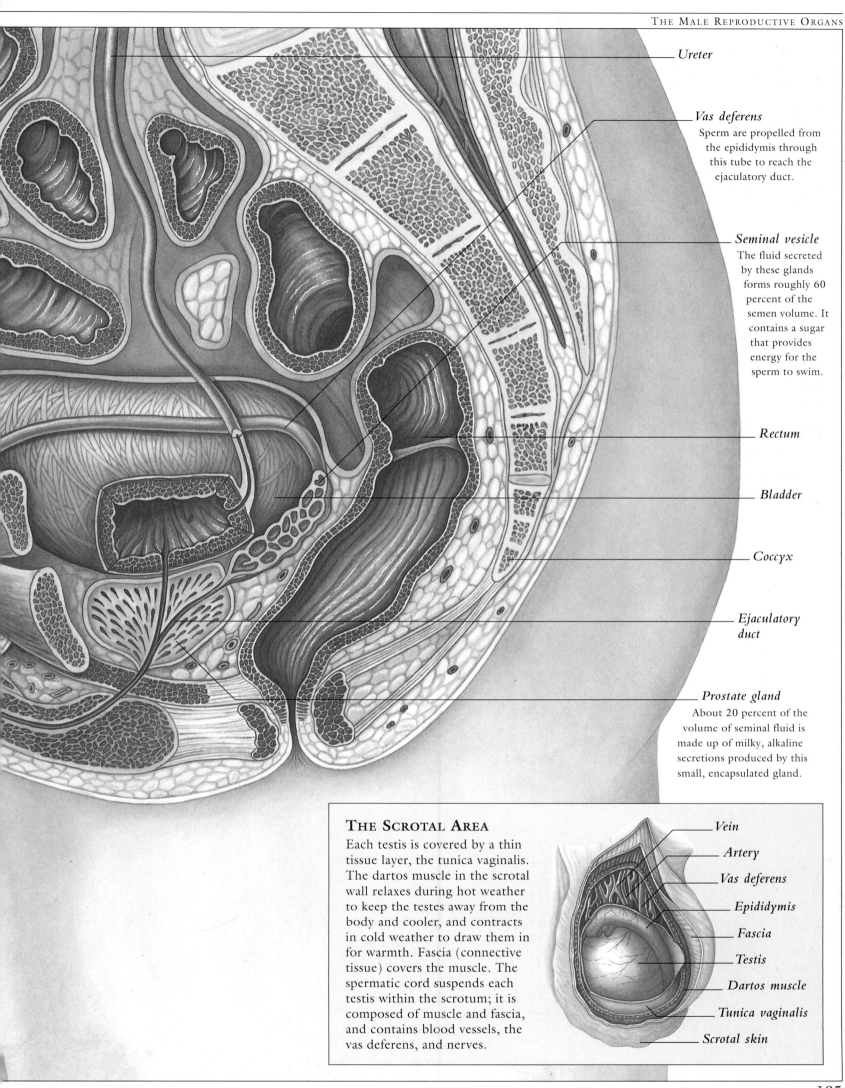

Ureter

Vas deferens
Sperm are propelled from
the epididymis through
this tube to reach the
ejaculatory duct.

Seminal vesicle
The fluid secreted
by these glands
forms roughly 60
percent of the
semen volume. It
contains a sugar
that provides
energy for the
sperm to swim.

Rectum

Bladder

Coccyx

Ejaculatory
duct

Prostate gland
About 20 percent of the
volume of seminal fluid is
made up of milky, alkaline
secretions produced by this
small, encapsulated gland.

THE SCROTAL AREA

Each testis is covered by a thin
tissue layer, the tunica vaginalis.
The dartos muscle in the scrotal
wall relaxes during hot weather
to keep the testes away from the
body and cooler, and contracts
in cold weather to draw them in
for warmth. Fascia (connective
tissue) covers the muscle. The
spermatic cord suspends each
testis within the scrotum; it is
composed of muscle and fascia,
and contains blood vessels, the
vas deferens, and nerves.

Vein

Artery

Vas deferens

Epididymis

Fascia

Testis

Dartos muscle

Tunica vaginalis

Scrotal skin

The FEMALE REPRODUCTIVE ORGANS

THE ESSENTIAL SEX GLANDS IN WOMEN are the two ovaries. At puberty, they begin to release female germ cells, known as ova, and synthesize the sex hormones, estrogen and progesterone, that affect the development of female sexual characteristics, such as general body shape, breast enlargement, and menstrual cycles. Each month an ovum is released and travels down one of the fallopian tubes to the uterus, a hollow organ in the center of the pelvis; if the egg is not fertilized, menstruation occurs.

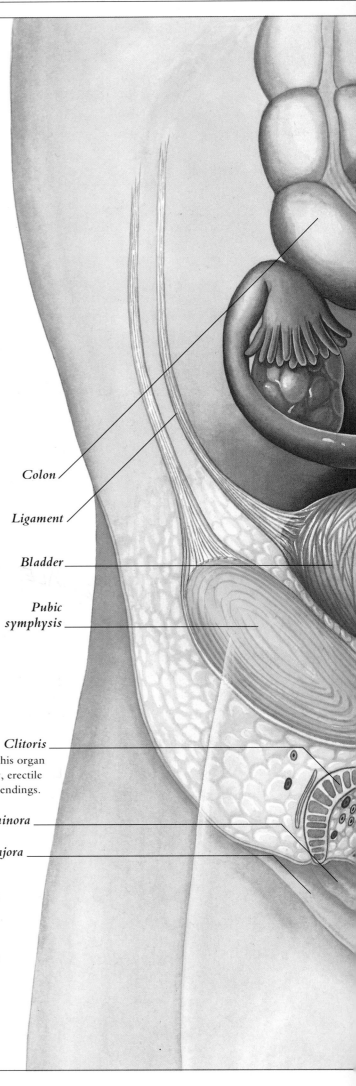

Colon

Ligament

Bladder

Pubic symphysis

LOCATION OF ORGANS

Female reproductive organs are located within the pelvic cavity, a broad, empty space between the encircling pelvic bones. This space is wider and shallower in a woman than in a man, permitting the uterus to enlarge during pregnancy as the fetus grows.

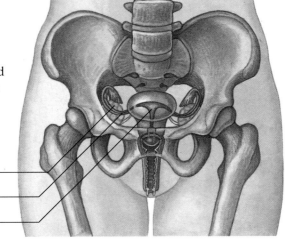

Ovary

Fallopian tube

Uterus

Clitoris
Like the penis, this organ contains spongy, erectile tissue and nerve endings.

Labia minora

Labia majora

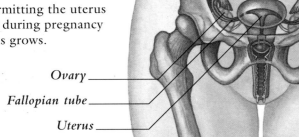

Urethral opening

Pubic symphysis

Vaginal opening

Clitoris

Vestibular bulb

Labia minora

Vestibular gland

Inguinal ligament

Muscle

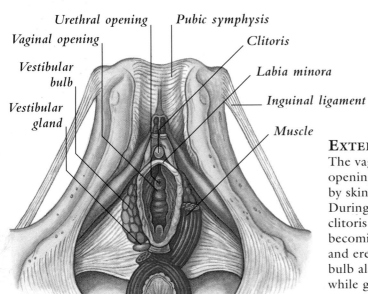

EXTERNAL GENITALS

The vaginal and urethral openings are surrounded by skin folds called labia. During sexual arousal, the clitoris swells with blood, becoming highly sensitive and erect. The vestibular bulb also becomes erect, while glands lubricate the mucous membranes.

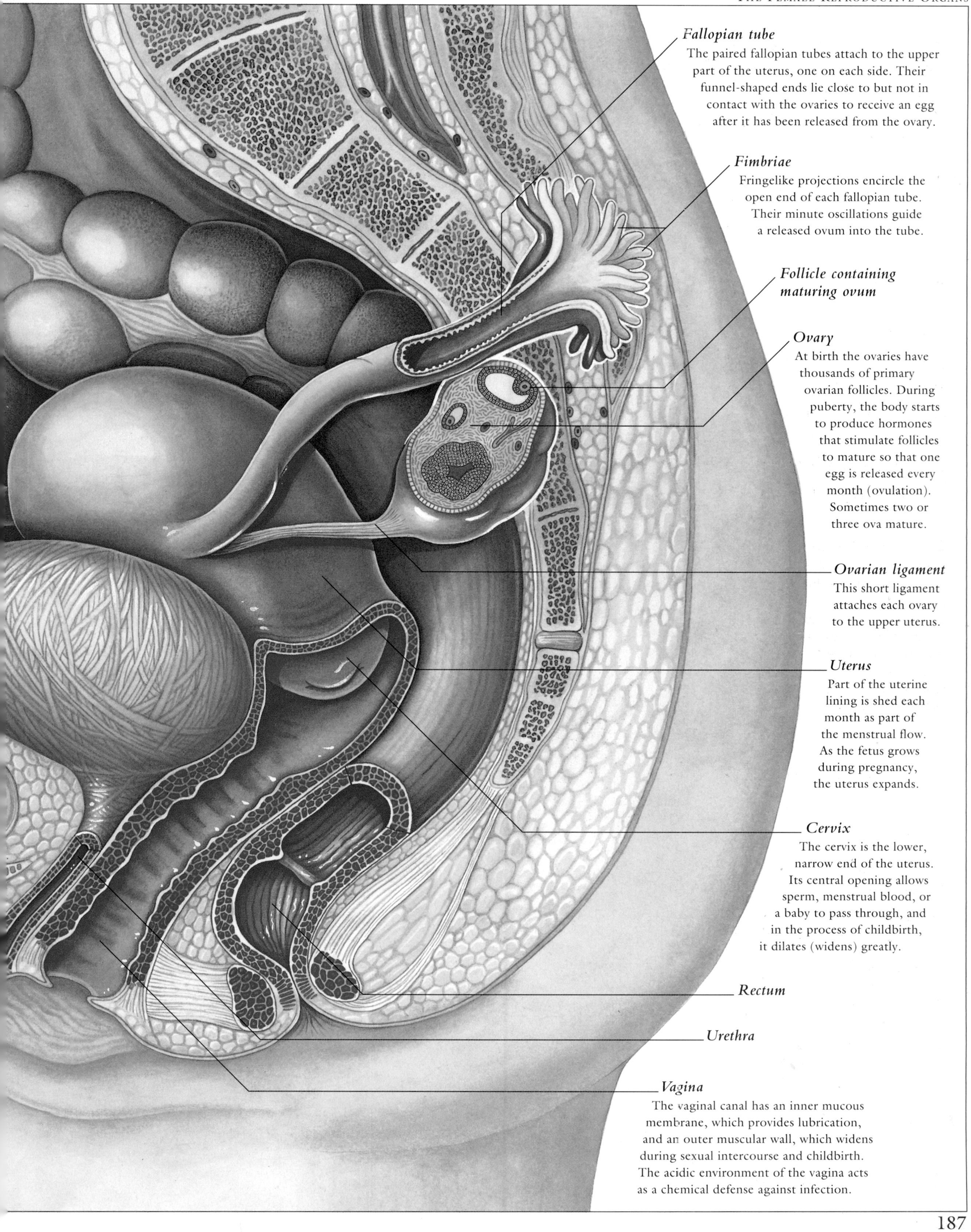

Fallopian tube
The paired fallopian tubes attach to the upper part of the uterus, one on each side. Their funnel-shaped ends lie close to but not in contact with the ovaries to receive an egg after it has been released from the ovary.

Fimbriae
Fringelike projections encircle the open end of each fallopian tube. Their minute oscillations guide a released ovum into the tube.

Follicle containing maturing ovum

Ovary
At birth the ovaries have thousands of primary ovarian follicles. During puberty, the body starts to produce hormones that stimulate follicles to mature so that one egg is released every month (ovulation). Sometimes two or three ova mature.

Ovarian ligament
This short ligament attaches each ovary to the upper uterus.

Uterus
Part of the uterine lining is shed each month as part of the menstrual flow. As the fetus grows during pregnancy, the uterus expands.

Cervix
The cervix is the lower, narrow end of the uterus. Its central opening allows sperm, menstrual blood, or a baby to pass through, and in the process of childbirth, it dilates (widens) greatly.

Rectum

Urethra

Vagina
The vaginal canal has an inner mucous membrane, which provides lubrication, and an outer muscular wall, which widens during sexual intercourse and childbirth. The acidic environment of the vagina acts as a chemical defense against infection.

BREAST DISORDERS

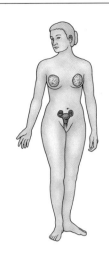

THE BREASTS ARE STRONGLY INFLUENCED by the sex hormones. Many women notice breast sensations that are related to hormonal changes of the menstrual cycle or during pregnancy. Common symptoms of disorders are pain, nipple discharge, and lumps. More than 80 percent of breast lumps are not cancerous but, if one is found, tests need to be done. The main tests are mammography to identify a suspicious area and either needle aspiration to withdraw fluid or cells, or a needle biopsy to obtain tissue for microscopic examination.

COMMON PROBLEMS

Many women experience cyclical changes in their breasts that cause pain. The symptoms often begin approximately a week before menstruation and disappear within a day of the beginning of the period. The pain can be quite unbearable. A number of remedies have been tried. The symptoms may be relieved by reducing the dietary intake of saturated fat, or by following certain drug treatments.

Gynecomastia

Gynecomastia is the enlargement of breast tissue on one or both sides in males. Possible causes include hormone disorders, side-effects of prescribed or illicit drugs, alcohol abuse, or cancer. This disorder is common in puberty to age 18.

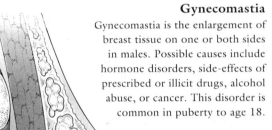

Fibroadenoma

Fibroadenomas are common noncancerous breast lumps. Occurring most frequently in women under age 30, they are painless, fibrous growths that may be removed under local or general anesthesia.

Cyst

Cysts are fluid-filled sacs within the breast. Usually noncancerous, many contain a clear fluid, which can be withdrawn through a needle, occasionally with the use of ultrasound guidance.

Fibrocystic disease

This disorder may cause pain and increased lumpiness in the last few days of each menstrual cycle. In most cases, it has no long-term effects.

Fatty tissue

Breast abscess

An abscess, or collection of pus, may develop after bacteria enter through a crack in the nipple. This may occur in women who are breastfeeding. The area of the abscess is red and painful; chills and fever may occur. Treatment includes antibiotic drugs and surgical drainage.

Abscess

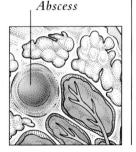

BREAST CANCER

About one in nine women develops breast cancer, and this disease is a leading cause of death. Risk factors are increasing age and whether a mother or a sister has also been affected. The outlook depends on the type of cancer and how far it has spread.

Signs

The primary sign of breast cancer is a lump. Other signs are a blood-stained discharge from the nipple, indrawing of the nipple, and sometimes dimpling of the skin over the breast (shown at right). In a small percentage of cases, the disease attacks both breasts.

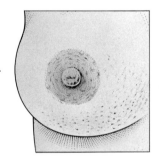

MAMMOGRAPHY

Mammography is a simple X-ray procedure that is employed to examine women for breast cancer. Most useful for women over 50, the technique enables cancers to be detected at an early stage, even before a tumor can be discovered by other means. Breast cancer can usually be treated more effectively if it is detected at an early stage.

Mammogram

On an X-ray film, a tumor usually appears as a dense mass, which occasionally contains flecks of calcium, and an irregular border, such as the one colored orange at left. A biopsy usually confirms whether a lump is cancerous.

X-RAY

The procedure

An X-ray exposure is made of each breast in turn. The breast is compressed between a cassette containing the film and a plastic cover that is attached to the X-ray machine.

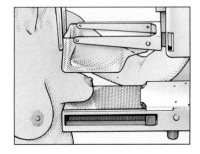

SURGICAL PROCEDURES

A range of operations can be performed to remove breast cancers. The procedure carried out depends on the size of the tumor and whether it has spread. A procedure called mammoplasty may be performed in order to reconstruct the removed breast and/or reduce the size of the unoperated breast.

LUMPECTOMY

In this procedure, the breast lump is removed with a small area of surrounding tissue. For localized breast cancers, it may offer as good a chance for survival as more extensive surgery.

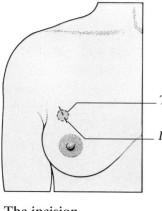

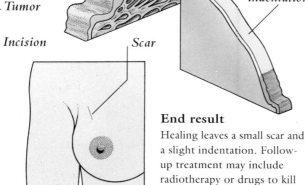

Tissue removed

Tumor

Tumor

Incision

Small indentation

Scar

The incision
A small incision is made over the lump. The tumor and a rim of surrounding tissue is removed, along with lymph nodes on the affected side.

End result
Healing leaves a small scar and a slight indentation. Follow-up treatment may include radiotherapy or drugs to kill any remaining cancer cells.

PARTIAL MASTECTOMY

Partial mastectomy is the surgical removal of part of the breast. In some cases, the cancer and a segment of the surrounding breast tissue are removed (segmental mastectomy). In other cases, the affected quarter of the breast is removed (quadrantectomy). The lymph nodes in the armpit are also usually removed.

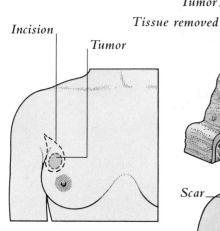

Incision

Tumor

Tumor

Tissue removed

Indentation

Scar

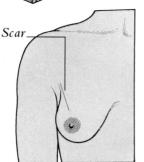

The incision
The affected section of breast tissue is removed along with the overlying skin. The lymph nodes from the armpit on the same side as the breast are also removed.

End result
Healing leaves a visible scar and a depression in the skin. The breast may be slightly smaller than before. Drugs and radiotherapy may also be part of the treatment.

SUBCUTANEOUS MASTECTOMY

In a subcutaneous mastectomy, all or most of the breast tissue is removed but the overlying nipple and skin are left intact. Subcutaneous mastectomy may be suitable as a preventive measure for women who are at high risk but do not have cancer. This technique is not considered a suitable operation for breast cancer by most surgeons.

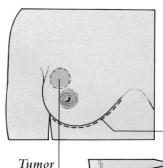

The incision
The incision is usually made underneath the breast and then the tissue is removed. Locating the incision here and placing an implant can restore a natural appearance.

Incision

Tumor

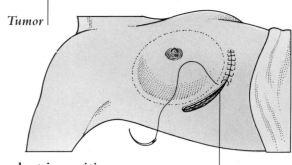

Implant in position
An implant may be inserted under the skin to replace the missing breast tissue and to restore the appearance of the breast.

Incision being closed

MODIFIED RADICAL MASTECTOMY

In this procedure, the entire breast and the lymph nodes in the armpit are removed. The breast can be reconstructed at the time of mastectomy or at a later date. A radical mastectomy, in which the pectoral (chest) muscles are also removed, is now regarded as disfiguring and is rarely performed.

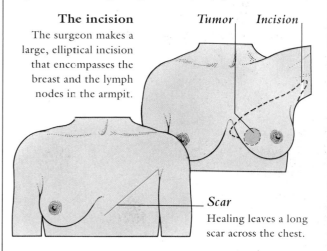

The incision
The surgeon makes a large, elliptical incision that encompasses the breast and the lymph nodes in the armpit.

Tumor   *Incision*

Scar
Healing leaves a long scar across the chest.

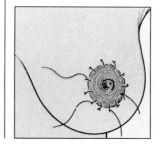

Breast reconstruction
Skin and underlying fat and muscle are taken from the abdomen or back, and used to reconstruct the breast. An implant may be inserted and the nipple is reconstructed.

UTERUS DISORDERS

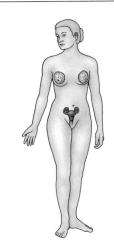

A WOMAN'S UTERUS UNDERGOES MORE CHANGES than any other organ in her body. From about age 13 up to about age 55, it sheds its lining during the monthly menstrual cycle. Normally only about the size of a woman's fist, it enlarges enormously during pregnancy. Apart from the complications that follow childbirth, such as a torn cervix, disorders that can affect the uterus include endometriosis, prolapse, fibroids, polyps, and cancer. When deemed necessary, these disorders can usually be relieved by drugs or surgery.

FIBROIDS

About one in five women over the age of 35 develops a fibroid, a noncancerous uterine tumor composed of fibrous tissue and muscle. Large fibroids may cause discomfort, heavier periods, and frequent urination; these may require surgery. Hypothalamic hormones may be injected before surgery to reduce the size of very large fibroids and avoid complications.

Subserosal fibroid
These are located just under the surface of the outer uterine wall.

Outer lining

Endometrium (mucosal inner lining)

Intramural fibroid

Uterine muscle

Stalk

Submucosal fibroid
This type of fibroid occurs under the endometrium and may cause bleeding.

Pedunculated fibroid
A stalk that grows from this type may protrude into the cervix, ulcerate, and bleed.

Cervix

TYPES OF FIBROID

Intramural fibroid

An intramural fibroid (seen at right), which develops within the muscular uterine wall, is the most common type of fibroid. As the fibroid grows, the uterus may enlarge and become deformed; this sometimes affects the woman's ability to sustain pregnancy.

LM

ENDOMETRIOSIS

According to one theory, the cause of this disorder is that endometrial fragments shed during menstruation back up the fallopian tubes and attach to organs in the pelvic cavity. Cysts may form, which bleed during each period. Pain may occur during menstrual periods and intercourse. Drugs to stop menstruation are given, or surgery is performed to treat complications of cysts or to remove the affected tubes, ovaries, and uterus.

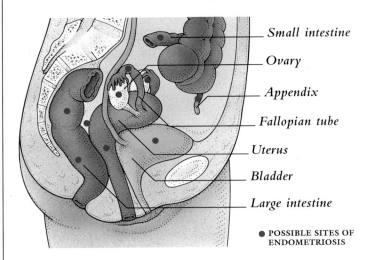

Small intestine

Ovary

Appendix

Fallopian tube

Uterus

Bladder

Large intestine

● POSSIBLE SITES OF ENDOMETRIOSIS

PROLAPSE OF UTERUS

Pregnancy and childbirth can stretch the ligaments that hold the uterus in place, especially in women who have had many children. Lax ligaments allow the uterus to sag, distorting the vagina and causing problems with bowel movements and urination. An operation may be performed to tighten the lax ligaments, or a pessary, a plastic device that provides support, may be inserted.

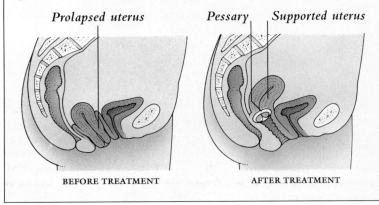

Prolapsed uterus

Pessary   *Supported uterus*

BEFORE TREATMENT

AFTER TREATMENT

CERVICAL CANCER

The cervix, or neck of the uterus, is an important site of cancer. In the cervical (Pap) smear test, cells are scraped from the surface of the cervix and smeared onto a glass slide. When examined under a microscope, the Pap smear may reveal abnormal changes, called dysplasia, before cancer develops or spreads. Treatment at an early stage can prevent progression.

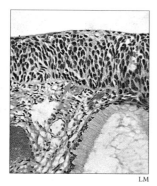

Noninvasive cancer
Severe dysplasia is shown by the abnormally large cells confined above the basement membrane (arrow), the lowest layer of the cervical lining.

LM

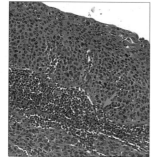

Invasive cervical cancer
Seen at right are tumor cells that have broken through the basement membrane (arrow) and spread from the cervical lining into deeper tissue.

LM

DIAGNOSIS AND TREATMENT

If a Pap smear reveals dysplasia, a colposcopy may follow, in which a magnifying viewing instrument with a strong light is used to examine the cervix. Biopsy samples are taken of suspicious areas. If necessary, a cone of cervical tissue may be removed. Abnormal clusters of cells may be removed either by freezing (cryocautery) or by laser treatment.

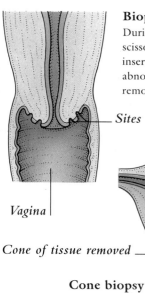

Biopsy
During colposcopy, a small, scissorlike instrument is inserted. Tiny samples of abnormal-looking tissue are removed for examination.

Sites of biopsy

Vagina

Cone of tissue removed

Cone biopsy
If colposcopy cannot precisely delineate the area of abnormal cells, a cone biopsy is done. A cone-shaped section of tissue is removed from the part of the cervix considered most likely to contain abnormal cells.

HYSTERECTOMY

Hysterectomy, removal of the uterus, is the most common operation performed on women in developed countries. It may be done to treat menometrorrhagia (heavy bleeding during and between periods), endometriosis, fibroids, and cancer of the cervix or body of the uterus.

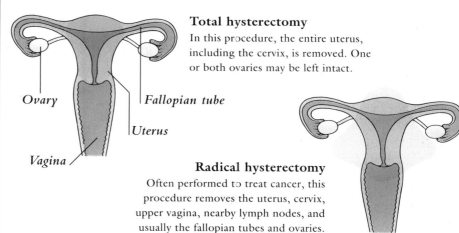

Total hysterectomy
In this procedure, the entire uterus, including the cervix, is removed. One or both ovaries may be left intact.

Ovary

Fallopian tube

Uterus

Vagina

Radical hysterectomy
Often performed to treat cancer, this procedure removes the uterus, cervix, upper vagina, nearby lymph nodes, and usually the fallopian tubes and ovaries.

OPERATION

ABDOMINAL HYSTERECTOMY

Removal of the uterus may be performed through an incision in the abdominal wall. A woman in good health is usually up within a week, and has made a full recovery in 4 to 6 weeks. Alternatively, a vaginal incision is often utilized, with laparoscopic assistance.

1 An incision is made in the abdomen either horizontally, parallel to the upper pubic hair line, or vertically between the navel and pubic hair.

Vertical incision

Horizontal incision

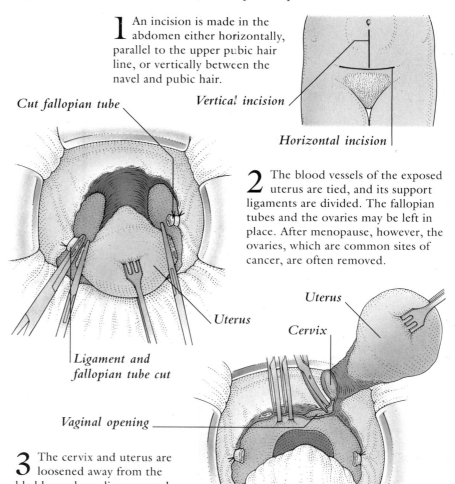

Cut fallopian tube

2 The blood vessels of the exposed uterus are tied, and its support ligaments are divided. The fallopian tubes and the ovaries may be left in place. After menopause, however, the ovaries, which are common sites of cancer, are often removed.

Uterus

Ligament and fallopian tube cut

Uterus

Cervix

Vaginal opening

3 The cervix and uterus are loosened away from the bladder and are disconnected from the wall of the vagina before being removed.

OVARY, TESTIS, *and* PROSTATE DISORDERS

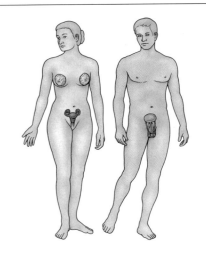

THE OVARIES, TESTES, AND PROSTATE GLAND are possible sites of cancers, nonmalignant tumors, and sexually transmitted infections. Diseases of the ovaries and testes can also be responsible for disorders elsewhere in the body related to the hormones they secrete. A few of the conditions that directly affect the organs themselves are described here.

OVARIAN DISORDERS

Although an ovarian disorder may follow an infection due to a virus or bacteria, a very common disorder is the presence of one or more cysts. These fluid-filled swellings may develop at any age. Many cysts do not cause any symptoms; however, some may be painful if they rupture. About 95 percent are benign.

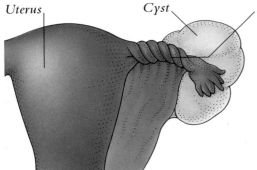

Uterus   Cyst   *Twisted fallopian tube and pedicle*

Twisted pedicle
The stalk, or pedicle, of a cyst may become twisted, cutting off the pedicle's blood supply and causing sudden, severe abdominal pain. Immediate surgery may be required.

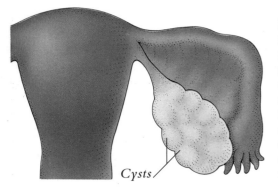

Polycystic ovaries
Multiple small cysts can form due to a hormone imbalance. Features of this syndrome include obesity and excessive body hair, and periods may be infrequent or absent. Infertility is often present, but is treated with drugs.

Cysts

LAPAROSCOPY
This technique involves the use of a fiberoptic viewing tube in order to examine the abdominal cavity directly. It is used to determine causes of pelvic pain and infertility. Laparoscopy is also used in the diagnosis and treament of a range of gynecological disorders, including ovarian cysts and endometriosis, as well as ovarian cancer.

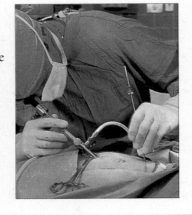

OVARIAN CANCER
Cancer of the ovary kills more women than cancers of the uterus and cervix together, essentially because it is rarely detected while still at a curable stage. It is most common in women over age 50 and those who are childless. Long-term use of the oral contraceptive pill may be protective against this type of cancer. Anticancer drugs may prolong survival, even in advanced tumors.

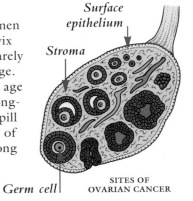

Surface epithelium

Stroma

Germ cell

SITES OF OVARIAN CANCER

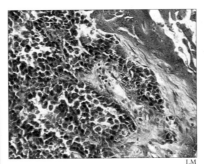

LM

Cancerous cells
Seen in the tissue section at left are darkly stained cancerous cells invading the normal tissue of the surface epithelium. This is the most common site for ovarian cancer.

DISORDERS OF THE SCROTUM

A common symptom of a scrotal disorder is swelling. Most swellings are painless and harmless, but should be checked. Swellings may be due to an injury or to a collection of fluid, sperm, or blood. Some swellings are associated with a fever, possibly due to an infection of the testis. Only rarely is a swelling a sign of cancer.

Hydrocele
A straw-colored fluid may accumulate in the space around the testis. Called a hydrocele, this type of swelling very commonly affects middle-aged men. It is usually soft and painless, and requires no treatment unless it becomes so large that it causes embarrassment or discomfort. In these cases, the fluid may be aspirated (drawn out) or the sac may be removed.

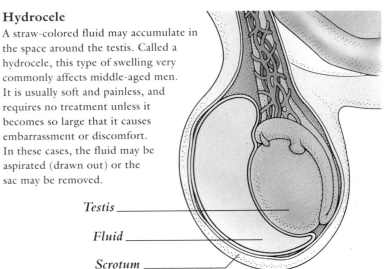

Testis

Fluid

Scrotum

TESTICULAR CANCER

Cancer of the testis is rare, but is most common in men under age 40, especially those who had an undescended testis. It first appears as a firm, usually painless swelling, often discovered during self-examination. Ultrasound helps confirm the diagnosis. Removal of the testis and chemotherapy usually cure the disease. A number of those who relapse may be salvaged by using high-dose chemotherapy associated with bone marrow rescue.

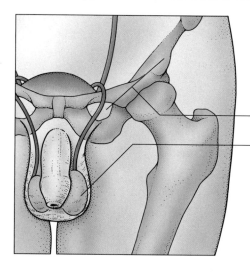

Removing the testis
A surgeon makes an incision in the groin to remove the affected testis. A synthetic implant may be inserted.

Site of incision

Testis to be removed

Testicular tumor
Seen on the right is a cross-section of an excised testis. The large, white area is a tumor. Removal of one testis does not usually affect subsequent fertility.

PROSTATE DISORDERS

The prostate gland, located at the base of the bladder, surrounds the urethra. The gland sometimes becomes swollen and inflamed as a result of bacterial infection, possibly sexually transmitted. An enlarged prostate is common in men aged over 50, as are symptomless cancerous changes. Prostate cancer is one of the main causes of death from cancer in men.

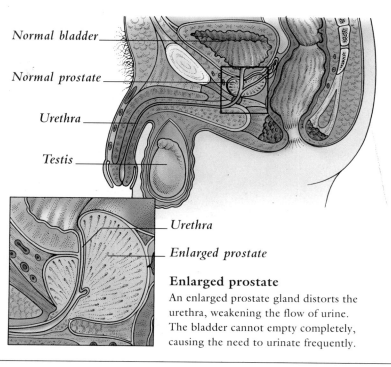

Normal bladder

Normal prostate

Urethra

Testis

Urethra

Enlarged prostate

Enlarged prostate
An enlarged prostate gland distorts the urethra, weakening the flow of urine. The bladder cannot empty completely, causing the need to urinate frequently.

TRANSURETHRAL PROSTATECTOMY

Two main methods are available to treat an enlarged prostate gland. One is transurethral prostatectomy, in which only part of the gland – the part that blocks the flow of urine – is removed. The procedure requires no surgical incision and a brief hospital stay.

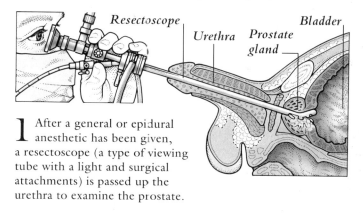

Resectoscope   Urethra   Prostate gland   Bladder

1 After a general or epidural anesthetic has been given, a resectoscope (a type of viewing tube with a light and surgical attachments) is passed up the urethra to examine the prostate.

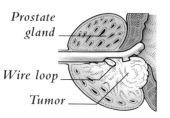

Prostate gland

Wire loop

Tumor

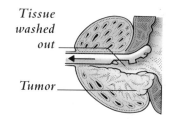

Tissue washed out

Tumor

2 A hot wire loop, a cutting instrument, or a laser is used to shave away as much of the tumor as is needed to restore normal urine flow.

3 Fragments of prostate tissue are washed away through the resectoscope. A catheter is left in the bladder for several days.

RADICAL RETROPUBIC PROSTATECTOMY

A more extensive procedure may be used to remove a greatly enlarged or cancerous prostate. In a number of cases, impotence and incontinence occur. Freezing the prostate for cancer is an experimental procedure.

1 The surgeon makes an incision low in the abdomen to the prostate gland. Lymph glands are biopsied to ensure that cancer has not spread.

Incision

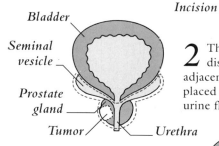

Bladder

Seminal vesicle

Prostate gland

Tumor   Urethra

2 The surgeon removes the diseased prostate gland and adjacent structures. A catheter is placed in the bladder to maintain urine flow during healing.

3 The bladder and urethra are reconnected. The catheter is usually removed 3 to 4 weeks later. Exercise strengthens the sphincter muscles in the pelvis.

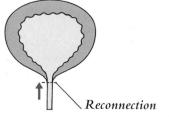

Reconnection

SEXUALLY TRANSMITTED DISEASES

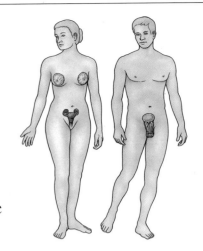

INFECTIONS SPREAD BY SEXUAL INTERCOURSE are called STDs. These are very common worldwide, especially among young adults. In developed countries, syphilis and gonorrhea declined after World War II, increased in the 1960s and '70s, declined in the 1980s, and are rising again in some areas. Some STDs increase the risk of transmitting HIV, the AIDS virus.

PELVIC INFLAMMATORY DISEASE

Pelvic inflammatory disease, PID, means infection of the upper female reproductive tract. The disorder is usually caused by chlamydial infection or gonorrhea. The diagnosis is based on the symptoms and on tests carried out on samples of cervical or vaginal discharge taken during a physical examination. PID is usually treated with a combination of antibiotics.

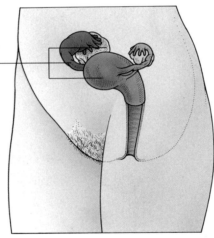

Inflamed ovary, fallopian tube, and uterus

Affected organs
Affected organs can include the cervix, uterus, fallopian tubes, and ovaries. These areas become inflamed and may cause various symptoms such as vaginal discharge, lower abdominal pain, fever, and painful intercourse.

COMPLICATIONS
Pelvic inflammatory disease is serious since it can cause severe damage and scarring in the reproductive tract. Worldwide, it is the most common cause of infertility. Other complications of PID include chronic pelvic pain and abnormal menstrual bleeding. Scarring in the fallopian tubes can block passage of a fertilized egg, increasing the risk of an ectopic pregnancy, in which the egg implants outside the uterine cavity.

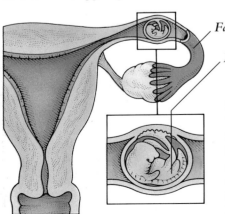

Fallopian tube

Embryo

Ectopic pregnancy
The most common site for an ectopic pregnancy is one of the fallopian tubes. The tube may rupture, resulting in peritonitis with severe lower abdominal pain, and vaginal blood loss. Internal bleeding can be severe.

NONGONOCOCCAL URETHRITIS

Nongonococcal, also known as nonspecific, urethritis is urethral inflammation caused by an infection other than gonorrhea. This infection is among the most prevalent of all STDs. Symptoms usually start 1 to 3 weeks after exposure. In women, the main symptom is vaginal discharge; effects in men are described below. Treatment consists of antibiotics such as tetracycline.

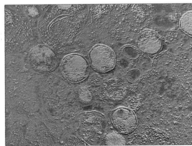

TEM x 29,100

A common cause
Nonspecific urethritis is most often caused by the bacterium *Chlamydia trachomatis* (see left). This lives and multiplies only inside human cells. Chlamydia also causes an STD called lymphogranuloma venereum and a persistent eye infection, trachoma, that is the most common cause of blindness in the Middle East.

Effects in men
As a result of inflammation of the urethra, men commonly experience discharge from the penis and pain when passing urine. Infection may spread to the epididymis, producing scrotal pain and swelling.

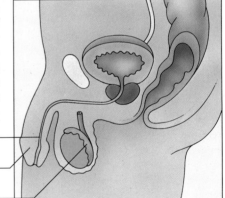

Urethra

Penis

Epididymis

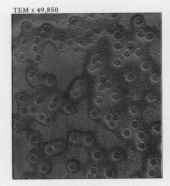

TEM x 49,850

Human papillomavirus

GENITAL WARTS
Warts on the genitals are due to human papillomavirus (HPV) infection. The virus also infects the anal region. HPV may cause cancer of the anus and cervix, so affected women should have regular Pap smears. Attempts to destroy genital warts can be done by freezing, burning, hot wire, laser, application of podophyllin, or injection of interferon.

GONORRHEA

Gonorrhea, a bacterial infection, causes discharge of pus from the penis or sometimes vagina, and pain on urination. The main sites of infection are the urethra and, in women, the cervix, from where organisms can spread to the uterus, fallopian tubes, and ovaries. The rectum can also be affected. A pregnant woman risks passing the infection to her baby during childbirth. Gonorrhea is treated by antibiotics, but in some parts of the world resistance has developed to most drugs.

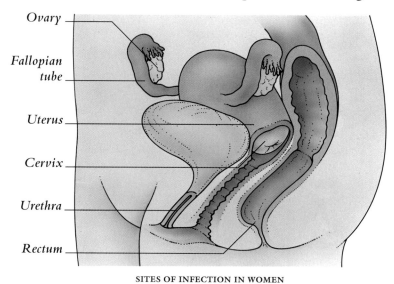

Ovary
Fallopian tube
Uterus
Cervix
Urethra
Rectum

SITES OF INFECTION IN WOMEN

SYPHILIS

The earliest symptom of syphilis is usually an ulcer (chancre) of the genitals. However, a chancre can also develop in the mouth or anus. A rash, mouth ulcers, and enlargement of the lymph nodes follow. Later effects include brain, heart, and bone disorders. A pregnant woman can pass the infection to her baby.

Infecting organism

The organism that causes syphilis is a spiral bacterium, *Treponema pallidum*. The bacteria are seen here as wavy and extended threads inside a testicular cell. Treatment with penicillin can cure syphilis if given early, but the effects of late stages of syphilis are irreversible.

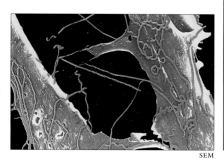

SEM

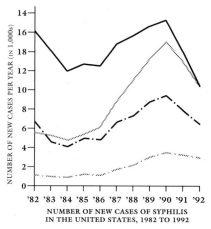

NUMBER OF NEW CASES PER YEAR (IN 1,000s)

'82 '83 '84 '85 '86 '87 '88 '89 '90 '91 '92
NUMBER OF NEW CASES OF SYPHILIS IN THE UNITED STATES, 1982 TO 1992

KEY

Males	20–34 ——	35+ — • —
Females	20–34 ～～	35+ ～ ⌃ ～

Decrease in syphilis

The number of syphilis cases is much higher in men than it is in women. The number of new cases in men decreased during the period from 1982 to 1986, followed by a sharp decline from 1990 to 1992, in part due to safe sex practices. The number of female cases showed a smaller decrease over the same period.

GENITAL HERPES

One of the most common STDs is genital herpes, and its reported incidence has increased over recent years in some countries. Caused by an organism known as herpes simplex virus, genital herpes tends to recur, The first episode is the most severe with subsequent occurrences decreasing in severity and frequency.

TEM x 120,000

Herpes simplex virus

Here two herpes simplex virus particles can be seen emerging from a host cell's nucleus into the surrounding cytoplasm. The herpes simplex virus also causes mouth ulcers and cold sores.

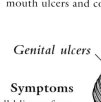

Genital ulcers

Symptoms

During an episode, crops of small blisters form on the penis or around the vagina, then develop into shallow, painful ulcers. The first attack of herpes may be accompanied by headache, fever, and pain in the groin, buttocks, and legs.

TREATMENT

There is no cure for genital herpes. Painkillers, such as aspirin, and warm baths can help relieve symptoms. The antiviral drug acyclovir can provide pain relief and speed healing in people during an attack. Prolonged use may reduce the number and frequency of occurrences.

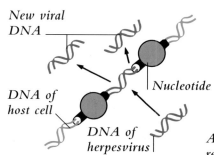

New viral DNA
DNA of host cell
Nucleotide
DNA of herpesvirus

Replication of herpesvirus

The herpesvirus can multiply only within a host cell. It makes new copies of itself by using the host's DNA (genetic material). It is then free to invade other cells.

Replication prevented

Acyclovir works by interfering with the process of viral replication. When the drug is inside herpes-infected cells, it is activated to resemble a nucleotide and blocks the synthesis of viral DNA.

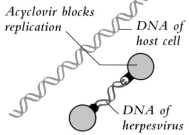

Acyclovir blocks replication
DNA of host cell
DNA of herpesvirus

PREVENTION OF STDs

To prevent STDs, sexually active people should ideally have a single sexual partner, always use a condom when engaged in penetrative sex, and avoid practices that could damage the delicate lining of the vagina or anus. People with symptoms of or being treated for an STD should abstain from sex; their partners should be checked for infection.

Condoms

Latex rubber condoms can provide an effective barrier against microorganisms, but they can burst and should always be used correctly.

INFERTILITY

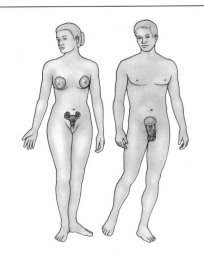

AS MANY AS ONE COUPLE IN SIX asks for medical advice about infertility. The numbers may be growing for social as well as medical reasons: many women delay childbearing and, after the age of 30, a women's fertility begins to decline. However, in only about one-third of infertile couples is the cause a disorder in the woman: in one-third of the cases, a disorder in the man is responsible, most commonly a low sperm count. About half of all couples who are treated for infertility eventually achieve a pregnancy.

ENDOMETRIOSIS

Worldwide, one of the most common causes of infertility is blockage of the fallopian tubes, which may occur as a result of untreated endometriosis. Normally, fertilization occurs when a sperm and ovum meet in the fallopian tube, but a blockage prevents this. Surgery may be needed.

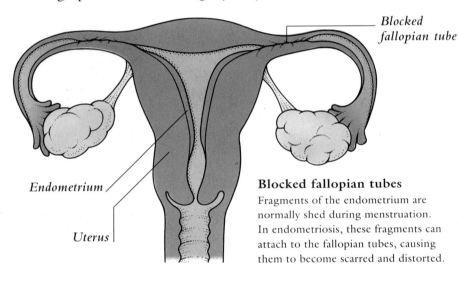

Blocked fallopian tube

Endometrium

Uterus

Blocked fallopian tubes
Fragments of the endometrium are normally shed during menstruation. In endometriosis, these fragments can attach to the fallopian tubes, causing them to become scarred and distorted.

ANTIBODIES TO SPERM

Sometimes infertility is due to the formation of antibodies to sperm by either the man or the woman. If a man has one-sided obstruction of sperm flow, his immune system may form antibodies to his own sperm. These antibodies immobilize sperm by making them stick together.

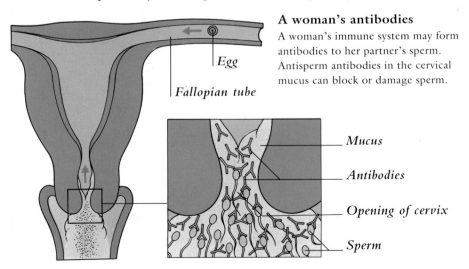

A woman's antibodies
A woman's immune system may form antibodies to her partner's sperm. Antisperm antibodies in the cervical mucus can block or damage sperm.

Egg

Fallopian tube

Mucus

Antibodies

Opening of cervix

Sperm

CAUSES OF FEMALE INFERTILITY

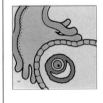

Egg release
Ovaries may fail to release mature eggs, or may release them at irregular intervals. The causes include hormone imbalance caused by either excessive weight loss or polycystic ovarian disease.

Blocked or damaged fallopian tubes
Narrowed or blocked tubes resulting from scarring from infection, endometriosis, or an ectopic pregnancy can prevent either fertilization or implantation.

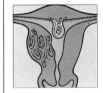

Abnormalities of the uterus
Structural abnormalities are a rare cause of infertility. The uterus may have been deformed since birth, or it may have been damaged by the formation of fibroids, by surgery, or by an infection.

Cervical problems
A hormone imbalance may result in a thick cervical mucus, which blocks sperm from traveling successfully through the woman's reproductive tract. Damage to the cervix may result in miscarriages.

CAUSES OF MALE INFERTILITY

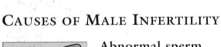

Abnormal sperm
An inadequate number of sperm is the most common cause of male infertility; sperm may also be misshapen or unable to move rapidly. These problems may be due to hormone imbalance, drugs, or illness.

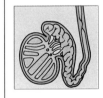

Difficult passage of sperm
Sperm must make their way through the tubal systems of the epididymis and the vas deferens before mixing with semen to be ejaculated. Blockage of any of these pathways may be a cause of infertility.

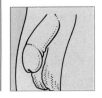

Ejaculation problems
Nerve conduction may be impaired by diabetes mellitus, poor blood flow, some medications, or a spinal cord disease. After a prostate operation semen may have been blocked or enter the bladder.

INFERTILITY TESTS

Tests are not usually considered until a couple has been trying to conceive for at least a year. Both partners are tested: the woman is examined to confirm that she is ovulating, and the man is checked to see that he is producing healthy sperm. More elaborate investigations may be needed.

SEMEN ANALYSIS

Semen is obtained by the man masturbating or is taken from his partner's vagina after sexual intercourse. About 20 percent of the millions of sperm produced daily are abnormal; only if a higher number are defective is infertility likely.

SEM x 2,500

Normal sperm

The number of sperm in the semen is counted, and their appearance and mobility are assessed. Normal sperm, such as those shown at left, are of a uniform size and shape, and also swim vigorously.

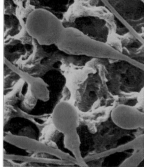

SEM x 2,500

Abnormal sperm

A semen analysis may show too few sperm, or sperm that are immobile or abnormally shaped. Shown at right are sperm of varying sizes with irregularly shaped bodies.

CHECKING THE FALLOPIAN TUBES

Once ovulation has been confirmed, the next step is to check whether the fallopian tubes are blocked by passing dye into the tubes via the uterus. Successful passage of the dye is confirmed with a laparoscope or, less often, by an X-ray.

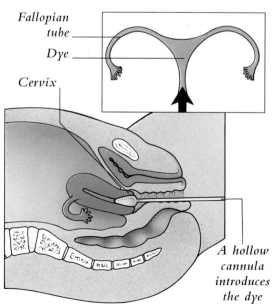

Fallopian tube
Dye
Cervix
A hollow cannula introduces the dye

INFERTILITY TREATMENT

If a woman is not ovulating normally, her overall general health may need improvement, especially if she is seriously underweight. She may also need to be treated with fertility drugs or hormones to stimulate the ovaries to form mature eggs and then release them. If the quality of a man's sperm is poor, treatment of the cause may be undertaken. Healthy sperm may be used to fertilize the ovum of his partner or that of a surrogate in an assisted conception technique. The couple may also consider donor insemination.

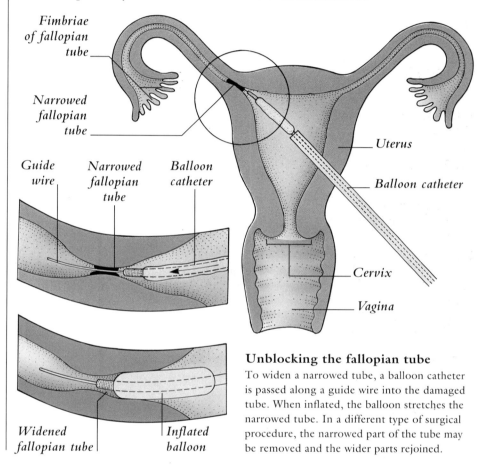

Fimbriae of fallopian tube
Narrowed fallopian tube
Guide wire
Narrowed fallopian tube
Balloon catheter
Uterus
Balloon catheter
Cervix
Vagina
Widened fallopian tube
Inflated balloon

Unblocking the fallopian tube

To widen a narrowed tube, a balloon catheter is passed along a guide wire into the damaged tube. When inflated, the balloon stretches the narrowed tube. In a different type of surgical procedure, the narrowed part of the tube may be removed and the wider parts rejoined.

IN VITRO FERTILIZATION (IVF)

IVF may be the best chance for a couple to conceive if the fallopian tubes are blocked, or if other forms of treatment have failed. The woman's eggs are collected, mixed with sperm, and incubated. The fertilized eggs are then transferred into the woman's uterus or tubes. Other assisted fertilization techniques exist to aid infertile couples.

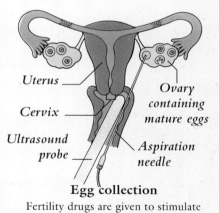

Uterus
Cervix
Ultrasound probe
Ovary containing mature eggs
Aspiration needle

Egg collection

Fertility drugs are given to stimulate several eggs to mature. These are collected using a laparoscope or a needle guided by an ultrasound probe.

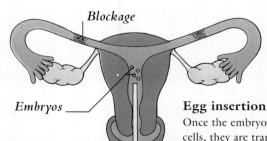

Blockage
Embryos
Catheter

Egg insertion

Once the embryos have grown to about eight cells, they are transferred into the uterus via the cervix. Several will be transferred to maximize the chances of successsful pregnancy.

CHAPTER TWELVE

C H A P T E R   1 2

The HUMAN LIFE CYCLE

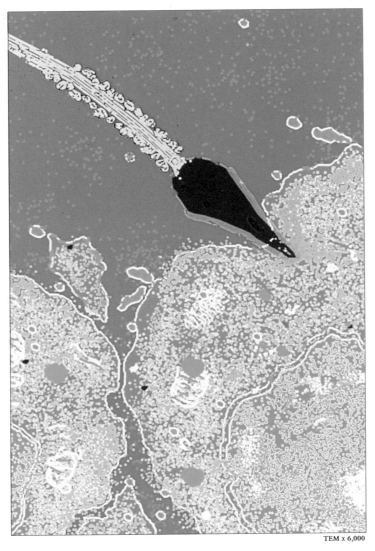

A human sperm
penetrates the outer
covering of an egg

TEM x 6,000

INTRODUCTION

After an egg has been fertilized, it divides and then differentiates into three layers of cells. These embryonic layers later develop into tissues and organs. Instructions for this development are held in the nucleus of the original cell by the genes, units that together make up 23 pairs of structures called chromosomes. Deciphering the genetic code is one of the most exciting ventures of modern biology: the human genome project has located the precise position of many thousands of genes. This has made it possible to give advice to families in which hereditary diseases are discovered. The replacement of faulty genes with healthy ones is likely to be an important issue in future decades. Other ways to cut the risks of some birth defects include immunizing women against rubella and giving them extra folic acid during the weeks before and after conception. Children born in the 1990s should live longer than those born previously. Most people who now live to age 80 or more stay in reasonably good health, although they are more susceptible to some diseases. The challenge that faces medicine in future is to ensure that better health and an enjoyable, active lifestyle go together with longer life.

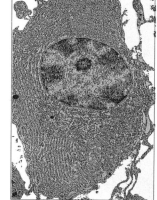

A cell nucleus – location of the chromosomes

TEM x 10,800

The ninth month of pregnancy

Chromosomes separating during sperm or egg formation

The EMBRYO

DURING ITS FIRST 8 WEEKS OF DEVELOPMENT, the unborn child is called an embryo. For the rest of the pregnancy, it is known as a fetus. The embryo develops from a cluster of cells formed by repeated division of the fertilized egg. Some of these cells form membranes to protect both the embryo and its placenta, which nourishes the embryo and removes its waste products.

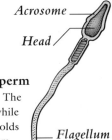

The sperm's journey
After each ejaculation, about 300 million sperm enter the cervix. Only about 300 of these will reach the fallopian tube, and only one will fertilize the ovum.

FERTILIZATION

Fertilization takes place high in the fallopian tube when the head of the sperm penetrates a mature ovum. After penetration occurs, the nuclei of the sperm and ovum, each of which contains 23 chromosomes, fuse to form the zygote. With its 46 chromosomes, the zygote starts to divide as it is travels down the fallopian tube.

Sperm fertilizing egg

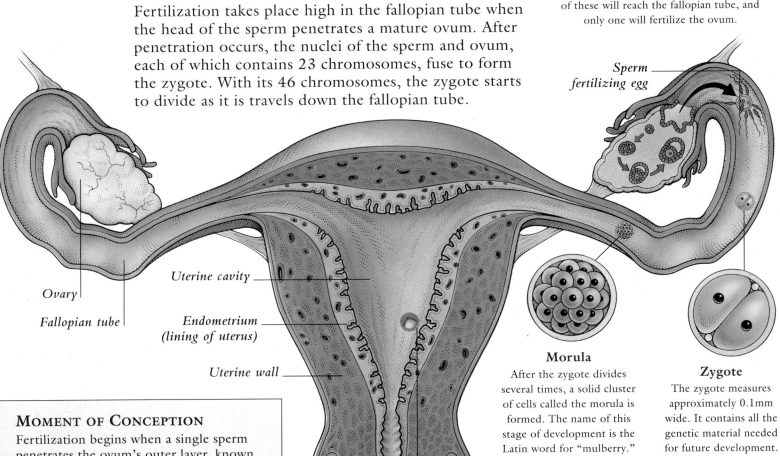

Ovary

Uterine cavity

Fallopian tube

Endometrium
(lining of uterus)

Uterine wall

Cervix

Vagina

Morula
After the zygote divides several times, a solid cluster of cells called the morula is formed. The name of this stage of development is the Latin word for "mulberry."

Zygote
The zygote measures approximately 0.1mm wide. It contains all the genetic material needed for future development.

Blastocyst
About 6 days after fertilization, the cell mass develops a hollow cavity. Now called a blastocyst, it is ready to become embedded in the endometrial tissue.

MOMENT OF CONCEPTION
Fertilization begins when a single sperm penetrates the ovum's outer layer, known as the corona radiata. After one sperm head penetrates the ovum, chemical changes that are triggered by enzymes prevent the entry of any other sperm. The sperm sheds its body and tail, while the head containing the nucleus and genetic material continues to move toward the ovum's nucleus.

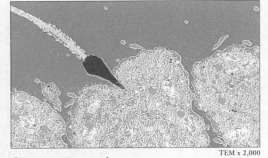

TEM x 2,000

Sperm penetrating egg

Acrosome

Head

Structure of sperm
Each sperm is about 0.5mm long. The head contains genetic material, while the acrosome capping the head holds enzymes that stop other sperm from penetrating the ovum. The whiplike tail propels sperm.

Flagellum
(tail)

IMPLANTATION AND EARLY DEVELOPMENT

Once the blastocyst forms, it floats freely within the uterine cavity for about 48 hours before drifting to a location in the endometrium. Part of the uterine lining thins and softens to facilitate the process of implantation. By about the tenth day after fertilization, the embryo is completely embedded in the uterine wall. If levels of estrogen and progesterone are too low, the endometrium may break down and cause a miscarriage.

1 The blastocyst is covered by an outer layer called the trophoblast. After the blastocyst becomes attached, specialized trophoblast cells secrete an enzyme that softens the tissue of the endometrium; other trophoblast cells burrow more deeply, eventually forming the nourishing placenta. The inner cell cluster in the blastocyst's fluid-filled cavity develops into the embryo.

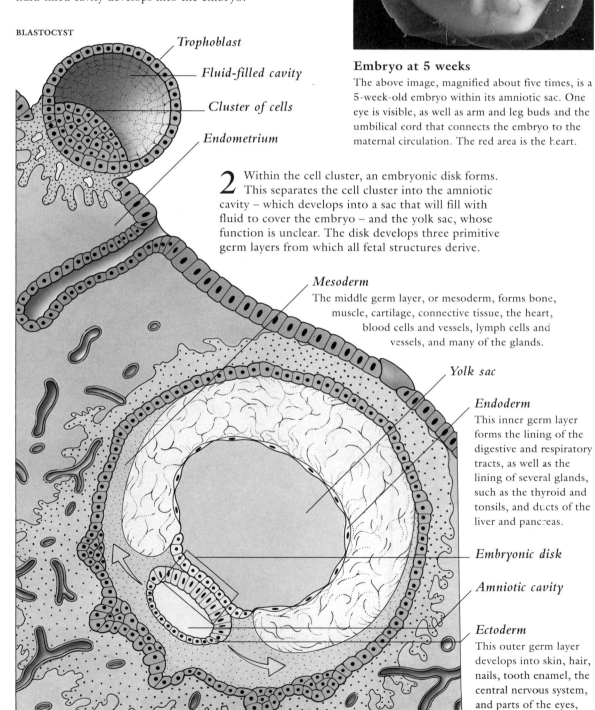

BLASTOCYST

Trophoblast

Fluid-filled cavity

Cluster of cells

Endometrium

Embryo at 5 weeks
The above image, magnified about five times, is a 5-week-old embryo within its amniotic sac. One eye is visible, as well as arm and leg buds and the umbilical cord that connects the embryo to the maternal circulation. The red area is the heart.

2 Within the cell cluster, an embryonic disk forms. This separates the cell cluster into the amniotic cavity – which develops into a sac that will fill with fluid to cover the embryo – and the yolk sac, whose function is unclear. The disk develops three primitive germ layers from which all fetal structures derive.

Mesoderm
The middle germ layer, or mesoderm, forms bone, muscle, cartilage, connective tissue, the heart, blood cells and vessels, lymph cells and vessels, and many of the glands.

Yolk sac

Endoderm
This inner germ layer forms the lining of the digestive and respiratory tracts, as well as the lining of several glands, such as the thyroid and tonsils, and ducts of the liver and pancreas.

Embryonic disk

Amniotic cavity

Ectoderm
This outer germ layer develops into skin, hair, nails, tooth enamel, the central nervous system, and parts of the eyes, ears, and nasal cavity.

THE GROWING EMBRYO

By the end of the third week, a neural tube has formed, which will become the spinal cord. Between the third and fourth weeks, the heart starts to beat, and the liver and lungs can be seen. By the eighth week, the embryo starts to "quicken," or move. It is now called a fetus.

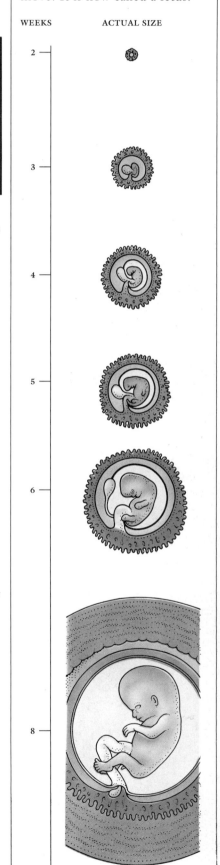

WEEKS   ACTUAL SIZE

2

3

4

5

6

8

FETAL DEVELOPMENT

FROM THE EIGHTH WEEK OF PREGNANCY until childbirth, the unborn baby, now called a fetus, develops inside a sac within the uterus. The sac is filled with a clear fluid that cushions the delicate fetus against injury; called amniotic fluid, it is swallowed by the fetus, absorbed into its bloodstream, and excreted as urine. Oxygen and nutrients needed by the fetus are supplied from maternal blood through the placenta.

Placenta

Amniotic fluid

LENGTH 1in (2.5cm)
WEIGHT 0.07oz (2g)

THE GROWING FETUS

The major body organs of the fetus develop in the early months of pregnancy. During this phase, the fetus is most vulnerable to infectious organisms, including the virus that causes rubella (German measles), as well as to alcohol and various other toxic substances. Later, the fetus increases in complexity and size. Around week 32, the fetus turns head-down and looks much as it will at birth.

8 weeks
The arms, legs, and major joints of the fetus are forming and it begins to move, although these movements will not be felt by the mother at this early stage. Toes and fingers are distinct but may still be joined by webs of skin. The fetal blood cells circulate within immature blood vessels.

12 weeks
Although its head is very large compared to the rest of its body, the fetus is recognizably a human being. Its major internal organs have developed, and tiny nails are growing on its fingers and toes. The external ears, the eyelids, and 32 permanent teeth buds have formed.

16 weeks
The fetus is growing rapidly and is able to move vigorously, although these movements are still not felt by the mother. External genital organs are visible, and a fine, downy hair, called lanugo hair, grows over its body.

LENGTH 3in (7.5cm)
WEIGHT 0.6oz (18g)

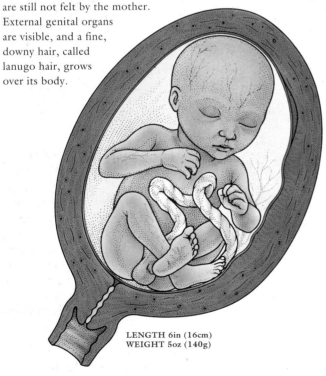

40 weeks
The fetus, now mature, is ready for life outside the uterus. Its skin is covered in a somewhat greasy, white substance, called vernix, to ease its passage down the birth canal. A baby that is born before 37 weeks is termed premature, and may need to be placed in an incubator.

LENGTH 6in (16cm)
WEIGHT 5oz (140g)

LENGTH 20in (51cm)
WEIGHT 7.5lb (3.4kg)

THE DEVELOPING PLACENTA

The placenta is a special organ that supplies the fetus with nutrients and oxygen, absorbs fetal waste products, and acts as a barrier against harmful substances. It derives from the trophoblast, the outer layer of the blastocyst (the mass of cells that implants in the uterine lining after fertilization). It begins to form as soon as implantation occurs and is well established by the tenth day. Hormones from the placenta help maintain the endometrium so the pregnancy continues.

1 Specialized cells of the embedded trophoblast extend into nearby uterine blood vessels. Blood from the mother flows from these blood vessels into spaces within the trophoblast.

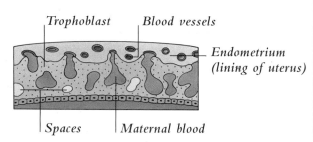

Trophoblast *Blood vessels* *Endometrium (lining of uterus)* *Spaces* *Maternal blood*

2 Other trophoblast cells extend fingerlike projections, known as chorionic villi, into the endometrium. These are surrounded by the spaces filled with maternal blood. Fetal blood vessels grow into the chorionic villi.

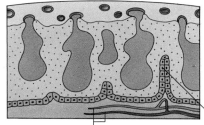

Fetal blood vessels *Chorionic villus*

3 Maternal and fetal blood do not make direct contact in the placenta; they are separated by a barrier of cells. Oxygen, nutrients, and antibodies cross the barrier to the fetus, and waste products pass back through the placenta.

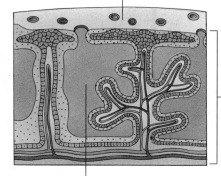

Endometrium *Maternal blood* *Placenta*

4 The placenta continues to develop as the fetus grows so that by the end of the pregnancy it is about 8in (20cm) wide and 1in (2.5cm) thick. It is attached to the center of the baby's abdomen by the umbilical cord.

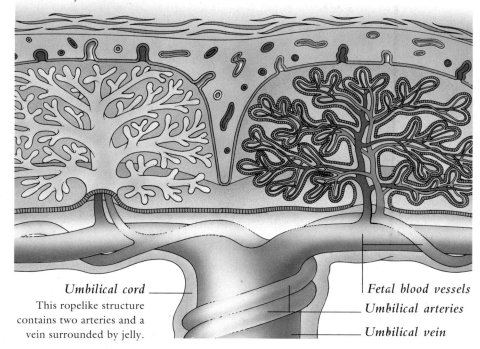

Umbilical cord
This ropelike structure contains two arteries and a vein surrounded by jelly.

Fetal blood vessels
Umbilical arteries
Umbilical vein

TRIMESTERS OF PREGNANCY

Pregnancy typically lasts 40 weeks from the first day of a woman's last menstrual period. By convention, the duration of pregnancy is divided into trimesters, each about 3 months long. During this time, a woman's body undergoes many changes to support the fetus and prepare for childbirth.

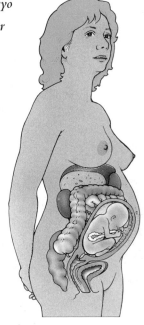

FIRST TRIMESTER

The woman's breasts become tender and begin to enlarge. Vaginal discharge sometimes increases, as does the need to urinate. Weight gain begins, and the areola surrounding the nipple darkens. Vomiting and nausea are common.

Areola
Stomach
Uterus
Embryo
Bladder

SECOND TRIMESTER

The woman begins to look noticeably pregnant as her uterus enlarges. Her heart rate increases as a result of circulatory changes. The fetus often begins to move at approximately 8 weeks, although most women feel the baby move only after 20 weeks of pregnancy.

THIRD TRIMESTER

The skin stretches over the abdomen, and very slight contractions are sometimes felt. The enlarged uterus presses on the bladder, which may cause slight incontinence. Fatigue is a common symptom, as is back pain, heartburn, or occasional breathlessness.

Compressed bladder

PRENATAL TESTS

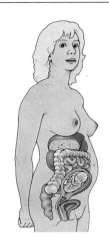

SOON AFTER A WOMAN KNOWS SHE IS PREGNANT, she should begin regular visits to a physician or midwife in order to check on her health and that of the growing fetus. Nowadays, prenatal care includes maternal tests, such as blood typing, blood count, tests for infections, and blood pressure readings, as well as tests to identify abnormalities in the developing fetus. Recognizing problems at an early stage is desirable because it may allow for treatment. If an abnormality is severe, the parents may seek counseling.

ULTRASOUND SCANNING

A clear view of the fetus is obtained by utilization of ultrasound scanning, which is a safe, reliable method of creating an image. Scans are done at 16 weeks in order to check on the fetus's growth and position, and the development of its body parts and internal organs, such as the heart and lungs.

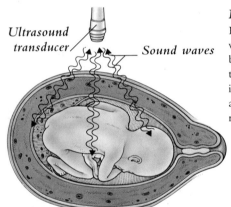

Ultrasound transducer — Sound waves

Method
High-frequency ultrasound waves penetrate body tissues but do not damage them. A transducer emits the waves as it is moved over the abdomen and detects them as they are reflected back from the fetus.

The image on the screen
Sound waves deciphered by a computer create an image of the moving fetus on the screen. Fetal size helps confirm its age. Its position enables a doctor to anticipate complications that may affect fetal development or delivery.

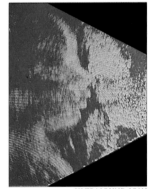

ULTRASOUND SCAN

ALPHAFETOPROTEIN TESTING
Alphafetoprotein is produced in the liver of the fetus, and then passes into the mother's bloodstream where it may be measured. A concentration above normal may suggest the possibility of twins or an abnormality such as spina bifida, in which vertebrae fail to close around the spinal cord.

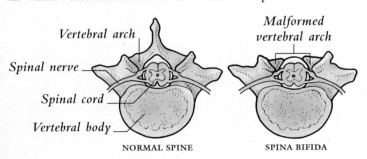

Vertebral arch

Malformed vertebral arch

Spinal nerve

Spinal cord

Vertebral body

NORMAL SPINE   SPINA BIFIDA

AMNIOCENTESIS

A membranous bag, the amniotic sac, surrounds and protects the fetus. Fluid taken from the sac contains chemicals, which are analyzed, and fetal cells, which are cultured and examined to reveal fetal chromosomes. Amniocentesis may be performed between 16 and 18 weeks of pregnancy; miscarriage is a rare risk.

Chromosome analysis
Fetal cells are grown in the laboratory and a chemical, colchicine, is added to stop division at a stage when they are most easily seen. The cells and the 23 pairs of chromosomes are checked. Four chromosomes are seen at right.

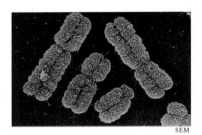

SEM

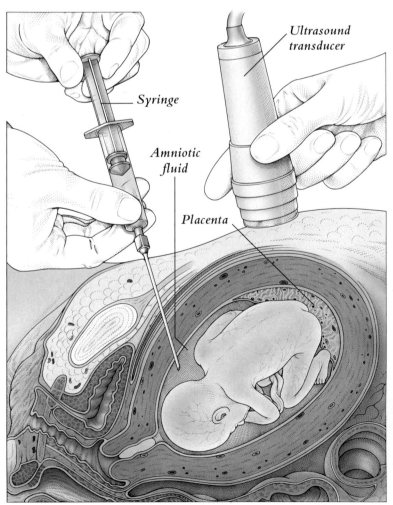

Ultrasound transducer

Syringe

Amniotic fluid

Placenta

CHORIONIC VILLUS SAMPLING

The chorion is the outermost of the two membrane layers of the amniotic sac surrounding the fetus. A sample of the villi, which are tiny projections from the chorion, may be removed as early as the eighth week of pregnancy. Fetal cells from this sample may be cultured for chromosomal analysis. Miscarriage or damage to the fingers or the hands occurs rarely.

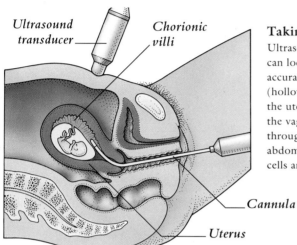

Ultrasound transducer *Chorionic villi*

Cannula

Uterus

Taking a sample
Ultrasound scanning can locate the fetus accurately. A cannula (hollow tube) enters the uterus either via the vagina or directly through the wall of the abdomen, and chorion cells are withdrawn.

1 Chorion cells are transferred to a culture solution in order to provide them with nutrients as they multiply. For chromosome analysis, cell division is artificially halted at the optimum stage for viewing them under a microscope.

Syringe

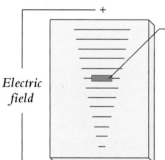

Piece of DNA for study

Electric field

2 For gene testing, DNA from the chorion cells is cut into fragments, using enzymes, and these are placed on a special gel. An electric current passed through the gel sorts the DNA strands by size. The strands are transferred to a membrane. A genetic probe is added.

3 The genetic probe consists of radioactively labeled DNA that will bind to DNA strands on the membrane if they contain a matching pattern. Binding makes dark bands appear on a film called an autoradiograph, which can be compared with reference patterns.

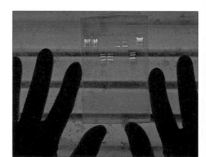

Autoradiograph analysis

Results of a genetic analysis
Shown at left are the results of genetic probes carried out on four boys at risk of Duchenne-type muscular dystrophy. The pattern at far left is normal, but the others are abnormal, with dark bands either missing or of a different size.

FETAL HEART MONITORING

During pregnancy and especially during labor, one of the most reliable measures of fetal health is the heart rate. An obstetrical nurse measures the heart rate and the contractions of the uterus. Continuous electronic heart monitoring during labor is usually reserved for babies who are thought to be at higher than average risk of developing complications.

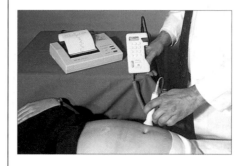

Doppler ultrasound
This technique, which may be used from about the 12th week of pregnancy, detects echoes of ultrasound that bounce off the fetus's heart; the waves are then converted into audible signals.

OTHER FETAL TESTS

If there is an indication that the fetus suffers from a blood disorder, a blood sample may be taken from the umbilical cord. If fetal problems cannot be diagnosed by other tests, fetoscopy, in which the fetus is viewed inside the uterus by means of an endoscope, may be done; the procedure carries a risk of miscarriage.

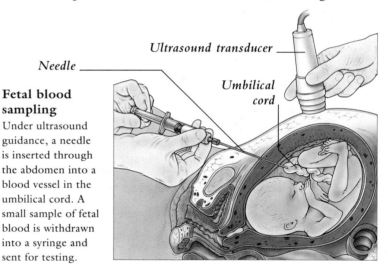

Ultrasound transducer

Needle

Umbilical cord

Fetal blood sampling
Under ultrasound guidance, a needle is inserted through the abdomen into a blood vessel in the umbilical cord. A small sample of fetal blood is withdrawn into a syringe and sent for testing.

SUMMARY OF PRENATAL TESTS

TIMING	PROCEDURE AND REASON
INITIAL VISIT	Tests usually include: blood type, blood count, tests for STDs. Urine tests check for protein and glucose. A full physical examination is made.
THROUGHOUT PREGNANCY	Blood pressure is measured and urine analyzed. The abdomen is palpated and fetal heartbeat checked. Any warning signs are treated.
9–10 WEEKS	DNA analysis of fetal cells obtained by chorionic villus sampling assesses fetuses at risk of single-gene disorders, such as muscular dystrophy.
16–18 WEEKS	Ultrasound tests check growth and development. Amniocentesis or chorionic villus sampling may be performed in high-risk pregnancies.

ONSET *of* LABOR

CHANGES OCCUR IN THE BODY during late pregnancy, signaling the approach of childbirth. The head of the fetus drops lower into the pelvis, and the expectant mother may experience weight loss. As labor begins, the mucus plug that seals off the cervix is expelled as a blood-stained discharge known as "show." Uterine contractions increase in strength and become more regular. The sac around the amniotic fluid ruptures, known as "breaking of the bag of waters."

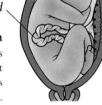

Twins

The presence of more than one fetus in the uterus is called a multiple pregnancy. Twins are relatively common; they occur in about one in every 80 pregnancies.

Frank breech

In this presentation, known as "frank" or "incomplete" breech, the baby's hips are flexed and the legs extend alongside the body. The feet lie beside the head.

Placenta

Umbilical cord

Complete breech

In "complete breech," the baby's legs are flexed at the knees and at the hips. This presentation occurs less commonly than frank breech.

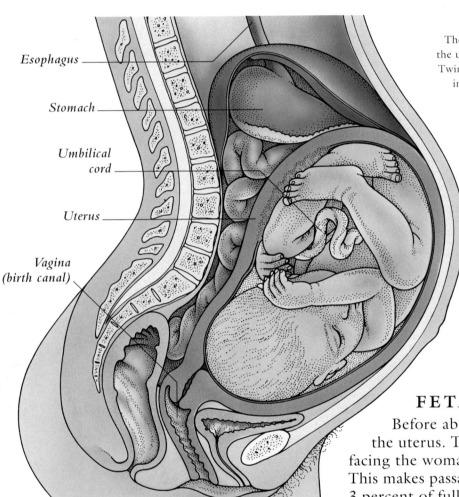

Esophagus

Stomach

Umbilical cord

Uterus

Vagina (birth canal)

FETAL POSITIONS

Before about 30 weeks, the fetus tends to turn in the uterus. The most usual position is head downward, facing the woman's back, with the neck flexed forward. This makes passage through the birth canal easiest. About 3 percent of full-term deliveries are breech, in which the baby's bottom is delivered before the head. Breech babies may be delivered by cesarean section. The incidence of breech delivery is much higher among premature babies.

CHANGES IN THE CERVIX

The cervix is a firm band of muscle and connective tissue that forms the lower end of the uterus. During late pregnancy, the cervix softens in an early phase of childbirth. Braxton-Hicks uterine contractions are painless and gently help thin the cervix so that it merges with the uterus's lower segment.

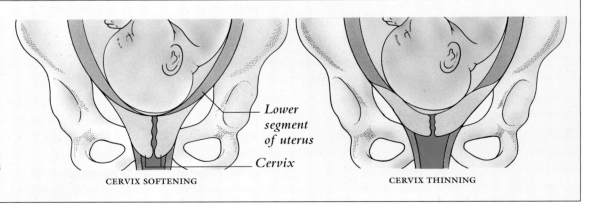

Lower segment of uterus

Cervix

CERVIX SOFTENING

CERVIX THINNING

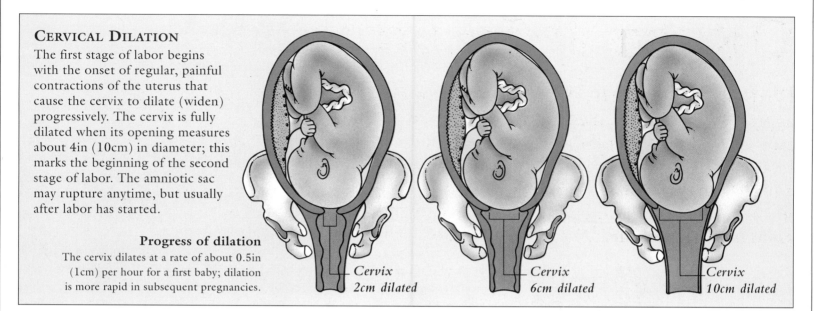

CERVICAL DILATION

The first stage of labor begins with the onset of regular, painful contractions of the uterus that cause the cervix to dilate (widen) progressively. The cervix is fully dilated when its opening measures about 4in (10cm) in diameter; this marks the beginning of the second stage of labor. The amniotic sac may rupture anytime, but usually after labor has started.

Progress of dilation

The cervix dilates at a rate of about 0.5in (1cm) per hour for a first baby; dilation is more rapid in subsequent pregnancies.

Cervix
2cm dilated

Cervix
6cm dilated

Cervix
10cm dilated

PELVIC SIZE AND SHAPE

The size and shape of the woman's pelvis are very important in determining the ease of childbirth. Any mismatch, also known as "disproportion," between the dimensions of the mother's pelvis and the baby's head can obstruct the progress of labor.

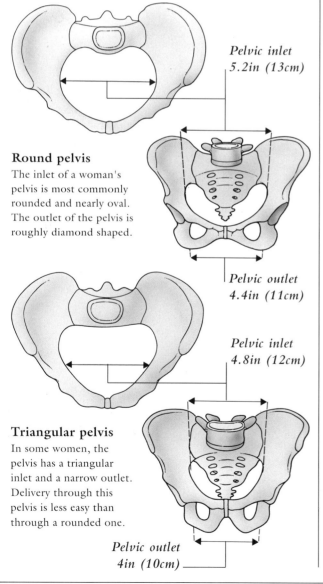

Pelvic inlet
5.2in (13cm)

Round pelvis

The inlet of a woman's pelvis is most commonly rounded and nearly oval. The outlet of the pelvis is roughly diamond shaped.

Pelvic outlet
4.4in (11cm)

Pelvic inlet
4.8in (12cm)

Triangular pelvis

In some women, the pelvis has a triangular inlet and a narrow outlet. Delivery through this pelvis is less easy than through a rounded one.

Pelvic outlet
4in (10cm)

ENGAGEMENT

During the last weeks of pregnancy, the baby's head descends into the cavity of the pelvis, a process called engagement. When this happens, many women feel the load "lightening" as the descent of the baby's head takes pressure off the diaphragm, making breathing easier. Engagement usually takes place at around 36 weeks during a first pregnancy. It may not happen until the onset of labor during second and subsequent pregnancies.

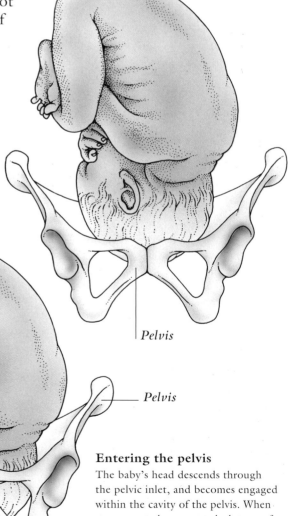

Before engagement

Toward the end of pregnancy and before the baby's head has engaged, the top of the uterus reaches up to the sternum, or breastbone. The widest section of the baby's head has not yet passed down through the inlet of the pelvis into the cavity.

Pelvis

Pelvis

Entering the pelvis

The baby's head descends through the pelvic inlet, and becomes engaged within the cavity of the pelvis. When engagement has occurred, the top of the uterus drops down and the baby's head then rests against the cervix.

DELIVERY *of the* BABY

DURING THE FIRST STAGE OF LABOR, the opening of the cervix gradually widens. In the second stage, the woman feels a strong urge to push with each contraction until the baby is born. The third stage follows, lasting from delivery of the baby until the placenta is delivered. Some women experience relatively little pain during childbirth, others a great deal.

PROGRESS OF LABOR

Labor progression is usually monitored by a physician or midwife. The second stage lasts about 50 minutes for the firstborn and about 20 minutes for subsequent babies. The third stage lasts about 5 minutes. Drugs are given to the mother to reduce the risk of bleeding after delivery.

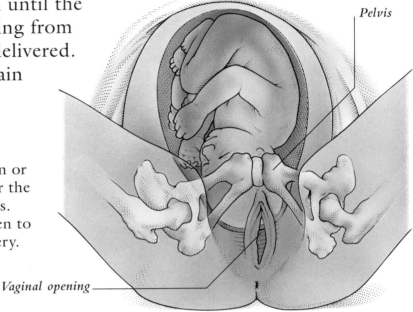

Pelvis

Vaginal opening

1 The baby's body rotates as it moves through the birth canal, while the muscles of the pelvic floor are pushed down. The perineum, which is the area around the vagina and the anus, bulges out, and the vaginal opening widens.

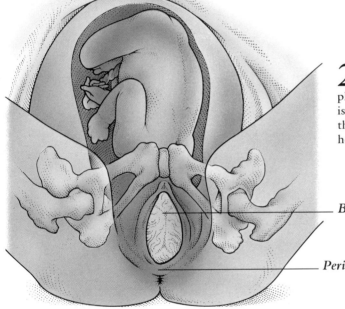

2 As the baby's head emerges, it is delivered slowly by the physician or midwife. If the head is expelled rapidly it may increase the risk of damage to the baby's head, or a tear in the perineum.

Baby's head

Perineum

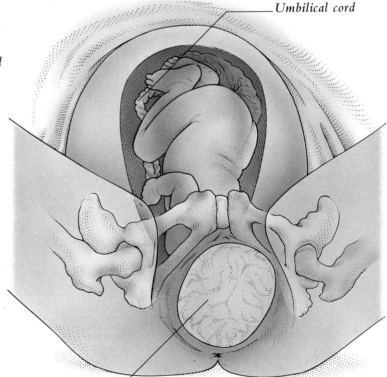

Umbilical cord

Baby's head

3 The baby's head usually emerges facing toward the mother's anus. The head then turns sideways as the shoulders move down into the pelvis, so that the alignment of the baby's body is restored.

FETAL MONITORING

During labor, the condition of the fetus is monitored by measuring fetal heart rate (normally between 120 and 160 beats per minute, or BPM). The rate normally decreases when each contraction begins and should return to normal quickly. Prolonged deceleration may indicate a problem.

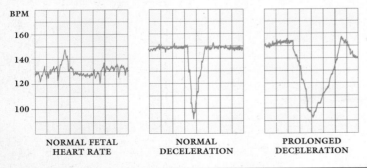

BPM

160

140

120

100

NORMAL FETAL HEART RATE

NORMAL DECELERATION

PROLONGED DECELERATION

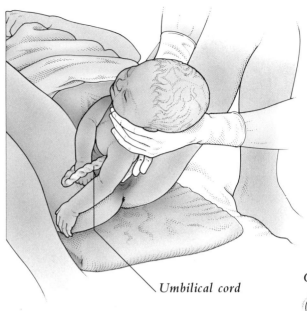

4 The physician or the midwife runs a finger along the baby's neck to check whether the umbilical cord is wrapped around it; if it is, the cord is slipped over the baby's head. Fluid is cleared from the baby's nose and mouth, then the shoulders are delivered.

Umbilical cord

5 The rest of the baby slides out easily with the next contraction. After the baby has been delivered, the umbilical cord is clamped in two places and is cut between the clamps. Cutting the cord does not hurt the baby.

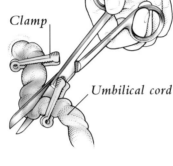

Clamp

Umbilical cord

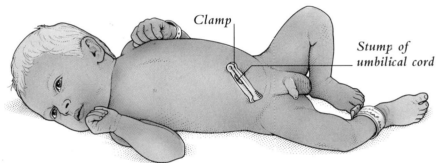

Clamp

Stump of umbilical cord

6 The condition of the newborn baby is evaluated rapidly. If all is well, the baby is immediately given to the mother to hold. The contractions continue, which cause the placenta to separate from the uterine wall.

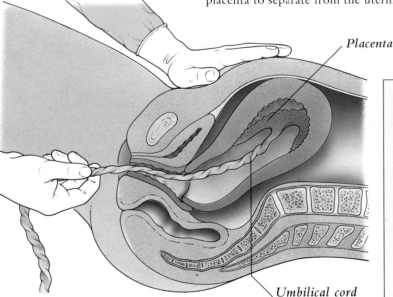

Placenta

Umbilical cord

7 When the placenta separates from the wall of the uterus, the physician or midwife pulls gently on the umbilical cord with one hand, pressing on the lower abdomen with the other. The placenta is eased out of the vagina.

PAIN RELIEF

Natural techniques, like relaxation and breathing, may help to relieve the pain of childbirth. Another safe method is a 50/50 mixture of nitrous oxide and air given via a handheld mask at the start of each contraction. Narcotic injections are effective but should not be given near the time of delivery.

Epidural anesthesia
A hollow needle is inserted into the epidural space of the spinal canal, as shown left. A soft tube is fed though the needle and the needle is removed. Doses of local anesthetic can be injected into the tube as required.

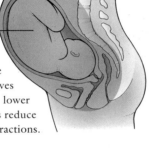

Area affected (shaded)

Area of numbness
An epidural provides effective pain relief by numbing the nerves that supply the pelvis and the lower abdomen. Epidural injections reduce awareness of uterine contractions.

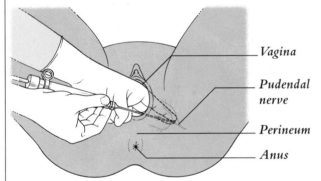

Vagina

Pudendal nerve

Perineum

Anus

Pudendal block
Injection of local anesthetic via the vagina into the pudendal nerve can abolish stretching pain during the second stage of labor. However, it does not relieve the pain of contractions.

MANAGING TWINS

Twins must be monitored closely during labor. After the first baby is delivered, the position of the second must be noted. If the second baby is lying in either a head-down or breech position and is in good condition, it can be delivered normally. If lying transversely, an obstetrician may be able to turn the baby longitudinally.

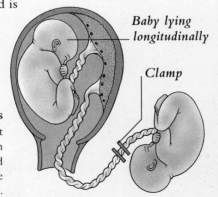

Baby lying longitudinally

Clamp

Delivery of twins
After delivery of the first twin, the cord is clamped in order to prevent the second twin from bleeding into the circulation of the first.

COMPLICATIONS *of* PREGNANCY *and* LABOR

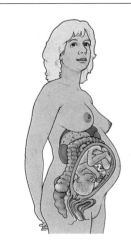

MOST WOMEN HAVE NORMAL PREGNANCIES and after 9 months they have a healthy baby with relative ease. The discomforts that are often experienced are not a threat to the well-being of either the mother or child. However, not all pregnancies are straightforward and problems may occur; these may pose a serious threat to the health of the mother or her baby, and sometimes both.

EARLY PROBLEMS

If the fertilized egg implants outside the main cavity of the uterus, the pregnancy is known as ectopic. The cause is not always known, although it occurs most often in women who have used an intrauterine device or who have had pelvic infections or a previous ectopic pregnancy. Symptoms are severe pain and bleeding.

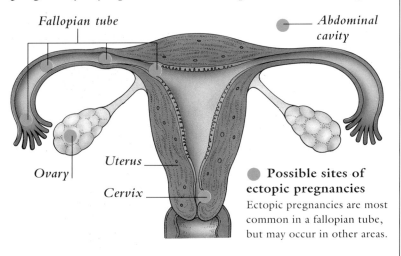

Fallopian tube

Abdominal cavity

Ovary

Uterus

Cervix

● **Possible sites of ectopic pregnancies**
Ectopic pregnancies are most common in a fallopian tube, but may occur in other areas.

MISCARRIAGE

A miscarriage is the loss of a fetus before week 20 of a pregnancy. About 20 percent of all pregnant women miscarry, and for many women this happens so early that they do not even know they are pregnant. The reason may be unknown, but common causes are fetal chromosomal abnormalities or developmental defects.

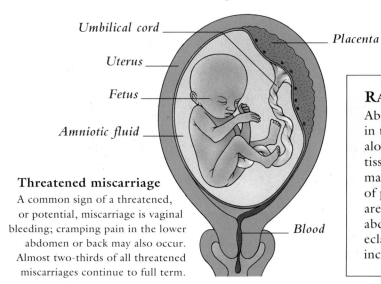

Umbilical cord

Placenta

Uterus

Fetus

Amniotic fluid

Threatened miscarriage
A common sign of a threatened, or potential, miscarriage is vaginal bleeding; cramping pain in the lower abdomen or back may also occur. Almost two-thirds of all threatened miscarriages continue to full term.

Blood

PLACENTAL PROBLEMS

A healthy placenta to nourish the fetus is essential for a normal pregnancy and a thriving baby. Very early in pregnancy, the placenta should develop in the upper wall of the uterus. Problems may occur if the placenta detaches or if it is abnormally low, which may lead to cervical obstruction, bleeding, or premature labor.

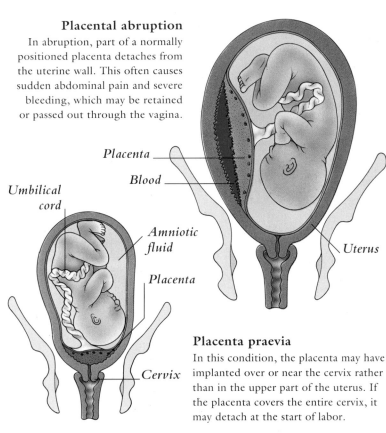

Placental abruption
In abruption, part of a normally positioned placenta detaches from the uterine wall. This often causes sudden abdominal pain and severe bleeding, which may be retained or passed out through the vagina.

Placenta

Blood

Umbilical cord

Amniotic fluid

Placenta

Uterus

Cervix

Placenta praevia
In this condition, the placenta may have implanted over or near the cervix rather than in the upper part of the uterus. If the placenta covers the entire cervix, it may detach at the start of labor.

RAISED BLOOD PRESSURE

Abnormally raised blood pressure in the second half of pregnancy, along with edema (fluid in the tissues) and protein in the urine, may indicate the serious condition of preeclampsia. Other symptoms are headaches, blurred vision, and abdominal pain. If left untreated, eclampsia may result, which may include life-threatening seizures.

Measuring blood pressure

ASSISTED DELIVERY

An assisted delivery may be necessary if labor is not progressing satisfactorily. Forceps are not used as commonly today as in the past, but they help deliver a baby quickly – especially if the baby is in distress or the mother is exhausted or has bled excessively. An alternative to forceps is vacuum extraction.

Forceps delivery

Obstetrical forceps consist of two curved metal blades designed to fit around the baby's head. The doctor pulls gently to guide the baby's head down into the vagina. Once the head has emerged, the forceps are removed and delivery can then continue normally.

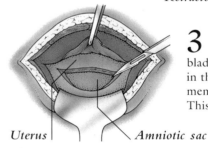

Forceps

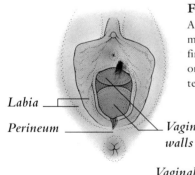

Plastic cup

Vacuum extraction

A vacuum pump machine attaches to a suction cup made of metal, rubber, or plastic. The cup is placed on the baby's head. The pump is turned on and, with each contraction, the doctor gently pulls the baby down the birth canal.

GENITAL TISSUE INJURIES

Injury to tissues of the genital tract is most common in women having their first child. The perineum, the area of tissue between the vagina and the anus, is most often torn. Tears vary in size from tiny splits to large, ragged tears. In rare cases the cervix is torn.

First-degree tear

A tear of the labia as well as the mucous membrane of the vagina is classed as a first-degree tear; it may heal on its own or need a few stitches. A second-degree tear extends into the perineal muscles.

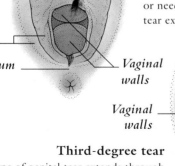

Labia

Perineum

Vaginal walls

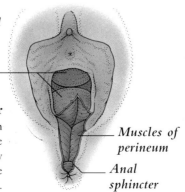

Vaginal walls

Third-degree tear

This type of genital tear extends through vaginal tissues, perineal muscles, and the anal sphincter. Each layer is individually sutured to maintain sphincter muscle strength and bowel movement control.

Muscles of perineum

Anal sphincter

CESAREAN SECTION

In a cesarean, the baby is delivered through an incision in the abdomen. The operation may be preplanned for multiple births or an abnormal fetal position, or if the mother has a vaginal infection or a scarred uterus as a result of previous cesareans. It is also performed as an emergency procedure during labor if the fetus becomes distressed. The operation takes about 40 to 60 minutes.

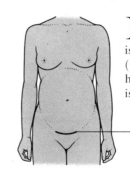

1 A general anesthetic or an epidural block is used. The woman's abdomen is cleaned with antiseptic, and a catheter (thin tube) is inserted in order to empty her bladder. Then a horizontal incision is made just below the pubic hairline.

Incision

Skin and fatty tissue

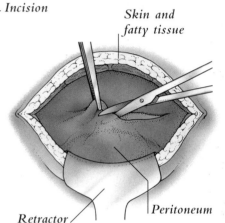

2 The surgeon makes a cut through the fatty tissues and muscles of the abdominal wall. A retractor holds back the tissues, and an incision is made in the peritoneum, the membrane lining the abdominal cavity.

Retractor

Peritoneum

3 The surgeon repositions the retractor to hold back the bladder, while making an incision in the lower uterus down to the membranes of the amniotic sac. This sac encloses the baby.

Uterus

Amniotic sac

4 The surgeon ruptures the amniotic sac and, placing a hand underneath the baby's head or buttocks, removes the baby. The cord is clamped and cut, and the placenta is removed.

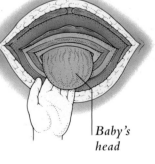

Baby's head

Stitches in uterus

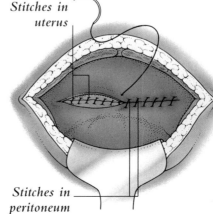

5 The uterus and abdominal layers are then closed with stitches. The incision in the skin is closed by metal clips (staples) or a long stitch. Stitches or staples are removed after 5 days, and the woman can go home.

Stitches in peritoneum

AFTER CHILDBIRTH

THE TIME FROM DELIVERY until most pregnancy changes have reverted to normal is called the postnatal period or puerperium. This lasts about 6 weeks, during which time minor discomforts such as genital soreness and constipation are common, but are usually overshadowed by the experience of having a new baby. Meanwhile, the newborn has to adapt to existence outside the protected environment of the uterus.

Fontanelles
Fontanelles are soft gaps between skull bones. By about 18 months, the bones close over these spots.

THE NEWBORN BABY

A full-term baby weighs, on average, 7.7lb (3.5kg) and measures 20in (51cm) in length. During the first few days, the baby loses up to 10 percent of its birthweight, but regains this by about the 10th day. At birth, the baby is usually covered with a greasy white substance called vernix, which provides protection within the uterus. The vernix is wiped away soon after birth.

Liver
Immaturity of the liver enzymes that break down the pigment bilirubin can cause temporary yellowing (jaundice).

Genitals
The external genitals of newborn boys and girls appear relatively large. Girls sometimes have a slight vaginal discharge.

Eyes
Newborn babies can see but tend to keep their eyes shut. The eyes are often grayish blue at first, but change color over the next few months.

Thymus
This lymph gland, which plays a role in the body's defenses, is large at birth but shrinks over the next several years.

Skin
Slight skin peeling can occur in the first week. Minor rashes and blemishes commonly disappear during the baby's early months.

Intestines
The first fecal material excreted by the baby is a thick, sticky, greenish black substance called meconium.

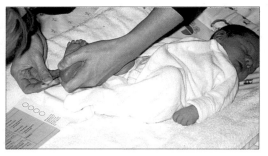

Heel-prick test
Within the first 10 days, a blood sample is obtained to test for phenylketonuria, which is a rare cause of mental retardation, and for thyroid deficiency. Blood may also be tested for Tay-Sachs disease, sickle cell disease, and other disorders.

APGAR SCORE

A test for new babies is the Apgar score. Breathing and heart rate, responses, muscle tone, and color are scored from 0 to 2 at 1 and 5 minutes after birth. A total score of 7 to 10 at 5 minutes is considered to be normal.

SIGN	SCORE: 0	SCORE: 1	SCORE: 2
HEART RATE	None	Below 100	Over 100
BREATHING RATE	None	Slow or irregular; weak cry	Regular; strong cry
MUSCLE TONE	Limp	Some bending of limbs	Active movements
REFLEX RESPONSES	None	Grimace or whimpering	Cry, sneeze, or cough
COLOR	Pale or blue	Blue extremities	Pink

CHANGES IN CIRCULATION

Because the fetus obtains oxygen and nutrients from the placenta, its circulatory system (illustrated below) differs from that of a baby at birth. Special features of the fetal circulation are: the foramen ovale, a hole that allows blood to flow from the right atrium to the left atrium; the ductus arteriosus, a channel that bypasses the lungs; and the ductus venosus, a liver bypass.

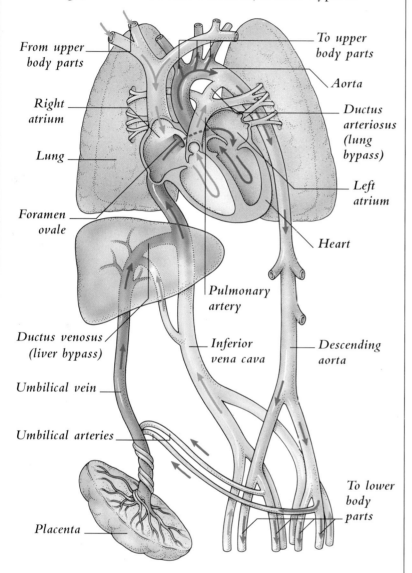

From upper body parts

Right atrium

Lung

Foramen ovale

Ductus venosus (liver bypass)

Umbilical vein

Umbilical arteries

Placenta

To upper body parts

Aorta

Ductus arteriosus (lung bypass)

Left atrium

Heart

Pulmonary artery

Inferior vena cava

Descending aorta

To lower body parts

CIRCULATION AT BIRTH

At birth, the lungs take over from the placenta. Lung blood flow increases and placental blood flow ceases. Pressure in the left heart chambers rises, causing the foramen ovale to shut. The umbilical vessels, ductus arteriosus, and ductus venosus close, forming ligaments.

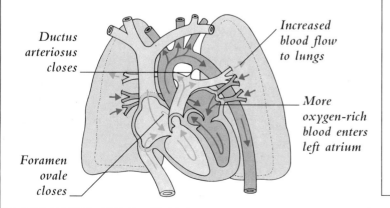

Ductus arteriosus closes

Foramen ovale closes

Increased blood flow to lungs

More oxygen-rich blood enters left atrium

THE PUERPERIUM

During the puerperium, the woman's genital tract gradually reverts back to its prepregnant state. As the placental site heals, tissue debris from the uterus is expelled in the form of a vaginal discharge called lochia. For the first week after childbirth, the lochia is bloodstained but then becomes cream colored. The vagina slowly shrinks to nearly its previous size.

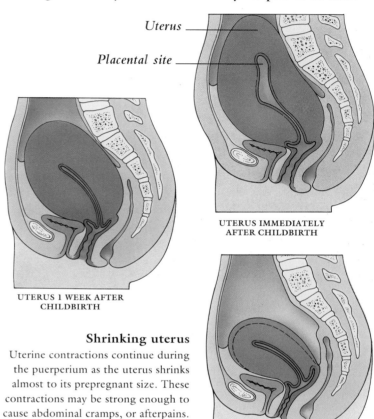

Uterus

Placental site

UTERUS IMMEDIATELY AFTER CHILDBIRTH

UTERUS 1 WEEK AFTER CHILDBIRTH

Shrinking uterus

Uterine contractions continue during the puerperium as the uterus shrinks almost to its prepregnant size. These contractions may be strong enough to cause abdominal cramps, or afterpains. Nursing triggers release of a hormone that may cause the uterus to contract.

UTERUS 6 WEEKS AFTER CHILDBIRTH

CERVIX OF NULLIPAROUS WOMAN

CERVIX AFTER GIVING BIRTH

The cervix

In nulliparous women (those who have never given birth) the cervical opening is nearly circular. Childbirth stretches and slightly tears the cervix. The cervical opening closes again, but does not regain its original appearance.

LACTATION

Breast size increases during pregnancy as glands develop for breastfeeding. Breast milk provides all the nourishment that a newborn baby requires, and also helps protect against infection. The baby is able to suck immediately after birth; this stimulates the release of oxytocin, a pituitary hormone that promotes milk flow and possibly uterine contraction.

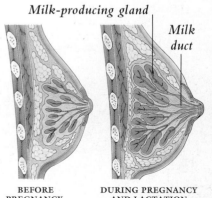

Milk-producing gland

Milk duct

BEFORE PREGNANCY

DURING PREGNANCY AND LACTATION

GROWTH *and* DEVELOPMENT

DURING THE EARLY YEARS, children develop basic physical skills such as walking and talking. While childhood progresses, agility improves and intellectual abilities increase. Growth is very rapid during infancy, then occurs fairly steadily until the growth rate speeds up again at puberty. During puberty, a child becomes an adult, physiologically capable of reproduction.

BONE GROWTH

Most of the long bones develop from cartilage by an orderly sequence of changes called ossification. The process starts before birth at zones called primary ossification centers in the bone shafts. After birth, secondary ossification centers develop near the bone ends. Growth ceases when ossification is complete.

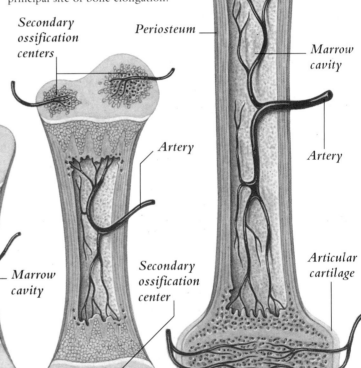

Epiphyseal line
The epiphyseal growth plate ossifies during adolescence or early adult life, forming a dense epiphyseal line.

Growing ends
The shaft, or diaphysis, of a long bone is separated from the growing end, or epiphysis, of the bone by a zone near the end of the bone called the epiphyseal plate. This plate is the principal site of bone elongation.

Epiphyseal line

Secondary ossification centers

Ossifying cartilage

Periosteum

Marrow cavity

Epiphyseal plate

Cartilage

Artery

Artery

Marrow cavity

Secondary ossification center

Articular cartilage

Diaphysis

Joint space

Epiphyseal plate

Epiphyses

NEWBORN

ABOUT 11 YEARS

Diaphysis

Bone

Epiphysis

Bone age
X-rays can reveal the maturity of a growing child because each bone ossifies at a predictable age and because cartilage, which is not as dense, shows up less clearly than bone does. The epiphyseal plates, other zones of cartilage growth, and joint spaces appear as gaps.

CHANGING BODY PROPORTIONS

Superimposing the body at different ages onto a grid divided into eight equal parts demonstrates the dramatic changes in body proportions that take place during childhood. In a newborn infant, the head is relatively large, representing about one-quarter of the baby's total length. As the child grows, the relative sizes of the head and trunk decrease and the limbs become longer. When final adult height is reached during adolescence, the head represents only about one-eighth of body length.

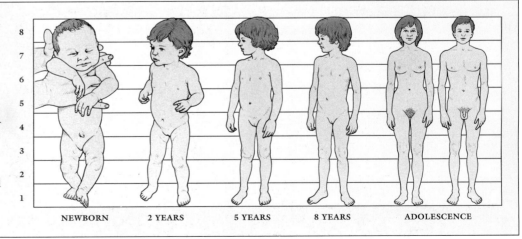

NEWBORN   2 YEARS   5 YEARS   8 YEARS   ADOLESCENCE

CHILD DEVELOPMENT

At birth, babies are capable of seeing, hearing, and responding with reflex actions. In early childhood, they acquire skills of body movement, manipulation, and social behavior, and develop language. Certain developmental "milestones," or well-recognized steps of development, occur at predictable ages. Each child, however, progresses at a slightly different rate.

Neonatal grasp reflex
Newborn babies can perform certain automatic movements that are known as primitive reflexes. One example is the grasping of an object that is put firmly in the palm. Primitive reflexes disappear after a few months.

Rooting reflex
Lightly stroking the baby's face near the corner of the mouth will prompt the baby to turn his or her head to that side and also to open the mouth. This rooting reflex helps the baby locate the mother's nipple to start feeding.

DENTAL DEVELOPMENT

The first set of teeth, known as primary or deciduous teeth, erupts through the gums in a set pattern from about 8 months into the third year. The primary teeth become loose and fall out as the second, or permanent, teeth push through the gums; this starts to happen at about the age of 6 years. The set of 32 permanent teeth is complete only when the third molars, also called the wisdom teeth, appear in the late teens or early twenties.

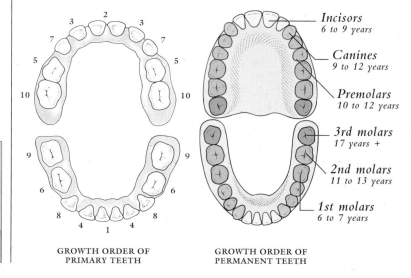

Incisors
6 to 9 years

Canines
9 to 12 years

Premolars
10 to 12 years

3rd molars
17 years +

2nd molars
11 to 13 years

1st molars
6 to 7 years

GROWTH ORDER OF PRIMARY TEETH

GROWTH ORDER OF PERMANENT TEETH

AGE	MOVEMENT		MANIPULATION		SOCIAL BEHAVIOR	
1 MONTH		Lies with head to one side. Sleeps most of time when not being fed or handled.		Hands are normally closed at rest but grasp onto finger when palm is touched.		Watches mother's nearby face intently. Starts smiling at about 5 or 6 weeks.
6 MONTHS		Sits with support. Holds head and back straight. Turns head to look around.		Uses whole hand to grasp objects in palm. Passes objects between hands.		Tests everything in mouth. Turns quickly to sound of familiar voice across room.
9 MONTHS		Attempts to crawl on all fours. Stands, holding onto support, for a few moments.		Grasps between thumb and index finger. Pokes at small object with index finger.		Holds bottle or cup. Holds and chews solids. Babbles. Shouts to attract attention.
12 MONTHS		Walks with one or both hands held. Walks around furniture stepping sideways.		Deliberately drops toys, one by one, to the ground and watches them fall.		Holds out arms and feet to be dressed. Understands some simple commands.
18 MONTHS		Can get up and down stairs with a helping hand or holding rail. Throws ball.		Can build a tower of three or four blocks. Scribbles on paper with pencil or crayon.		Uses spoon well. Indicates need for toilet. Uses some words; understands many.
2 YEARS		Runs around with ease. Can open doors. Kicks a ball without overbalancing.		Turns pages of book one at a time. Can build a tower six or seven blocks high.		Puts on shoes and socks. Makes simple sentences. Asks for food and drink.
3 YEARS		Can ride a tricycle and walk on tiptoe. Uses alternating feet to walk upstairs.		Can copy lines and circles. Able to copy a bridge made from three blocks.		Plays with others. Understands idea of sharing. Tries to neaten up. Uses fork.
4 YEARS		Able to hop on one foot, and to run on tiptoe. Can climb up trees and ladders.		Copies some letters, such as X, V, H, T, and O. Can draw a man and a house.		Can dress and undress. Speech grammatical and completely intelligible.
5 YEARS		Able to skip on alternate feet. Runs lightly on toes. Can dance well to music.		Copies squares, triangles, many letters. Writes a few letters without prompting.		Washes and dries face. Uses knife. Knows birthday. Can act out stories in detail.

PUBERTY

At puberty, hormonal changes stimulate physical growth, alterations in behavior, and the maturation of sex organs so that reproduction can occur. These changes are triggered when gonadotrophin-releasing hormone (GnRH) from the hypothalamus acts on the anterior pituitary gland.

HORMONES IN GIRLS

The pituitary gland releases follicle-stimulating hormone (FSH) and luteinizing hormone (LH), which stimulate the ovary to release eggs each month and to produce the female sex hormones.

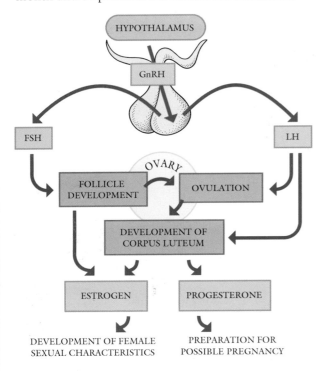

PUBERTAL DEVELOPMENT

The physical changes of puberty start at about age 10 or 11 years in girls, and about 12 or 13 years in boys, and sexual maturation is usually complete within about 3 or 4 years. In both sexes, puberty is accompanied by a rapid growth spurt and an increase in weight, and by emotional and attitudinal changes. The growth spurt in boys begins later than it does in girls, but boys have a longer period of steady growth, and thus usually attain a final greater adult height.

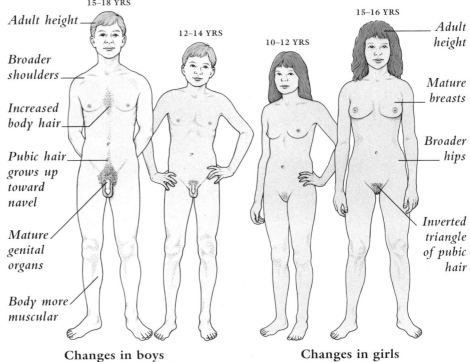

Changes in boys
The genital organs enlarge, pubic and underarm hair appear, facial and body hair increases, and muscle bulk increases. The voice deepens as the larynx enlarge.

Changes in girls
Budding of the breasts is followed by the growth of pubic and underarm hair, and by menstruation (which may be irregular at first). Fat is deposited around the hips.

HORMONES IN BOYS

FSH and LH from the pituitary gland prompt cells of the testes to increase their secretion of testosterone, the male sex hormone, and also to start producing spermatozoa, or sperm.

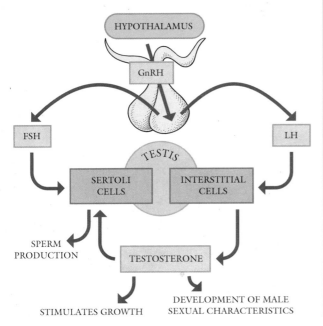

SPERM PRODUCTION

The production of sperm takes place within the seminiferous tubules of the testes. Sperm develop through a complex series of events from cells called spermatogonia, which change first into spermatocytes, and then into spermatids. As the spermatids mature into sperm, they move away from supporting cells, called Sertoli cells, into the central cavity of the seminiferous tubule.

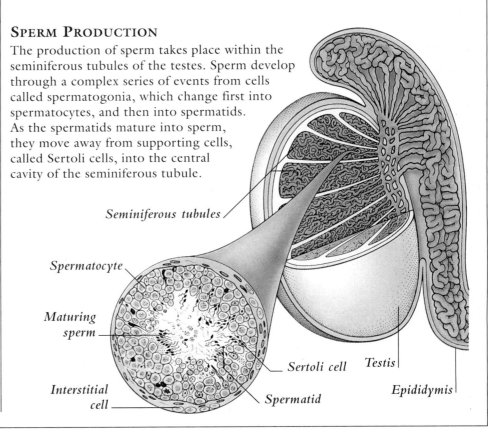

MENSTRUAL CYCLE

The principal sign that a girl has become sexually mature and is capable of reproduction is the onset of menstruation, a period of cyclical bleeding from the vagina. During each menstrual cycle, one or the other of the ovaries releases an egg, or ovum. Unless fertilization takes place, the uterine lining is shed about 2 weeks later during menstruation. The menstrual cycle is regulated by several hormones secreted by the pituitary gland and the ovaries.

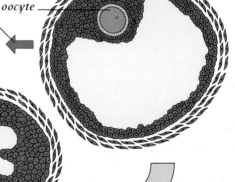

Menstruation

DAYS

28-day cycle

The standard cycle is 28 days, but can range from 23 to 35 days. Bleeding lasts an average of 5 days, although this can vary from 1 to 8 days. In the standard cycle, ovulation occurs around day 14.

1 FSH acts on the ovary to stimulate the growth of primary follicles (containing primary oocytes). Usually, only one follicle reaches full maturity during each cycle.

FSH

Primary oocyte

2 The developing oocyte enlarges, and the cells of the primary follicle multiply so that they form several layers around the oocyte.

Primary follicle

Layers of follicular cells

Primary oocyte

Secondary follicle

Fluid-filled cavity

3 As the follicle enlarges, a fluid-filled cavity forms and cells are pushed toward the rim of the follicle and around the oocyte. The structure is now called a secondary follicle.

4 The mature follicle bulges toward the surface of the ovary and increases production of the hormone estrogen.

Corpus luteum

Progesterone

Corpus luteum

Estrogen

Secondary oocyte

Estrogen

7 If fertilization does not take place, the corpus luteum breaks down during the second week after ovulation.

LH

6 After ovulation, the ruptured follicle develops into a structure called the corpus luteum, which secretes progesterone and estrogen.

Mature egg (secondary oocyte)

5 A surge of LH from the pituitary gland causes the mature follicle to rupture and, release the egg from the ovary; this process is called ovulation.

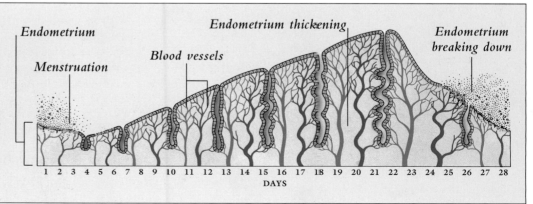

SEM x 72

Ovulation

A mature egg (red) is released from the surface of the ovary at ovulation. The egg is surrounded by cells and fluid from the ruptured follicle.

CHANGES IN THE UTERUS

At the start of each menstrual cycle the uterine lining, or endometrium, is shed during menstruation. After each period of bleeding, the endometrium regrows to prepare the uterus for nurturing a fertilized egg and subsequent pregnancy. But if fertilization does not occur, the endometrium again breaks down and is shed, and the cycle repeats itself unless or until the woman becomes pregant.

Endometrium

Menstruation

Blood vessels

Endometrium thickening

Endometrium breaking down

1 2 3 4 5 6 7 8 9 10 11 12 13 14 15 16 17 18 19 20 21 22 23 24 25 26 27 28
DAYS

The AGING PROCESS

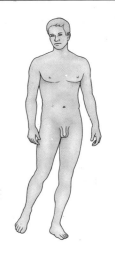

ALL LIVING CREATURES HAVE A NATURAL LIFESPAN. For humans this is around 85 years, although some people live much longer. The function of the brain, the muscles and joints, the eyes and ears declines with age, although normally the changes are small until well past age 60. To maintain vigor, people should maintain their normal weight, exercise regularly, and avoid overconsumption of addictive substances such as tobacco and alcohol. Regular health checkups allow early detection of problems, when they are most effectively treated.

TISSUE CHANGES

Connective tissue, which consists mainly of collagen and elastin, is the body's main structural material. It forms the bulk of tendons and ligaments and provides a framework for bones and muscles. As the body ages, tissues lose elasticity; the collagen fibers thicken and become stiff so that the muscles and joints are less flexible, and the skin becomes wrinkled.

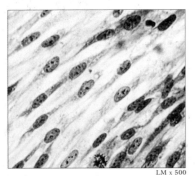

LM x 500

Young connective tissue
Dark, oval nuclei are arranged in a regular, pavementlike pattern throughout the structure, and pigment is evenly distributed.

LM x 500

Aging connective tissue
Tissue in older people has fewer cells; the irregular pattern makes the structure less flexible and less resilient after even minor injury.

SKIN

Older skin is thinner and more fragile, and the deep layers contain less elastic tissue. Blood vessels are also less elastic, so that even minor injuries can cause bruising. The skin may be mottled with small, flat brown areas called lentigines (from the Latin word for lentils).

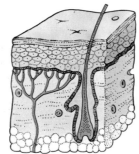

Young skin
A thick outer layer and a large number of elastic fibers in the deeper layers help maintain the smoothness of young skin.

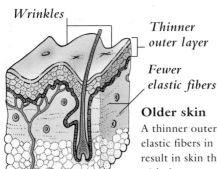

Wrinkles

Thinner outer layer

Fewer elastic fibers

Older skin
A thinner outer layer and fewer elastic fibers in the deeper layers result in skin that appears loose, with deeper creases and wrinkles.

THE NERVOUS SYSTEM

Brain cells begin to decrease in number after age 20, but this process accelerates with advancing age. Blood circulation to the brain may decrease, and memory and other mental functions become impaired. Nevertheless, many people remain mentally alert well past age 80.

HEARING

Aging may cause a loss of sensitivity to sounds, with the result that speech becomes difficult to follow. Hearing tests are recommended for older people if they notice a hearing problem; a hearing aid may help them understand speech more clearly.

Deteriorating structures
Hearing loss in older people may be due to degeneration of the cochlea; repeated or prolonged exposure to loud noises hastens deterioration.

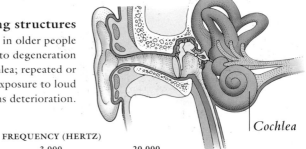

Cochlea

FREQUENCY (HERTZ)

| 400 | 3,000 | 20,000 |

HEARING LOSS (DECIBELS)

10
20
30
40
50

AGE 10
AGE 20

AGE 70   AGE 50   AGE 30

Hearing loss
As people get older, some hearing loss is inevitable. High-pitched sounds are the first that are difficult to detect; eventually all frequencies are affected.

VISION

With age, vision may become impaired by structural changes that affect the eyes' ability to focus on nearby objects. Vision is sometimes affected by degeneration of the macula, the central area of the retina, or by a cataract (clouding of the lens).

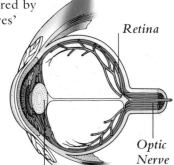

Retina

Optic Nerve

Lens

Changes in the lens
Loss of elastic tissue stiffens the lens so that it cannot change shape to create a clear image on the retina.

BONES, MUSCLES, AND JOINTS

The main, bulky structures of the body are affected by aging in several ways. Bones become thinner and more brittle as a result of osteoporosis, the loss of collagen reduces muscle bulk and strength, and loss of cartilage makes joints painful and stiff.

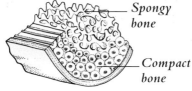

Spongy bone

Compact bone

Young bone
This type of bone has a thick, strong outer layer of dense, compact bone and an inner core of soft, spongy bone that is rich in blood vessels.

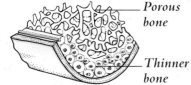

Porous bone

Thinner bone

Osteoporotic bone
Bone in older people has a thinner, outer layer that lacks strength. Inner spongy bone is more porous, and has fewer blood vessels and less calcium.

HEART AND CIRCULATION

The narrowing of the arteries by atherosclerosis and a constantly elevated blood pressure increase demands on the heart. As a consequence, the heart becomes less efficient. Heart valves become stiff, and the electrical conduction system that maintains a regular heart rate and rhythm frequently becomes faulty.

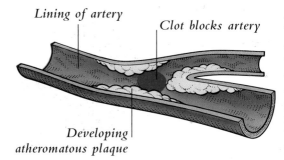

Lining of artery

Clot blocks artery

Developing atheromatous plaque

Atherosclerosis
The lining of arteries, or intima, thickens with cholesterol-rich deposits that then form atheromatous plaques. Blood clots may form and restrict blood flow if these plaques ulcerate.

LIVER AND KIDNEY FUNCTION

In youth, organs such as the liver and kidneys have a greater capacity than needed; they easily compensate for most types of damage. With age, earlier damage to the organs may permit a minor illness to cause failure. Hypertension, atherosclerosis, excessive alcohol, and a sedentary lifestyle can hasten natural decline.

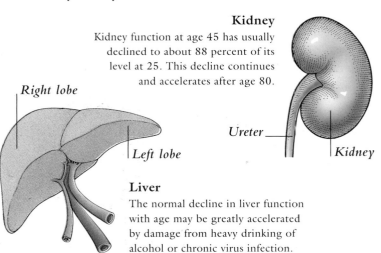

Kidney
Kidney function at age 45 has usually declined to about 88 percent of its level at 25. This decline continues and accelerates after age 80.

Right lobe

Left lobe

Ureter

Kidney

Liver
The normal decline in liver function with age may be greatly accelerated by damage from heavy drinking of alcohol or chronic virus infection.

MENOPAUSE

A gradual loss of ovarian function in women over several years causes variable symptoms, mostly due to lack of estrogen. Menstruation ceases, and some women have hot flashes, night sweats, a thinned and dry vagina that may cause discomfort during sexual intercourse, and urinary symptoms. Emotional strain may be aggravated by hormonal changes.

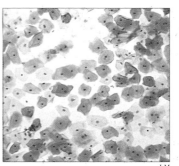

Premenopausal vaginal cells
Before menopause, the lining of the vagina is thick and well lubricated. A Pap smear, as here, usually shows many large cells with small nuclei.

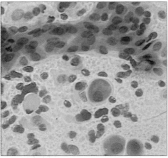

Postmenopausal vaginal cells
Declining levels of estrogen cause the vaginal lining to become thinner. The smear at right show fewer cells, which clump together, and larger nuclei.

HORMONE REPLACEMENT THERAPY

For many women, menopausal symptoms can be reversed by regular treatment with replacement hormones. These may be given as tablets, injections, implants, skin patches, creams, or suppositories. The choice of medication and the method of delivery depend on whether the woman has had a hysterectomy, her general health, and her symptoms.

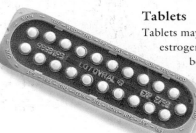

Tablets
Tablets may consist of estrogen alone or estrogen and progesterone. They may be taken every day or for 21 to 25 days every month.

Skin patches
Patches deliver estrogen through the skin into the bloodstream. They are usually applied to the abdomen, in a different position, every 3 to 4 days.

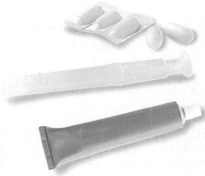

Creams and suppositories
To relieve dryness, an estrogen-containing cream can be applied to the vaginal walls. Rectal or vaginal suppositories that contain estrogen allow it to be absorbed into the bloodstream.

INHERITANCE

MOST OF THE TRAITS OR FUNCTIONS passed on in genes are basic, such as the timing and development of the nervous system and all the organs. Genes also pass on more complex traits concerning psychological and other features inherited from parents, including susceptibility to diseases or athletic ability. Genetics is the study of how genes function and how their alteration (mutation) produces inherited abnormalities.

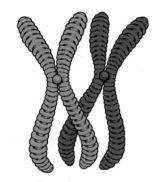

TEM x 10,800

Location of genes
Genes, which are tiny segments of chromosomes, are contained in the cell's nucleus (center of image).

MEIOSIS: GERM CELL FORMATION

During meiosis, genes from both parents are exchanged so that each sperm or egg cell has a unique genetic mix. The cells then divide in two stages to produce four new cells: each cell has 23 chromosomes, half the number in all other human cells. When the sperm and ovum, known as the germ cells, fuse during fertilization, the embryo has a complete set of 46 chromosomes, of which half come from each parent.

1 Meiosis starts with duplication of each member of the 23 pairs of chromosomes (four pairs are seen here). Each of these "doubled-up" chromosomes, which is X-shaped, lines up with its partner.

2 As pairs of chromosomes entwine, they exchange a random selection of DNA. Like shuffling a deck of cards, this process, known as crossing-over, combines genes in a way that will never be exactly repeated.

Crossing-over
During this process, pairs of homologous chromosomes exchange corresponding genes (those located at the same point on each chromosome).

KEY

Father's gene
Mother's gene

Homologous chromosomes
The chromosomes in a pair are similar but not identical, and are called homologous.

3 The matching (homologous) pairs of chromosomes line up in the middle of the cell. Threads form a structure called the spindle between the poles of the cell.

Spindle threads

A new nuclear membrane forms in each new cell

Cells splitting

4 The threads of the spindle pull each of the "doubled-up" chromosomes in a pair to opposite sides. The cell begins to divide into two separate cells.

5 Each new cell, complete with a new nuclear membrane, now has a "doubled-up" chromosome from each of the 23 pairs.

Spindle threads

Individual chromosomes

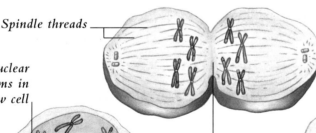

6 Spindle threads form and the chromosomes line up at the center of the cell. They separate to form individual chromosomes, which are pulled to opposite sides.

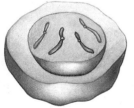   

7 The two cells divide again. After division, each of the four new cells contains a unique set of 23 chromosomes that contain DNA from the original cell's 46 chromosomes.

THE ROLE OF GENES

For most genes, the mix received from parents makes no difference since both genes code for the same chemical processes to occur. For other functions or characteristics, two or more genes "compete" for the same quality, such as hair color or height. The genes a person receives help determine his or her individual characteristics.

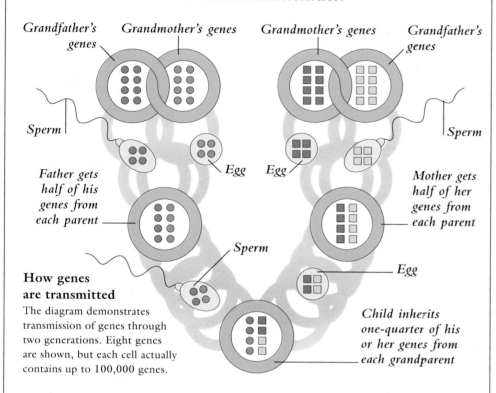

Grandfather's genes *Grandmother's genes*   *Grandmother's genes* *Grandfather's genes*

Sperm *Egg*   *Egg* *Sperm*

Father gets half of his genes from each parent   *Mother gets half of her genes from each parent*

Sperm

Egg

How genes are transmitted
The diagram demonstrates transmission of genes through two generations. Eight genes are shown, but each cell actually contains up to 100,000 genes.

Child inherits one-quarter of his or her genes from each grandparent

PATTERNS OF HEREDITY

Sex and some other traits are determined by the 23rd pair of chromosomes. The other 22 pairs carry genes for most other traits. Some are determined by a single gene, such as eye color, but most, such as intelligence, involve several genes on different chromosomes. Environmental factors also modify some characteristics, such as height.

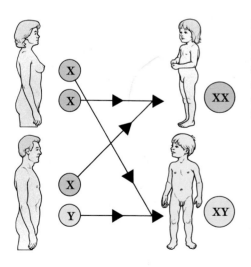

Sex determination
Embryos with two X chromosomes – one from each parent – develop into females; those with a Y chromosome from the father and an X chromosome from the mother develop into males. The Y chromosome is much smaller than the X chromosome.

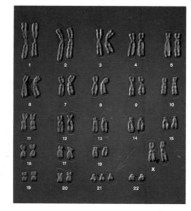

Chromosomal abnormalities
If germ cells with damaged chromosomes or the wrong number take part in fertilization, a miscarriage usually results, or a child may be handicapped. Down's syndrome results from an extra chromosome 21 (see right).

DOMINANT AND RECESSIVE GENES

When several different genes are found at the same position on a pair of chromosomes, the cell takes instructions from only one, the dominant gene. Its effects block those of the other, recessive gene. For a recessive characteristic to have an effect, a person must have two copies of the recessive gene.

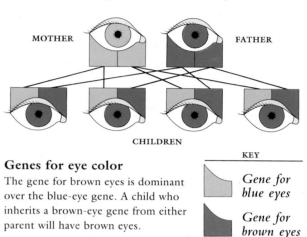

MOTHER   FATHER

CHILDREN

Genes for eye color
The gene for brown eyes is dominant over the blue-eye gene. A child who inherits a brown-eye gene from either parent will have brown eyes.

KEY
Gene for blue eyes
Gene for brown eyes

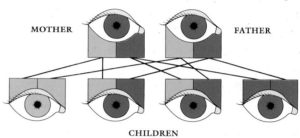

MOTHER   FATHER

CHILDREN

Blue or brown eyes?
If both parents have one gene for brown eyes and one for blue (known as heterozygous), each child has a 1 in 4 chance of being blue-eyed and a 3 in 4 chance of being brown-eyed.

SEX-LINKED INHERITANCE

Several important diseases, including hemophilia, are due to defective genes on the X chromosome. These genes are recessive, so a woman who inherits one normal and one diseased gene will often appear healthy but may carry the disease, while a man with one defective X chromosome develops the disease.

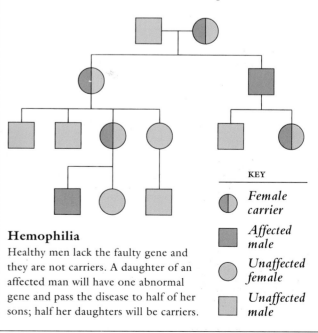

KEY
Female carrier
Affected male
Unaffected female
Unaffected male

Hemophilia
Healthy men lack the faulty gene and they are not carriers. A daughter of an affected man will have one abnormal gene and pass the disease to half of her sons; half her daughters will be carriers.

GLOSSARY

Text terms in *boldface italics* refer to other items in the glossary.

A

Abscess
A walled cavity containing pus, surrounded by inflamed or dying tissue.

Accommodation
The process by which the eyes adjust to focus on nearby or distant objects.

Acoustic neuroma
A *tumor* on the *nerve* that connects the ear and brain.

Acquired immune deficiency syndrome
(AIDS) An infection with the *human immunodeficiency virus* (HIV) that is spread by sexual intercourse or infected blood. AIDS results in the loss of resistance to infections and some *cancers*.

Acute
A medical condition that begins abruptly and may last a short time. Contrasts with *chronic* conditions.

Adenoids
Clusters of *lymphoid tissue* on each side of the back of the upper part of the throat.

Allergen
Any substance causing an allergic reaction in a person previously exposed to it.

Alveolus
See illustration above right.

Alzheimer's disease
A progressive *dementia* due to loss of nerve cells in the brain affecting more than 10 percent of people over 65.

Amniocentesis
The process of withdrawing a sample of fluid from the *uterus* to obtain information about the health and genetic composition of the *fetus*.

Anemia
A condition in which the amount of *hemoglobin* in the blood is reduced.

Aneurysm
A swelling of an *artery* caused by damage to or weakness in the vessel wall.

Angina
Pain or tightness in the center of the chest brought on by exertion, caused by an inadequate blood supply to the heart muscle.

Angiography
See illustration below right.

Angioplasty
Any process used to widen the bore of an *artery* that is narrowed by disease. See also *Balloon angioplasty*.

Antibody
A soluble protein that attaches to body invaders, such as *bacteria*, and helps destroy them.

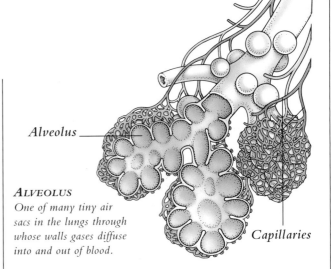

Alveolus

Capillaries

ALVEOLUS
One of many tiny air sacs in the lungs through whose walls gases diffuse into and out of blood.

Anticoagulant
A drug used to limit any tendency for blood to clot within *arteries* or *veins*.

Aorta
See illustration below left.

Aortic valve
A triple-cusped valve at the origin of the *aorta* that allows blood to leave the left ventricle of the heart but prevents backward flow.

Appendix
The wormlike structure attached to the initial part of the large intestine. It has no known function.

Aqueous humor
The fluid filling the front chamber of the eye between the back of the *cornea* and the front of the iris and *lens*.

Arrhythmia
An irregular heartbeat, either fast or slow, due to a defect in the pathways that control contractions of the heart.

Arteriole
A small terminal branch of an *artery* leading to even smaller *capillaries*, which make the link to the *veins*.

Artery
An elastic, muscular-walled tube that transports blood away from the heart to all other parts of the body.

Arthritis
Inflammation in a joint, causing varying degrees of pain, swelling, redness, and restriction of movement.

Articulation
A joint, or the way in which jointed parts are connected.

Asthma
A disease featuring variable narrowing of the air tubes so that breathing becomes intermittently difficult.

Atherosclerosis
A degenerative disease of the *arteries* in which raised plaques of fatty material limit blood flow and permit local *blood clotting*.

Atrial fibrillation
A disorder in which the *atria* beat very rapidly.

Atrial septal defect
A hole in the wall (the septum) between the upper two chambers of the heart.

Atrium
One of the two thin-walled upper chambers of the heart.

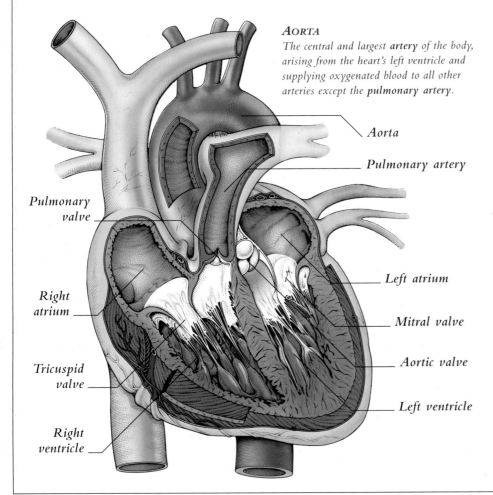

AORTA
*The central and largest **artery** of the body, arising from the heart's left ventricle and supplying oxygenated blood to all other arteries except the **pulmonary artery**.*

Aorta

Pulmonary artery

Pulmonary valve

Right atrium

Tricuspid valve

Right ventricle

Left atrium

Mitral valve

Aortic valve

Left ventricle

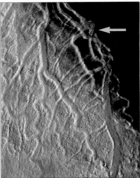

ANGIOGRAM

ANGIOGRAPHY
*A method of imaging blood vessels in which X-rays are taken after a **contrast medium** has been injected. Shown above is a narrowed **coronary** artery (arrow).*

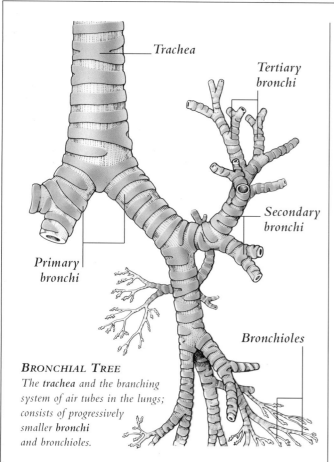

BRONCHIAL TREE
The trachea and the branching system of air tubes in the lungs; consists of progressively smaller bronchi and bronchioles.

Labels: Trachea; Tertiary bronchi; Secondary bronchi; Primary bronchi; Bronchioles

Autoimmune disease
A disease caused by a defect in the immune system, which fails to recognize body tissues as "self."

Autonomic nervous system
(ANS) The portion of the nervous system that controls unconscious functions, such as heartbeat and breathing.

Axon
The long, fiberlike process of a nerve cell that conducts nerve impulses to or from the cell body; bundles of many axons form *nerves*.

B

Bacterium
A type of microorganism consisting of one cell. Only a few of the many species of bacteria cause disease.

Balloon angioplasty
The use of a catheter with an inflatable balloon tip to widen an *artery*.

Basal ganglia
Paired masses of nerve cell bodies, or nuclei, lying deep in the brain; concerned with control of movement.

Benign
Mild and with no tendency to spread; contrasts with a *malignant* condition.

Beta-blocker
A drug that blocks the action of epinephrine, thus slowing the pulse and reducing blood pressure.

Bile
A greenish brown fluid from the *liver* and concentrated and stored in the *gallbladder*; helps the digestion of fats.

Biliary system
The network of bile vessels formed by the ducts from the *liver* and the *gallbladder*, and the gallbladder itself.

Biopsy
A sample of tissue from any part of the body that is suspected of disease, taken for *microscopic* examination.

Blood clot
A meshwork of the protein *fibrin*, *platelets*, and blood cells that forms when a blood vessel is damaged.

Boil
An inflamed, pus-filled area of skin, which is usually an infected *hair follicle*.

Bolus
A chewed-up quantity of food ready to be swallowed; also, a drug rapidly injected into the bloodstream.

Bone marrow
The fatty tissue within bone cavities, which may be red or yellow. Red bone marrow produces *red blood cells*.

Bradycardia
A slow heart rate. This is normal in athletes but may signal disorders in others.

Brain stem
The lower part of the brain; houses the centers that control vital functions such as breathing and heartbeat.

Breech delivery
A buttock-first birth; carries a slightly higher risk to the fetus than a head-first birth.

Bronchial tree
See illustration at left.

Bronchitis
Inflammation of the lining of the breathing tubes, resulting in a cough that produces large amounts of sputum (phlegm).

Bronchus
One of the larger air tubes in the lungs. Each lung has a main bronchus that divides into smaller branches.

CANCER
A localized growth from uncontrolled cell reproduction, which can spread (if untreated) to other parts of the body.

Labels: Epidermis of the skin; Cluster of cancer cells; Lymph vessel

C

Calcium-channel blocker
A drug that limits movement of dissolved calcium across cell membranes; it is used to treat high blood pressure and *arrhythmias*.

Cancer
See illustration above.

Capillary
One of the tiny blood vessels that link the smallest *arteries* and smallest *veins*.

Carcinoma
A *cancer* of either the inner or the outer surface layer (epithelium). Carcinomas commonly occur in the skin, linings of the air tubes, large intestine, breast, *prostate gland*, and *uterus*.

CARPAL TUNNEL SYNDROME
Numbness and pain in the thumb and middle fingers. It results from pressure on the median nerve where it passes through the gap under a ligament in front of the wrist.

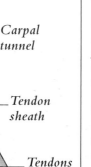

Labels: Carpal tunnel; Tendon sheath; Tendons; Median nerve; Ligament

Carpal tunnel syndrome
See illustration below.

Cartilage
A tough, fibrous connective tissue, also known as gristle.

Central nervous system
(CNS) The brain and spinal cord; receives and analyzes sensory data, and then initiates a response.

Cerebellum
The region of the brain located behind the *brain stem*. It is concerned with balance and the control of fine movement.

Cerebrospinal fluid
A watery fluid that bathes the brain and spinal cord.

Cerebrum
The largest part of the brain, which is made up of two hemispheres. It contains the nerve centers for thought, personality, the senses, and voluntary movement.

Chlamydia
Small bacteria causing the eye disease trachoma and *pelvic inflammatory disease*.

Cholecystitis
An inflammation of the *gallbladder*; commonly the result of obstructed outflow of *bile* by a *gallstone*.

Cholecystography
X-ray of the *gallbladder* after a *contrast medium* has been introduced into it.

Cholestasis
A slowing or cessation of the flow of *bile* in the *liver*.

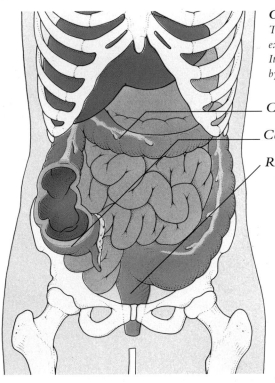

COLON
The part of the large intestine that extends from the cecum to the rectum. Its main function is to conserve water by absorbing it from the bowel contents.

Colon

Cecum

Rectum

D

Defibrillation
A strong pulse of electric current applied to the heart to restore its normal rhythm.

Dementia
The loss of mental powers and *memory* as well as the ability to care for oneself; dementia is often a result of degenerative brain disease.

Dermis
See illustration below.

Dialysis
The basis of artificial *kidney* machines, which separate dissolved substances and permit waste excretion and preservation of nutrients.

Diaphragm
The dome-shaped muscular sheet separating the chest from the abdomen. When the muscle contracts, the dome flattens, increasing the volume of the chest.

Diastole
The period in the heart cycle when the ventricles are relaxed and the heart is filling with blood.

Digestive system
The mouth, *pharynx*, *esophagus*, stomach, and intestines. Associated organs are the *pancreas, liver*, and *gallbladder* and their ducts.

Diverticular disease
See illustration above right.

Dopamine
A chemical messenger in the brain that is involved in the control of body movement.

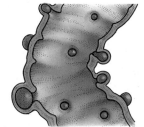

DIVERTICULAR DISEASE
The presence of diverticula, small sacs that are created by protrusion of the intestine's inner lining through the wall.

Down's syndrome
A genetic disorder in which a person's cells contain an extra *chromosome* 21.

Duodenum
The C-shaped first part of the small intestine, into which the stomach empties. Ducts from the *gallbladder*, the *liver*, and the *pancreas* all enter the duodenum.

Dura mater
A tough membrane, the outer layer of the *meninges*, which cover the brain and spinal cord. It lies over the arachnoid and pia mater and adheres closely to the skull.

E

Eardrum
The membrane separating the outer ear from the *middle ear* that vibrates in response to sound.

Ectopic pregnancy
Implantation of a fertilized egg in any site other than the uterine lining.

Chorionic villus sampling
Removal of a small piece of tissue from the *placenta* for *chromosome* or *gene* analysis; allows for early detection of fetal abnormalities.

Chromosome
See illustration below.

Chronic
Persistent medical conditions lasting over 6 months that may result in a long-term change; contrasts with *acute*.

Cirrhosis
Replacement of *liver* tissue by fine fibrous tissue, which results in hardening and impaired function; may be caused by excessive alcohol consumption or infection.

Cochlea
The coiled structure in the inner ear that contains the organ of Corti, which converts sound vibrations into nerve impulses for transmission to the brain.

Collagen
The body's most important structural protein, present in bones, *tendons*, *ligaments*, and other connective tissues. Collagen fibrils are twisted into bundles called fibers.

Colon
See illustration above left.

Congenital
Present at birth. Congenital disorders may be hereditary or may result from diseases or injuries that occur during fetal life or the birth itself.

Contrast medium
A substance through which X-rays are unable to pass.

Cornea
The transparent dome at the front of the eyeball that is the eye's main focusing lens.

Coronary
A term meaning "crown." Refers to the *arteries* that encircle and supply the heart with blood.

Corpus callosum
The wide, curved band of about 20 million nerve fibers that connects the two hemispheres of the *cerebrum*.

Corticosteroid
A drug that simulates the natural steroid *hormones* of the outer zone (cortex) of the adrenal glands.

Cranial nerves
The 12 pairs of *nerves* that emerge from the brain and *brain stem*. They include the nerves for smell, sight, eye movement, facial movement and sensation, hearing, taste, and head movement.

CROHN'S DISEASE
A inflammatory disease that affects the gastrointestinal tract. Symptoms may include pain, fever, and diarrhea.

Crohn's disease
See illustration above.

Cyst
A walled cavity, which is usually spherical, filled with secreted fluid or semisolid matter. Usually *benign*.

Cystadenoma
A harmless, *cyst*-like growth of glandular tissue.

Cystitis
Inflammation of the urinary bladder usually caused by an infection. Produces frequent, painful urination and, in some cases, incontinence.

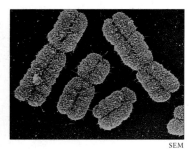

CHROMOSOME
A threadlike structure, present in all nucleated body cells, that carries the genetic code for the formation of the body. A normal human body cell has 46 chromosomes arranged in 23 pairs.

SEM

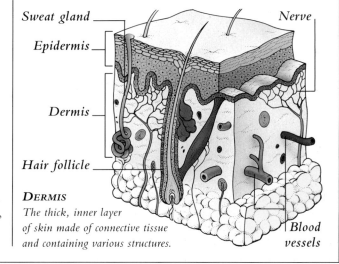

Sweat gland

Epidermis

Nerve

Dermis

Hair follicle

DERMIS
The thick, inner layer of skin made of connective tissue and containing various structures.

Blood vessels

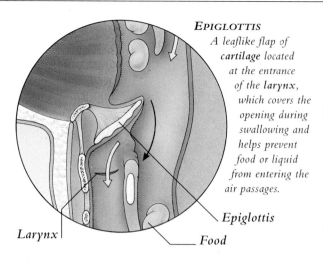

EPIGLOTTIS
*A leaflike flap of
cartilage located
at the entrance
of the **larynx**,
which covers the
opening during
swallowing and
helps prevent
food or liquid
from entering the
air passages.*

Epiglottis

Larynx

Food

Electrocardiography
Recording and study of the
heart's electrical changes.

Electroencephalography
Recording and study of the
brain's electrical signals.

Embolus
Any material, such as *blood
clots*, air bubbles, *bone
marrow*, fat, or *tumor* cells,
carried in the bloodstream.

Embryo
The developing baby from
conception until the eighth
week of pregnancy.

Endocarditis
An inflammation that affects
either the inner lining of the
heart wall or a heart valve.

Endorphin
A morphinelike substance
produced by the body in
times of pain and stress.

Enzyme
A protein that accelerates a
chemical reaction by acting
as a catalyst.

Epidermis
The outer layer of the skin;
its cells become flatter and
scalier toward the surface.

Epiglottis
See illustration above.

Epilepsy
A disorder featuring episodes
of unregulated electrical
discharge throughout the
brain or in a specific area.

Esophagitis
An inflammation that affects
the *esophagus*, often caused
by the reflux of stomach
acid into the esophagus.

Estrogen
A *sex hormone* that prepares
the uterine lining for an
implanted fertilized egg and
stimulates the development
of a female's secondary
sexual characteristics.

Eustachian tube
The tube that connects the
back of the nose to the
cavity of the *middle ear* and
equalizes air pressure.

F

Fallopian tube
One of two tubes along
which an *ovum* travels to
the *uterus*, after release from
an *ovary*; the most common
site of an *ectopic pregnancy*.

FERTILIZATION
*The union of a sperm and an
egg, after sexual intercourse or
artificial insemination, or in a
laboratory test tube.*

Fertilization
See illustration above.

Fetus
See illustration at right.

Fever
A body temperature that
registers above 98.6°F
(37°C), measured in the
mouth, or 99.8°F (37.7°C),
measured in the rectum.

Fiberoptics
The transmission of images
through bundles of flexible,
glass or plastic threads.
Some types of endoscope
use fiberoptic transmission
to view directly and treat
structures that are located
far within the human body.

Fibrin
An insoluble protein that is
converted from the blood
protein fibrinogen to form a
fibrous network – a stage in
the creation of a *blood clot*.

Fibroid
A *benign* tumor of fibrous
and muscular tissue growing
in the wall of the *uterus*,
usually in women over 30.
Fibroids are often multiple
and may cause symptoms.

Fibrosis
An overgrowth of scar or
connective tissue that is
formed as the body's natural
healing response to any
wound or burn. Fibrous
tissue may modify an organ's
structure, and thereby
impair its effectiveness.

Fistula
An abnormal channel that
lies between any part of the
interior of the body and
the surface of the skin, or
between two internal organs.

G

Gallbladder
The small, fig-shaped bag
lying under the *liver*, into
which *bile* secreted by the
liver passes to be stored.

Gallstone
An oval or faceted mass of
cholesterol, calcium, and
bile pigment that forms in
the *gallbladder*. Gallstones
vary in size and are more
common in women than
they are in men.

Gastric juice
A mixture produced by the
cells of the stomach that
contains hydrochloric acid
and digestive *enzymes*.

Gastritis
An inflammation of the
stomach lining from any
cause, including infection
or alcohol.

Gastrointestinal tract
The muscular tube that
extends from the mouth,
through the *pharynx*,
esophagus, stomach, and
small and large intestines,
to the rectum.

Gene
A distinct section of a
chromosome that is the basic
unit of inheritance. Each
gene consists of a segment
of deoxyribonucleic acid
(DNA) containing the code
that governs the production
of a specific protein.

Glaucoma
An abnormal rise in the
pressure of the fluids within
the eye that, if it is left
untreated, causes internal
damage to the eye that may
result in blindness.

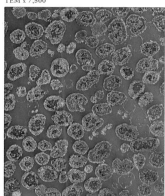

TEM x 7,500

GONORRHEA
*A sexually transmitted
disease that may cause
pelvic inflammation in
women and narrowing of
the urine outlet tube in
men. If untreated, the
disease may spread to other
parts of the body. Shown
at left is a colony of the
bacteria causing the disease,
Neisseria gonorrhoeae.*

Glial cell
A nerve cell that provides
support for *neurons*.

Glue ear
A disorder in which sticky
fluid accumulates in the
middle ear and impedes the
movement of the *ossicles*.

Gonorrhea
See illustration above.

Gout
A metabolic disorder
causing attacks of *arthritis*,
usually in a single joint.

Gray matter
The regions of the brain
and spinal cord that are
composed mainly of *neuron*
cell bodies as opposed to
their projecting fibers,
which form white matter.

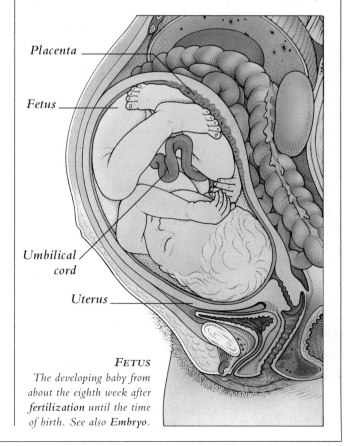

Placenta

Fetus

*Umbilical
cord*

Uterus

FETUS
*The developing baby from
about the eighth week after
fertilization until the time
of birth. See also **Embryo**.*

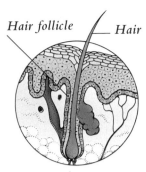

Hair follicle — Hair

HAIR FOLLICLE
A pit on the surface of the skin from which a hair grows.

H

Hair follicle
See illustration above.

Heart-lung machine
A pump and oxygenator that performs the functions of the heart and the lungs during cardiac operations.

Heart valve
One of four structures of the heart that allow passage of blood in one direction only.

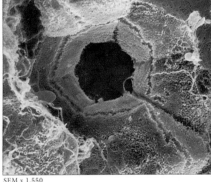

SEM x 1,550

Hematoma
Accumulated blood within any part of the body, caused by a torn blood vessel.

Hemiplegia
Paralysis of one-half of the body, from damage to the motor areas in the brain, or to the nerve tracts that connect these motor areas to the spinal cord.

Hemoglobin
The protein in *red blood cells* that combines with oxygen, carrying it from the lungs throughout the body.

Hemophilia
An inherited bleeding disorder caused by shortage of a clotting protein.

Hemorrhage
The escape of blood from a blood vessel, usually as a result of an injury.

Hemorrhoids
Ballooning of veins in the lining of the anus (external hemorrhoids) or in the lower part of the rectum (internal hemorrhoids).

Hepatitis
Inflammation of the *liver*, usually as a result of a viral infection, alcohol, or toxic substances. Symptoms include *fever* and *jaundice*.

Hepatocyte
See illustration below.

Hiatal hernia
Sliding upward of part of the stomach through the opening in the *diaphragm*.

Hippocampus
A structure in the brain concerned with learning and long-term *memory*.

HEPATOCYTE
A type of liver cell with many functions. Seen here in brown are hepatocytes surrounding a channel filled with blood.

Homeostasis
Active processes by which an organism maintains a constant internal state.

Hormone
A chemical released by the endocrine glands and some tissues. Hormones act on receptor sites in other areas.

Human immunodeficiency virus
(HIV) The virus that causes AIDS and destroys cells of the immune system, thereby undermining its efficiency.

Hypothalamus
A small structure located at the base of the brain where the nervous and hormonal systems of the body interact.

I–K

Ileum
The final segment of the small intestine, the location where the absorption of nutrients is completed.

Immune deficiency
Any failure of the function of the immune system from causes such as AIDS, *cancer* treatment, or aging.

Immunosuppressant
A drug that interferes with the production and activity of certain *lymphocytes*.

Interferon
A protein produced by cells to defend against viral infections and some *cancers*.

In vitro fertilization
Fertilization of *ova* in a laboratory container by the addition of sperm; the resulting embryos are placed in the woman's *uterus*.

Irritable colon syndrome
Recurrent gas, abdominal discomfort, and alternating constipation and diarrhea.

Jaundice
A yellowing of the skin and whites of the eyes that is due to deposition of *bile* pigment. Jaundice results from altered *liver* function.

Kaposi's sarcoma
A slow-growing *tumor* of blood vessels that affects people with AIDS. Scattered bluish brown nodules occur on the skin and internally.

Left lobe

Portal vein

Right lobe

Hepatic artery

Hepatic duct

LIVER
The large organ in the upper right abdomen that performs vital chemical functions, including processing of nutrients from the intestines; synthesis of sugars, proteins, and fats; detoxification of poisons; and conversion of waste to urea.

Kidney
One of two bean-shaped structures in back of the abdominal cavity that filter blood and remove wastes.

Killer T cells
White blood cells that can destroy damaged, infected, or *malignant* body cells.

L

Laparoscopy
The visual inspection of the interior of the abdomen, through a narrow optical and illuminating device, and often using a video camera.

Larynx
The structure in the neck at the top of the *trachea*, known as the voice box, that contains the vocal cords.

Lens
The internal lens of the eye, also called the crystalline lens; it adjusts fine focus by alterations in curvature. The outer lens is the *cornea*.

Leukemia
A group of blood disorders in which malignant *white blood cells* grow in the bone marrow and invade organs elsewhere in the body.

Ligament
See illustration below.

Limbic system
A collection of structures in the brain that plays a role in the automatic (involuntary) body functions, emotions, and the sense of smell.

Liver
See illustration above.

LIGAMENT
A band of tissue consisting of collagen – a tough, fibrous, elastic protein. Ligaments support bones, mainly in and around joints.

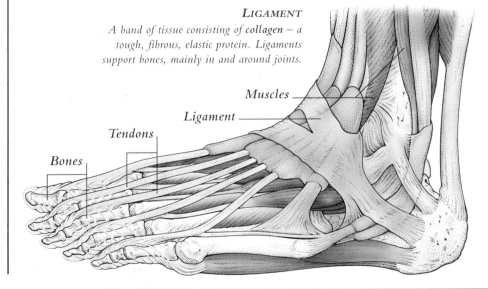

Muscles

Ligament

Tendons

Bones

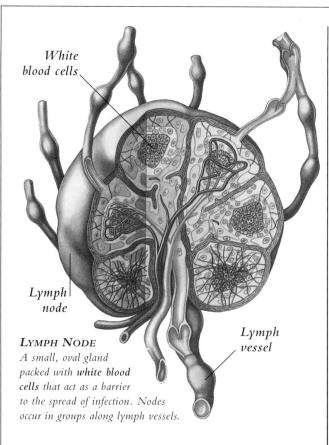

White blood cells

Lymph node

Lymph vessel

LYMPH NODE
A small, oval gland packed with white blood cells that act as a barrier to the spread of infection. Nodes occur in groups along lymph vessels.

Lobe
A rounded projection or subdivision forming part of a larger structure such as the brain, lung, or *liver*.

Lymphatic system
An extensive network of transparent lymph vessels and *lymph nodes*. It returns excess tissue fluid to the circulation and helps combat infections and *cancer* cells.

Lymph node
See illustration above.

Lymphocyte
Small *white blood cell* that is a part of the immune system; it protects against *virus* infections and *cancer*.

Lymphoid tissue
A tissue rich in *lymphocytes* found in *lymph nodes*, the *spleen*, intestines, and *tonsils*.

M

Macula
Any small, flat, colored spot on the skin; also the central region of the *retina*.

Malignant
Refers to a cancerous *tumor* that may spread throughout the body, causing death; contrasts with *benign*.

Mammography
The X-ray screening of the breasts, using low-radiation X-rays; used to detect breast *cancer* at an early stage.

Mastectomy
Surgical removal of part or all of the breast. It is usually performed to treat breast *cancer* and is often followed by radiotherapy.

Mastitis
Inflammation of the breast, usually resulting from an infection acquired during breastfeeding. *Bacteria* enter through cracks in the nipples. Symptoms include *fever*, and hardening and tenderness of the breast.

Medulla
The inner part of an organ, such as the *kidneys*, or the adrenal glands. Also refers to the part of the *brain stem* lying immediately above the start of the spinal cord, just in front of the *cerebellum*.

Meiosis
The stage in the formation of sperm and egg cells when chromosomal material is randomly redistributed and the number of *chromosomes* is reduced to 23 instead of the usual 46 found in the other body cells.

Memory
The data store for recent and remote experience. Short-term memory stores are small and the contents are soon lost. Long-term memory stores are larger.

Meninges
The three membrane layers around the brain and spinal cord; the pia mater on the inside, the arachnoid, and the *dura mater* next to the skull.

Meningitis
An inflammation of the *meninges*, sometimes as a result of a *virus* infection.

Meniscectomy
Surgical removal of a torn or displaced *cartilage (meniscus)* from the knee joint; usually carried out with the use of a *fiberoptic* viewing tube, which is inserted into the joint, and a TV monitor.

Meniscus
A crescent-shaped pad of *cartilage* found in the knee and some other joints.

Menopause
The end of the reproductive period in a woman, when the *ovaries* have ceased their production of eggs and menstruation has stopped.

Metabolism
The sum of all the physical and chemical processes that take place in the body.

Metastasis
The spread or transfer of any disease, but especially *cancer*, from its original site to another site, where the disease process continues.

Microscopy
See illustration below.

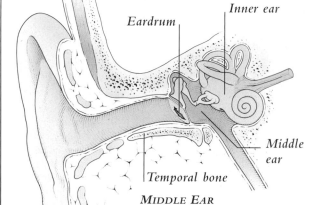

Eardrum

Inner ear

Middle ear

Temporal bone

MIDDLE EAR
The cleft within the temporal bone between the eardrum and the outer wall of the inner ear.

Middle ear
See illustration above.

Migraine
The effects of narrowing and then widening of some of the *arteries* of the scalp and brain, usually on one side. The symptoms include visual disturbances, nausea, and severe headache.

Miscarriage
A spontaneous ending of a pregnancy before the *fetus* is mature enough to survive outside the *uterus*.

Mitochondrion
A cell organelle containing genetic material; it is also involved in the production of energy for cell functions.

Mitosis
The process by which a cell nucleus divides to produce two daughter cells, each of which has the identical genetic makeup of the parent cell.

Mitral valve
The valve that lies between the left *atrium* and the left *ventricle* of the heart.

Mole
Any birthmark, pigmented spot, growth, or *congenital* blemish, whether flat, raised, and/or hairy, on the skin.

Motor cortex
The part of the surface layer of each hemisphere of the *cerebrum* in which voluntary movement is initiated. The motor cortex can be mapped into areas that are linked to particular parts of the body.

Motor neuron
A nerve cell that carries the impulses to muscles that cause its movement.

Motor neuron disease
A rare disorder in which the motor *neurons* suffer a progressive destruction, resulting in a corresponding loss of movement.

Mucocele
A *cyst*-like abnormal sac filled with mucus that arises from a *mucous membrane*.

Mucolytic drug
A drug that makes sputum (phlegm) less sticky and thus easier to cough up.

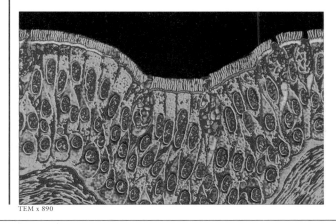

TEM x 890

MICROSCOPY
Examination by a microscope, usually to make a diagnosis. Simple techniques use focused light rays and magnifying lenses; in order to achieve higher magnifications, beams of electrons are used. Shown here is a type of electron microscope picture of the tissue lining the trachea.

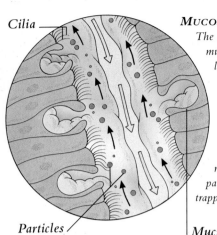

Cilia

Particles

MUCOUS MEMBRANE
The soft, skinlike, mucus-secreting layer lining the tubes and cavities of the body. Shown here is the mucous membrane that lines the respiratory tract. Its tiny hairs (cilia) remove foreign particles that are trapped in the mucus.

Mucus-secreting cell

PACEMAKER
An electronic device implanted in the chest that delivers short electric pulses via implanted electrodes to regulate heartbeat.

Mucous membrane
See illustration above.

Muscular dystrophy
One of several hereditary muscle disorders featuring gradual, progressive muscle degeneration and weakening.

Myocardium
The special muscle of the heart, in which the fibers make up a network that can contract spontaneously.

Myofibril
Cylindrical element within muscle cells (fibers) that consists of thinner filaments, which move to produce muscle contraction.

N

Nephron
The *kidney's* filtering and tubular system, consisting of a filtration capsule, the glomerulus, and a series of tubules that reabsorbs or excretes water and wastes to control fluid balance.

Nerve
The filamentous projections of individual *neurons* (nerve cells) held together by a fibrous sheath. Nerves carry electrical impulses to and from the brain and spinal cord and other body parts.

Neuron
See illustration below.

Nociceptor
A *nerve* ending responding to painful stimuli.

Noninvasive
Any medical procedure that does not involve penetration of the skin or an entry into the body through any of the natural openings.

O

Olfactory nerve
One of two *nerves* of smell that run from the olfactory bulb in the roof of the nose directly into the underside of the brain.

Optic nerve
One of the two nerves of vision. Each has about one million nerve fibers running from the *retina* to the brain, carrying visual stimuli.

Ossicle
One of three tiny bones of the *middle ear* that convey vibrations from the eardrum to the inner ear.

Ossification
The process of formation, renewal, and repair of bone. Most bones in the body develop from *cartilage*.

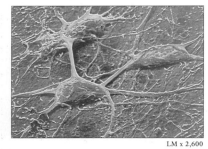

NEURON
A single nerve cell whose function is to transmit electrical impulses. Shown at left are three neurons found in the cerebral cortex, the outer layer of the brain.

LM x 2,600

Osteoarthritis
A degenerative joint disease that features damage to the *cartilage*-covered, weight-bearing surfaces in the joint.

Osteomalacia
Bone softening caused by defective mineralization, which usually results from poor calcium absorption that is due to a deficiency of vitamin D.

Osteon
The rod-shaped unit, also called a haversian system, that is the building block of cortical (hard) bone.

Osteoporosis
Loss of bone substance due to bone being reabsorbed faster than it is formed. The bones become brittle and easily fractured.

Osteosarcoma
A highly *malignant* form of bone *cancer* that mainly affects adolescents. It often develops near the knee.

Osteosclerosis
Increased bone density that may result from a severe injury, *osteoarthritis*, or osteomyelitis. It is detected on an X-ray film as an area of extreme whiteness.

Otitis media
Inflammation in the *middle ear* cavity, often caused by infection that has spread from the nose or throat.

Otosclerosis
A hereditary bone disease affecting the *middle ear*, in which the foot of the inner *ossicle* becomes fused to the surrounding bone.

Ovary
One of two structures, lying on each side of the *uterus*, that store primary ovarian follicles, release the mature *ova*, and produce the female *sex hormones* (estrogen and progesterone).

Ovulation
See illustration at right.

Ovum
The egg cell; if *fertilization* occurs, the ovum develops into an *embryo*.

P

Pacemaker
See illustration at right.

Paget's disease
A disease that causes bone to become weaker, thicker, and distorted.

Pancreas
A gland behind the stomach that secretes the digestive *enzymes* and *hormones* that regulate glucose levels.

Paralysis
Loss of movement of part of the body due to a nerve or a muscle disorder.

Paraplegia
Paralysis of the lower limbs, usually from injury or disease to the spinal cord or brain.

Parasite
An organism that lives in or on another organism (the host), and derives benefit at the host's expense.

Parasympathetic nervous system
One of the two divisions of the *autonomic nervous system*; it maintains and restores energy, for example by slowing the heart rate.

Parathyroid glands
Two pairs of glands located behind the thyroid gland that help control the level of calcium in blood.

Parietal
A term referring to the wall of a body cavity.

Parkinson's disease
A neurological disorder that features involuntary tremor, muscle rigidity, slowness of movements, tottering steps, and small handwriting. The intellect is not affected.

Parotid glands
The largest pair of *salivary glands* that are situated, one on each side, over the angles of the jaw just below and in front of the ears.

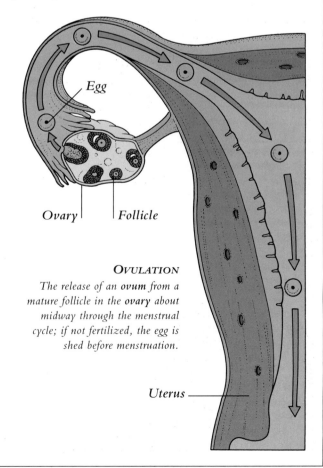

Egg

Ovary | Follicle

OVULATION
The release of an ovum from a mature follicle in the ovary about midway through the menstrual cycle; if not fertilized, the egg is shed before menstruation.

Uterus

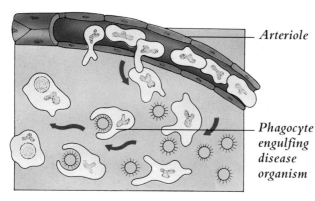

PHAGOCYTE
An amoeba-like cell of the immune system that can surround, kill, and digest disease organisms and scavenge foreign material.

Arteriole

Phagocyte engulfing disease organism

Pelvic inflammatory disease
Infection of the reproductive organs of the female. The cause may be unknown, but it often occurs following a sexually transmitted disease.

Pelvis
The basinlike ring of bones to which the lower *spine* is attached and with which the thigh bones articulate. The term is also used to refer to the soft tissue contents.

Peptic ulcer
The local destruction of the lining of the *esophagus*, stomach, or *duodenum* from the effects of the *bacterium Helicobacter pylori*, stomach acid, and digestive *enzymes*.

Pericarditis
An inflammation of the membranous *pericardium* that surrounds the heart. It may cause pain and the accumulation of fluid, called a pericardial effusion.

Pericardium
The tough, fibrous sac, with two layers, that encloses the heart as well as the roots of the major blood vessels emerging from it.

Periosteum
A tough tissue coating all bone surfaces except joints, from which new bone can be formed; it contains blood vessels and *nerves*.

Peripheral nervous system
All the *nerves* with their coverings that fan out from the brain and spinal cord, linking them with the rest of the body; consists of *cranial nerves* and *spinal nerves*.

Peristalsis
A coordinated succession of contractions and relaxations of the muscular wall of a tubular structure, such as the intestines. It causes the contents to move along.

Peritoneum
A double-layered membrane that lines the inner wall of the abdomen; it covers the organs in the abdomen and secretes fluid that lubricates their movement.

Peritonitis
An inflammation of the *peritoneum* due to *bacteria*, *bile*, pancreatic enzymes, or chemicals; sometimes the cause may be unknown.

Phagocyte
See illustration at left.

Pharynx
See illustration at right.

Pituitary gland
A gland hanging from the underside of the brain. It secretes *hormones* that control many other glands in the body and is regulated by the *hypothalamus*.

Placenta
See illustration below.

Plasma
The fluid part of blood from which all cells have been removed; contains proteins, salts, and various nutrients.

Platelet
A fragment of large cells, called megakaryocytes, that is present in large numbers in the blood and necessary for normal *blood clotting*.

Pleura
A double-layered membrane, the inner layer of which covers the lung and the outer layer of which lines the chest cavity. A layer of fluid lubricates and enables movement between the two.

Pleural effusion
Accumulation of excessive fluid between the layers of the *pleura*, which separates them and compresses the underlying lung.

Pleurisy
Inflammation of the *pleura*, usually from a lung infection such as *pneumonia*; may lead to adhesion between the pleural membranes.

Plexus
A network of interwoven *nerves* or blood vessels.

Pneumoconiosis
Any lung-scarring disorder due to inhalation of mineral dust; scarring causes the lungs to be less efficient in supplying oxygen to blood.

Pneumocystis pneumonia
A lung infection with the microorganism *Pneumocystis carinii*; occurs mainly in *immune deficiency disorders*.

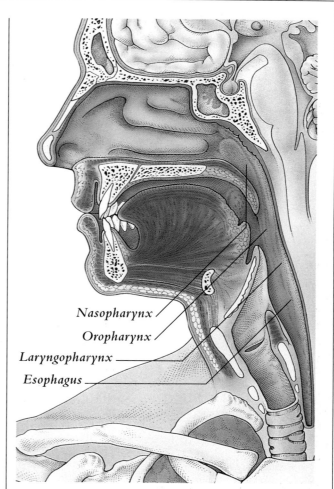

Nasopharynx
Oropharynx
Laryngopharynx
Esophagus

PHARYNX
The passage leading down from the back of the nose and the mouth to the esophagus; it consists of the nasopharynx, the oropharynx, and the laryngopharynx.

Pneumonia
Inflammation of the smaller air passages and *alveoli* of the lungs due to infection or contact with inhaled irritants or toxic material.

Pneumothorax
The presence of air in the space between the *pleura*, causing the lung to collapse.

Primary
A term describing a disorder that has originated in the affected structure.

Progesterone
A *sex hormone* secreted by the *ovaries* and *placenta* that allows the uterus to receive and retain a fertilized egg.

Prostaglandin
One of a group of fatty acids made naturally in the body that act much like *hormones*.

Prostate gland
The structure at the base of the bladder that secretes some of the fluid in semen.

Prosthesis
See illustration at right

Psoriasis
A common skin disease that features thickened patches of red, inflamed skin.

Pulmonary artery
The *artery* that conveys deoxygenated blood from the right *ventricle* of the heart to the lungs to be reoxygenated.

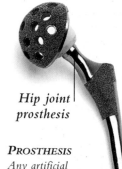

Hip joint prosthesis

PROSTHESIS
Any artificial replacement for a part of the body, internal or external, whether its purpose is functional or cosmetic.

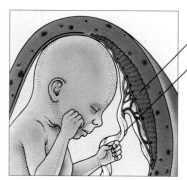

Placenta
Umbilical cord

PLACENTA
The disk-shaped organ that forms in the uterus during pregnancy. It links the blood supplies of the mother and baby via the umbilical cord.

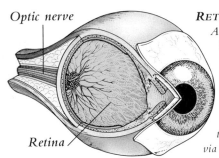

Optic nerve

Retina

RETINA
A light sensitive layer lining the inside of the back of the eye that converts optical images to nerve impulses, which travel to the brain via the optic nerve.

Pulse
Rhythmic expansion and contraction of an *artery* as blood is forced through it.

Pus
A yellowish or green fluid that forms at the site of a bacterial infection; contains *bacteria*, dead *white blood cells*, and damaged tissue.

Q–R

Quadriplegia
Paralysis of both arms, both legs, and the trunk, usually caused by severe spinal cord damage in the neck region.

Red blood cells
Biconcave, disk-shaped cells, without nuclei, that contain *hemoglobin*. There are four to five million red blood cells in a milliliter of blood.

Respiration
The process by which oxygen is conveyed to body cells and carbon dioxide is removed from the cells.

Reticular formation
Nerve cells scattered throughout the *brain stem* that are concerned with alertness and direction of attention to external events.

Retina
See illustration above.

Rheumatoid arthritis
A disorder that causes joint deformity and destruction.

Rubella
A mild viral infection, also known as German measles; if it affects a woman in early pregnancy, it can cause serious harm to the *fetus*.

S

Saccharide
The basic unit that makes up carbohydrates.

Saliva
A watery fluid secreted into the mouth by the *salivary glands* to aid in chewing, tasting, and digestion.

Salivary glands
See illustration below.

Sarcoma
A *cancer* that arises from connective tissue (such as bone), muscle, fibrous tissue, or blood vessels.

Sciatica
Pain caused by pressure on the sciatic nerve; felt in the buttock and back of thigh.

Secondary
A term describing a disorder that follows or results from another disorder (called the *primary* disorder).

Septal defect
An abnormal opening in the central heart wall that allows blood to flow from the right side to the left or vice versa.

Sex hormones
Steroid substances that bring about the development of bodily sex characteristics. Sex hormones also regulate sperm and egg production and the menstrual cycle.

Sinoatrial node
A cluster of specialized muscle cells in the right *atrium* that acts as the heart's natural pacemaker.

Sinus bradycardia
An abnormally slow, but regular, heart rate resulting from a low rate of pacing by the *sinoatrial node*.

Sphincter
A muscle ring, or local thickening of the muscle coat, surrounding an opening in the body.

Spinal fusion
A surgical operation to fuse the bodies of two or more adjacent *vertebrae* in order to stabilize the *spine*.

Spinal nerves
The 31 pairs of combined motor and sensory *nerves* that emerge from and enter the *spinal cord*.

Spine
See illustration at right.

Spleen
A lymphatic organ on the left of the abdominal cavity that removes and destroys worn-out *red blood cells* and helps fight infection.

Stapedectomy
An operation to relieve deafness due to *otosclerosis*.

Steroid drugs
Drugs that simulate the actions of the natural *corticosteroids* or the *sex hormones* of the body.

Stroke
Damage to the brain by deprivation of its full blood supply or leakage of blood from a ruptured vessel; may impair movement, sensation, vision, speech, or intellect.

Subarachnoid hemorrhage
Bleeding from a ruptured *artery* or *aneurysm* lying under the arachnoid layer of the *meninges*.

Subdural hemorrhage
Bleeding between the *dura mater* and the arachnoid layers of the *meninges*.

Sublingual glands
The pair of *salivary glands* in the floor of the mouth.

Submandibular glands
The pair of *salivary glands* that lie immediately under the jawbone near its angle.

Suture
A surgical stitch used to close a wound or incision.

Sympathetic nervous system
One of the two divisions of the *autonomic nervous system*; prepares the body for action, for example by constricting the intestinal and skin blood vessels, widening eye pupils, and increasing heart rate.

Synapse
The junction between two nerve cells, or between a nerve cell and a muscle fiber or a gland. Chemical messengers are passed across a synapse to produce a response in a target cell.

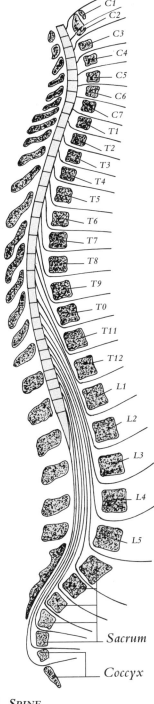

C1
C2
C3
C4
C5
C6
C7
T1
T2
T3
T4
T5
T6
T7
T8
T9
T0
T11
T12
L1
L2
L3
L4
L5

Sacrum

Coccyx

SPINE
The column of 33 ringlike bones, called vertebrae, that divides into seven cervical, 12 thoracic, five lumbar vertebrae, and the fused vertebrae of the sacrum and coccyx.

Synovial joint
A mobile joint lined with a membrane that produces a clear, lubricating fluid.

Syphilis
A sexually transmitted or *congenital* infection that, if untreated, passes through three stages and can involve serious damage to the nervous system. Congenital syphilis is now very rare.

Parotid gland

Sublingual gland

Submandibular gland

SALIVARY GLANDS
Three pairs of glands – the parotid, sublingual, and submandibular glands – that secret saliva via ducts into the mouth.

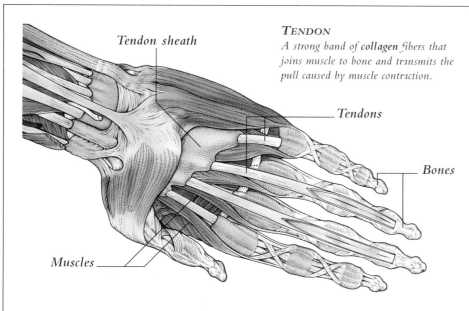

Tendon sheath

Tendons

Bones

Muscles

TENDON
A strong band of collagen fibers that joins muscle to bone and transmits the pull caused by muscle contraction.

T

Taste bud
A spherical nest of receptor cells found mainly on the tongue; each responds most strongly to a sweet, salty, sour, or bitter flavor.

Tendinitis
Inflammation of a *tendon*, causing pain and tenderness, usually from injury.

Tendon
See illustration above.

Tenosynovitis
Inflammation of the inner lining of a *tendon* sheath, usually from excessive friction due to overuse.

Testis
One of a pair of the sperm- and *hormone*-producing sex glands, in the scrotum.

Testosterone
The principal *sex hormone* produced in the *testis* and in small amounts in the adrenal cortex and *ovary*.

Thalamus
A mass of *gray matter* that lies deep within the brain. It receives and coordinates sensory information.

Thorax
The part of the trunk between the neck and the abdomen that contains the heart and the lungs.

Thrombolytic drug
A drug that dissolves *blood clots* and restores blood flow in blocked *arteries*.

Thrombus
A *blood clot* that is usually a result of damage to the blood vessel lining.

Tonsils
Oval masses of *lymphoid tissue* on the back of the throat on either side of the soft palate; help protect against childhood infections.

Trachea
The windpipe. A muscular tube lined with *mucous membrane* and reinforced by about 20 rings of *cartilage*.

Transient ischemic attack
(TIA) A "mini-*stroke*" that passes completely in 24 hours. An attack can imply danger of a full stroke.

Tumor
A *benign* or *malignant* swelling, especially a mass of cells resulting from uncontrolled multiplication.

U

Umbilical cord
The structure that connects the *placenta* and the *fetus*; provides the immunological, nutritional, and hormonal link with the mother.

Urea
A waste product of the breakdown of proteins and the nitrogen-containing component of *urine*.

Urethra
See illustration at left.

Urethritis
Inflammation of the lining of the *urethra*, which is usually caused by a sexually transmitted disease.

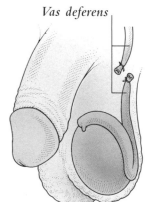

Bladder

Urethra

URETHRA
The tube that carries urine from the bladder to the exterior; much longer in the male than the female.

Urinary tract
The system that forms and excretes *urine*; made up of the *kidneys*, ureters, bladder, and *urethra*.

Urine
Fluid containing wastes produced by the *kidneys*.

Uterus
The hollow muscular organ in which the *fetus* grows and is nourished until birth.

V

Vagina
The passage from the *uterus* to the external genitals that stretches during sexual intercourse and childbirth.

Vagus nerves
The tenth pair of *cranial nerves*; help control automatic functions such as heartbeat and digestion.

Vas deferens
One of a pair of tubes that lead from the *testis* carrying sperm, which mix with fluid before entering the *urethra*.

Vasectomy
See illustration above right.

Vein
A thin-walled blood vessel that returns blood at low pressure to the heart.

Vena cava
One of two large *veins*, the superior and inferior, that empty into the right *atrium*.

Ventricle
One of the two thick-walled lower chambers of the heart; also one of four chambers in the brain that are filled with cerebrospinal fluid.

Vertebra
One of the 33 bones of the vertebral column (*spine*).

Virus
See illustration at right.

Vocal cord
One of two sheets of *mucous membrane* stretched across the inside of the *larynx* that vibrate to produce voice sounds when air passes between them.

Vas deferens

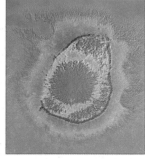

VASECTOMY
A surgical procedure for male sterilization in which each vas deferens is cut and tied.

W–Z

Wart
A contagious, harmless skin growth that is caused by the human papillomavirus.

White blood cell
Any of the colorless blood cells that play various roles in the immune system.

X chromosome
A sex *chromosome*. Body cells of females have two X chromosomes.

Y chromosome
A sex *chromosome*. Its presence is necessary for the development of male characteristics. Body cells of males have one Y and one *X chromosome*.

Zygote
The cell produced when an egg is fertilized by a sperm; contains genetic material for a new person.

TEM x 117,000

VIRUS
A small infectious agent capable of invading and damaging body cells and reproducing within them. Shown here is a Herpes simplex virus; herpes can cause mouth or genital infections.

INDEX

PICTURE CREDITS

c center; l left ; r right; b bottom; t top; bla bottom left above; cla center left above; bra bottom right above

Biophoto Associates: 11br, 13tr, 14tl, 21tr, 37bla, bl, 57c, cr, 72tr, 77crb, 101tr, 104tr, 109bra, br, 168bla, bl, 181tl, 190bl, 197cl, 214cl. Courtesy Dr. Leonard Hayflick, University of California, San Francisco, School of Medicine: 218cl, cla. HNE Healthcare: 205tr; Life Science Images/Ron Boardman: 37tl, 40t. Living Technology: 171bl. National Medical Slide Bank: 45bl, 131tl, 193c. Northwick Park Hospital: 170c, bc, 173bla, bl. Institute of Orthopaedics, University College, London: 34tr. Barry Richards, St. Petersburg, Florida: 114bl. Reynolds Medical Ltd.: 116bl. Audio Visual Department, St. Mary's Hospital Medical School: 164cl, 188c. Science Photo Library: 34tc, cr, 81crb, 107t, 131cl, 138bl, 166br, 191tl, cl, 192cr, 219tc, cr; John Durham: 2tr, 13br, 26; Professors Motta, Correr & Nottola/University "La Sapienza," Rome: 2bc, 17tr, 23cr, 94;

Bill Longcore: 3bc, 16c, 19tl; Professor P. Motta, Department of Anatomy, University "La Sapienza," Rome: 3cr, 14bl, 23bl, 33tl, bl, 49, 54cl, 134c, 151tr, 160tl, 161tl, 162tr, 174, 179bl, 226cl; Alfred Pasieka: 3cl, tc, 9tl, 199cl, 220tr, 227bc; Secchi-Lecaque/Roussel-UCLAF/CNRI: 6br, 65br, 87br, 228bl; Sandoz/D. Zagury/Petit Format: 7tl, 127bl; CNRI: 9bl, crb, cla, 12tr, 14br, 27cl, 39bc, 45cl, 62tr, 69bc, 84bl, 108br, 110bl, 123cl, 128cl, 130c, 140tr, 142bc, 161br, 167cl, 171tl, 172tl, 175cl, 180br, 183cl, 194tr, br, 195tr, 197tl, 204cl, 221bc, 222br; Mehau Kulyk: 9clb, 36tc, 84br; GCa/CNRI: 9tr, 37tr, 83tl; Tim Beddow: 9cra, 81bc, br; Petit Format/Nestle/Steiner: 9br, 68tr; J.C. Revy: 11tr, 58; Francis Leroy, Biocosmos: 11tl, 122; Don Fawcett: 12tl, 64tr, 100; Hossler/Custom Medical Stock Photo: 12bl; Astrid & Hanns Frieder Michler: 13cl, 86tr, 145tr, 182; Andrew Syred: 13bl; M.I. Walker: 14tr, 195tr; Dr. Gopal Murti: 20bl; David Scharf: 22tr, 135cl, 147bl; Biology Media: 23br; Dr. P. Marazzi: 25cl, clb, cr; Scott Camazine: 33br, 82br; Eric

Grave: 33cl; GJLP/CNRI: 35cl, 47bl; BSIP, LECA: 35bl; Biophoto Associates: 37cl, 66bl, 144bc, 204cr, 209tr, 224bl; Department of Clinical Radiology, Salisbury District Hospital: 44br; Princess Margaret Rose Orthopaedic Hospital: 46tr, 47cr; Manfred Kage: 59tr, 63bc, 179cr; Hank Morgan: 63tr; Petit Format/C. Edelmann: 66br; Dr. M. Phelps & Dr. John Mazziotta/ Neurology: 79bla; Dr. John Mazziotta: 79bl; Professor P. Motta/A. Caggiati/University "La Sapienza," Rome: 89tl; Dr. G. Oran Bredberg: 89bl; Omikron: 90tr; Stanford Eye Clinic: 91bc; Professor Tony Wright, Institute of Laryngology & Otology: 92tc; GEC Research/Hammersmith Hospital Medical School: 92bc; Western Ophthalmic Hospital: 93c; Alexander Tsiaras: 93bl; Martin Dohrn/Royal College of Surgeons: 105c; Lungagrafix: 105cr; Frieder Michler: 107bl; Philippe Plailly: 107br; Cardiothoracic Centre, Freeman Hospital, Newcastle-upon-Tyne: 113c, c; Adam Hart-Davis: 117bl; NIBSC: 131bl, 133tr, 231br; Jim Stevenson: 132tr; Barry Dowsett: 143br,

164br; Morendun Animal Health: 148bl, 172cl, 225tr; Cecil H. Fox: 150; John Burbidge: 166tc; King's College School of Medicine: 188bra; SIU: 192bl; NIH: 195cl; Petit Format/Nestle: 201tc; Petit Format/CSI: 200bl; Peter Menzel: 205bc; Mark Clarke: 212bl; Mark Clarke and Chris Priest: 210br; Professor P. Motta & J. van Blerkom: 217cr. C. James Webb: 154tr.

Every effort has been made to trace the copyright holders, and Dorling Kindersley apologizes in advance in case an omission has occurred. If any omission does come to light, the company will be pleased to insert the appropriate acknowledgment in any subsequent editions of this book.

ACKNOWLEDGMENTS

Additional editorial help: Edda Bohnsack, Edward Bunting, Will Hodgkinson, Deslie Lawrence, Sarah Miller, Seán O'Connell, and Michael Williams. Additional medical and scientific help: Philip Fulford, Editor, *Journal of Bone and Joint Surgery*.